T0122399

# NEUROMETHODS

*Series Editor*
**Wolfgang Walz**
**University of Saskatchewan**
**Saskatoon, SK, Canada**

For further volumes:
http://www.springer.com/series/7657

*Neuromethods* publishes cutting-edge methods and protocols in all areas of neuroscience as well as translational neurological and mental research. Each volume in the series offers tested laboratory protocols, step-by-step methods for reproducible lab experiments and addresses methodological controversies and pitfalls in order to aid neuroscientists in experimentation. *Neuromethods* focuses on traditional and emerging topics with wide-ranging implications to brain function, such as electrophysiology, neuroimaging, behavioral analysis, genomics, neurodegeneration, translational research and clinical trials. *Neuromethods* provides investigators and trainees with highly useful compendiums of key strategies and approaches for successful research in animal and human brain function including translational "bench to bedside" approaches to mental and neurological diseases.

# Methods for Preclinical Research in Addiction

Edited by

## María A. Aguilar

*Unit of Research 'Neurobehavioral Mechanisms and Endophenotypes of Addictive Behavior,' Department of Psychobiology, Faculty of Psychology, University of Valencia, Valencia, Spain*

 Humana Press

*Editor*
María A. Aguilar
Unit of Research 'Neurobehavioral
Mechanisms and Endophenotypes
of Addictive Behavior'
Department of Psychobiology
Faculty of Psychology
University of Valencia
Valencia, Spain

ISSN 0893-2336                    ISSN 1940-6045   (electronic)
Neuromethods
ISBN 978-1-0716-1750-2            ISBN 978-1-0716-1748-9   (eBook)
https://doi.org/10.1007/978-1-0716-1748-9

This Humana imprint is published by the registered company Springer Science+Business Media, LLC part of Springer Nature.
The registered company address is: 1 New York Plaza, New York, NY 10004, U.S.A.

# Preface to the Series

Experimental life sciences have two basic foundations: concepts and tools. The *Neuro-methods* series focuses on the tools and techniques unique to the investigation of the nervous system and excitable cells. It will not, however, shortchange the concept side of things as care has been taken to integrate these tools within the context of the concepts and questions under investigation. In this way, the series is unique in that it not only collects protocols but also includes theoretical background information and critiques which led to the methods and their development. Thus it gives the reader a better understanding of the origin of the techniques and their potential future development. The *Neuromethods* publishing program strikes a balance between recent and exciting developments like those concerning new animal models of disease, imaging, in vivo methods, and more established techniques, including, for example, immunocytochemistry and electrophysiological technologies. New trainees in neurosciences still need a sound footing in these older methods in order to apply a critical approach to their results.

Under the guidance of its founders, Alan Boulton and Glen Baker, the *Neuromethods* series has been a success since its first volume published through Humana Press in 1985. The series continues to flourish through many changes over the years. It is now published under the umbrella of Springer Protocols. While methods involving brain research have changed a lot since the series started, the publishing environment and technology have changed even more radically. Neuromethods has the distinct layout and style of the Springer Protocols program, designed specifically for readability and ease of reference in a laboratory setting.

The careful application of methods is potentially the most important step in the process of scientific inquiry. In the past, new methodologies led the way in developing new disciplines in the biological and medical sciences. For example, Physiology emerged out of Anatomy in the nineteenth century by harnessing new methods based on the newly discovered phenomenon of electricity. Nowadays, the relationships between disciplines and methods are more complex. Methods are now widely shared between disciplines and research areas. New developments in electronic publishing make it possible for scientists that encounter new methods to quickly find sources of information electronically. The design of individual volumes and chapters in this series takes this new access technology into account. Springer Protocols makes it possible to download single protocols separately. In addition, Springer makes its print-on-demand technology available globally. A print copy can therefore be acquired quickly and for a competitive price anywhere in the world.

*Saskatoon, SK, Canada*                                                      *Wolfgang Walz*

# Preface

The use of animal models to study addictive behavior has brought new knowledge of the neurobiological substrates of this disorder and facilitated the development of preventive and therapeutic tools. Currently, there are a great number of experimental paradigms used in preclinical research about addictive behavior. This book presents a wide perspective of the animal models that are most widely employed in the study of the individual and environmental factors involved in the development of addiction, the consequences of chronic drug exposure, and recently developed behavioral and biological procedures for the treatment of addictive disorders. The book is organized in three parts to cover these aspects in connection, mainly, with different drugs of abuse, but also with pathological gambling, as an example of behavioral addictions.

A main challenge for addiction research is to identify the factors that predispose individuals to addiction. Thus, the first part of this book focuses on the individual variables that confer enhanced vulnerability to addiction to nicotine, alcohol, cocaine, psychostimulants, or gambling. This part begins with the chapter of Chellian, Wilson, Behnood-Rod, and Bruijnzeel describing the methodology of *Intracranial Self-Stimulation* (ICSS) and how this paradigm can be used to evaluate the influence of sex in the propensity to develop addiction to nicotine and other drugs of abuse. The authors offer a detailed description of electrode implantation and the operant procedure for ICSS, nicotine administration via osmotic minipump implantation, and the methodology for evaluating the somatic withdrawal signs that characterize nicotine abstinence. In Chapter 2, Monleón, Redolat, Duque, Mesa-Gresa, and Vinader-Caerols highlight the importance of adolescence as a period of enhanced vulnerability to the impact of alcohol abuse. They offer a detailed explanation of the three main animal models of adolescent binge drinking: *Drinking in the Dark, Two-Bottle Choice*, and *Chronic-Intermittent Ethanol Administration*. In addition, the authors describe a battery of behavioral tests which are used to evaluate the effects of binge drinking on different kinds of memory, including the *Novel Object Recognition* task, *Inhibitory Avoidance*, and the *Morris Water Maze*. In Chapter 3, Arenas, Pujante-Gil, and Manzanedo describe the most common sensorimotor-gating index, the *Prepulse Inhibition* (PPI) of the startle reflex. The authors report the relationship between PPI impairment and cocaine addiction and provide guidelines on how to evaluate PPI in rodents and humans in order to identify subjects with a PPI deficit, which may be a biomarker of vulnerability to drug addiction. Chapter 4 addresses the influence of the novelty-seeking trait on the effects of different drugs of abuse (Calpe-López, Martínez-Caballero, García-Pardo, and Aguilar). We offer a detailed description of the *Hole-board test*, the *Conditioned Place Preference* paradigm, and different types of drug pre-treatments during the adolescent period and report on evidence that a high novelty-seeking phenotype in the hole-board is a marker of enhanced susceptibility to drug addiction. In the last chapter of this part Rygula, Hinchcliffe, and Noworyta describe a behavioral protocol to model pathological gambling (the *Rat Slot Machine Task*, *rSLM*) and to study the association between an enduring behavioral trait of animals—cognitive judgment bias—and vulnerability to the development of pathological gambling. The authors detail the screening and segregation of rats possessing "pessimistic" and "optimistic" traits by means of the *Ambiguous-Cue Interpretation test* and the subsequent assessment of gambling-like behavior (decision-making under risk in the rSLM).

The second part of the book is devoted to the environmental variables that contribute to vulnerability to drug addiction. In Chapter 6 Allain, Ndiaye, and Samaha focus on the importance of extended drug access for the escalation of *Drug Self-Administration*. They describe the protocols and methods of cocaine self-administration and describe in detail different patterns of drug access (short, long, intermittent, limited, unlimited, etc.). In Chapter 7, Borruto, Domi, Soverchia, Domi, Li, and Cannella also address the self-administration paradigm to focus on models of relapse to psychostimulant-seeking induced by environmental drug-associated stimuli. The authors explain the methodology of the *Extinction/Reinstatement of cocaine self-administration* and provide an overview and technical details of several models of relapse (drug-contingent and discriminative stimuli-induced reinstatement, context-induced reinstatement, the ABA renewal protocol, and the incubation of cue-induced psychostimulant craving). In Chapter 8, we address the role of exposure to stress on vulnerability to drug addiction (García-Pardo, De la Rubia, Calpe-López, Martínez-Caballero, and Aguilar). Besides providing a detailed description of the main protocols of *Social Defeat Stress*, we summarize the short- and long-term effects of social defeat on the rewarding properties of several drugs of abuse. This part ends with Chapter 9, by Stairs, Hanson, and Kellerman, who explain the procedures of *Environmental Enrichment* (cages, toy changes, feeding, etc.) and *Impoverished or Social Control Conditions* and discuss evidence of the protective role of environmental enrichment on the effects of a wide variety of drugs of abuse in different behavioral paradigms (locomotor activity, place preference, self-administration, drug discrimination, drug consumption, etc.).

The last part of the book deals with the consequences of chronic drug consumption and new approaches to the treatment of addiction. People diagnosed with substance use disorder (SUD) experience other psychiatric conditions and, frequently, post-traumatic stress disorder (PTSD). In Chapter 10, Wilkinson, Blount, Knackstedt, and Schwendt describe a novel animal model that uses *Predator Odor Stress* to study the comorbidity between SUD and PTSD. The authors describe the procedures used to evaluate anxiety (*Elevated Plus Maze* and *Light/Dark Box*) and hypervigilance (*Acoustic Startle Response*) in animals in response to predator odor stress and detail the protocols of acquisition, extinction, and *cue-induced reinstatement* of cocaine self-administration. Chapter 11 deals with the cognitive impairments frequently observed in chronic drug users. Mañas Padilla, Ávila Gámiz, Gil Rodríguez, Sánchez Salido, Santín, and Castilla-Ortega offer a detailed description of the materials and methodologies employed to study the long-term impact of cocaine exposure on working and reference memory. The authors describe the *"passive" model of chronic cocaine* administration and three protocols for assessing the cognitive consequences of such exposure to cocaine: the *Continuous Spontaneous Alternation Task* (to evaluate spatial working memory in a Y-maze), the *Novel Object Recognition Task*, and the *Novel Place Recognition Task* (to evaluate reference memory for objects and places, respectively). The last two chapters of the book address the issue of new and future treatments for addictive disorders. In Chapter 12, Dumontoy, Etievant, Van Schuerbeek, and Van Waes describe an animal model of *Transcranial Direct Current Stimulation* (tDCS), a neuromodulation technique that has recently attracted interest as a treatment for addiction and other psychiatric disorders. The authors explain the methodology of different variants of the tDCS protocol (on anesthetized, awake restrained, and freely moving mice), the main problems related with the technique, and the advantages and disadvantages of the different protocols of stimulation. Finally, in the last chapter of the book, Milton describes a protocol based on

*Manipulating Reconsolidation* to erase drug memories, as a new avenue for the treatment of addiction. She explains relevant aspects of the paradigm, such as the selection of the amnestic agent and the type of memory to target for reconsolidation blockade, the timing between memory reactivation and administration of amnestic agent, and the choice of memory reactivation protocol and memory retention tests.

The book provides a clear indication of how animal models of addiction have advanced and evolved to increase their validity and usefulness in the study of the factors involved in vulnerability to addictive disorders. The preclinical models described here have not only contributed to unraveling the neurobiological mechanisms underlying the individual and environmental variables that confer an enhanced risk of addiction, but will also undoubtedly help to develop new behavioral and pharmacological strategies to prevent and treat addictive disorders.

*Valencia, Spain*                                                   *María A. Aguilar*

The original version of this Frontmatter was revised. The correction to this chapter is available at https://doi. org/10.1007/978-1-0716-1748-9_14

# Contents

# Contributors

MARÍA A. AGUILAR • *Unit of Research 'Neurobehavioral Mechanisms and Endophenotypes of Addictive Behavior,' Department of Psychobiology, Faculty of Psychology, University of Valencia, Valencia, Spain; Faculty of Nursing, Catholic University of Valencia, Valencia, Spain*

FLORENCE ALLAIN • *Department of Pharmacology and Physiology, Faculty of Medicine, Université de Montréal, Montreal, QC, Canada*

M. CARMEN ARENAS • *Department of Psychobiology, Faculty of Psychology, University of Valencia, Valencia, Spain*

FABIOLA ÁVILA-GÁMIZ • *Institute of Biomedical Research in Malaga-IBIMA, University of Malaga, Malaga, Spain; Department of Psychobiology and Methodology of the Behavioral Sciences, University of Malaga, Malaga, Spain*

AZIN BEHNOOD-ROD • *Department of Psychiatry, University of Florida, Gainesville, FL, USA*

HARRISON BLOUNT • *Psychology Department, Behavioral and Cognitive Neuroscience Program, University of Florida, Gainesville, FL, USA*

ANNA MARIA BORRUTO • *School of Pharmacy, Pharmacology Unit, University of Camerino, Camerino, Italy*

ADRIAAN W. BRUIJNZEEL • *Department of Psychiatry, Center for Addiction Research and Education, University of Florida, Gainesville, FL, USA*

CLAUDIA CALPE-LÓPEZ • *Unit of Research 'Neurobehavioral Mechanisms and Endophenotypes of Addictive Behavior,' Department of Psychobiology, Faculty of Psychology, University of Valencia, Valencia, Spain; Faculty of Nursing, Catholic University of Valencia, Valencia, Spain*

NAZZARENO CANNELLA • *School of Pharmacy, Pharmacology Unit, University of Camerino, Camerino, Italy*

ESTELA CASTILLA-ORTEGA • *Institute of Biomedical Research in Malaga-IBIMA, University of Malaga, Malaga, Spain; Departamento de Psicobiología y Metodología de las CC, Facultad de Psicología, Universidad de Málaga, Malaga, Spain; Department of Psychobiology and Methodology of the Behavioral Sciences, University of Malaga, Malaga, Spain*

RANJITHKUMAR CHELLIAN • *Department of Psychiatry, University of Florida, Gainesville, FL, USA*

JOSÉ ENRIQUE DE LA RUBIA-ORTÍ • *Faculty of Nursing, Catholic University of Valencia, Valencia, Spain*

ANA DOMI • *School of Pharmacy, Pharmacology Unit, University of Camerino, Camerino, Italy*

ESI DOMI • *Center for Social and Affective Neuroscience, IKE, Linköping University, Linköping, Sweden*

STÉPHANIE DUMONTOY • *Laboratoire de Recherches Intégratives en Neurosciences et Psychologie Cognitive, Université Bourgogne Franche-Comté, Besançon, France*

ARÁNZAZU DUQUE • *Universidad Internacional de Valencia, Valencia, Spain*

ADELINE ETIEVANT • *Laboratoire de Recherches Intégratives en Neurosciences et Psychologie Cognitive, Université Bourgogne Franche-Comté, Besançon, France*

MARÍA PILAR GARCÍA-PARDO • *Department of Psychology and Sociology, Faculty of Social Sciences, University of Zaragoza (Campus Teruel), Teruel, Spain*

SARA GIL-RODRÍGUEZ • *Institute of Biomedical Research in Malaga-IBIMA, University of Malaga, Malaga, Spain; Department of Psychobiology and Methodology of the Behavioral Sciences, University of Malaga, Malaga, Spain*

TAENA HANSON • *Department of Psychological Science, Creighton University, Omaha, NE, USA*

JUSTYNA K. HINCHCLIFFE • *Affective Cognitive Neuroscience Laboratory, Department of Pharmacology, Maj Institute of Pharmacology Polish Academy of Sciences, Krakow, Poland; School of Physiology, Pharmacology and Neuroscience, University of Bristol, Bristol, UK*

KENDALL KELLERMAN • *Department of Psychological Science, Creighton University, Omaha, NE, USA*

LORI KNACKSTEDT • *Psychology Department, Behavioral and Cognitive Neuroscience Program, University of Florida, Gainesville, FL, USA; Center for Addiction Research and Education (CARE), University of Florida, Gainesville, FL, USA*

HONGWU LI • *School of Chemical Engineering, Changchun University, Changchun, Jilin, China*

M. CARMEN MAÑAS-PADILLA • *Institute of Biomedical Research in Malaga-IBIMA, University of Malaga, Malaga, Spain; Departamento de Psicobiología y Metodología de las CC, Facultad de Psicología, Universidad de Málaga, Malaga, Spain; Department of Psychobiology and Methodology of the Behavioral Sciences, University of Malaga, Malaga, Spain*

CARMEN MANZANEDO • *Department of Psychobiology, Faculty of Psychology, University of Valencia, Valencia, Spain*

M. ÁNGELES MARTÍNEZ-CABALLERO • *Unit of Research 'Neurobehavioral Mechanisms and Endophenotypes of Addictive Behavior,' Department of Psychobiology, Faculty of Psychology, University of Valencia, Valencia, Spain; Faculty of Nursing, Catholic University of Valencia, Valencia, Spain*

PATRICIA MESA-GRESA • *Department of Psychobiology, Universitat de València, Valencia, Spain*

AMY L. MILTON • *Department of Psychology, University of Cambridge, Cambridge, UK*

SANTIAGO MONLEÓN • *Department of Psychobiology, Faculty of Psychology, University of Valencia, Valencia, Spain*

NDEYE AISSATOU NDIAYE • *Department of Neurosciences, Faculty of Medicine, Université de Montréal, Montreal, QC, Canada*

KAROLINA NOWORYTA • *Affective Cognitive Neuroscience Laboratory, Department of Pharmacology, Maj Institute of Pharmacology Polish Academy of Sciences, Krakow, Poland*

SERGIO PUJANTE-GIL • *Department of Psychobiology, Faculty of Psychology, University of Valencia, Valencia, Spain*

ROSA REDOLAT • *Department of Psychobiology, Universitat de València, Valencia, Spain*

RAFAL RYGULA • *Affective Cognitive Neuroscience Laboratory, Department of Pharmacology, Maj Institute of Pharmacology Polish Academy of Sciences, Krakow, Poland*

ANNE-NOËL SAMAHA • *Department of Pharmacology and Physiology, Faculty of Medicine, Université de Montréal, Montreal, QC, Canada; Groupe de recherche sur le système nerveux central, Faculty of Medicine, Université de Montréal, Montreal, QC, Canada; Department of Pharmacology and Physiology, CNS Research Group, Faculty of Medicine, Université de Montréal, Montreal, QC, Canada*

LOURDES SÁNCHEZ-SALIDO • *Institute of Biomedical Research in Malaga-IBIMA, University of Malaga, Malaga, Spain; Mental Health Department, Regional University Hospital of Malaga, Malaga, Spain*

LUIS J. SANTÍN • *Institute of Biomedical Research in Malaga-IBIMA, University of Malaga, Malaga, Spain; Department of Psychobiology and Methodology of the Behavioral Sciences, University of Malaga, Malaga, Spain*

MAREK SCHWENDT • *Psychology Department, Behavioral and Cognitive Neuroscience Program, University of Florida, Gainesville, FL, USA; Center for Addiction Research and Education (CARE), University of Florida, Gainesville, FL, USA*

LAURA SOVERCHIA • *School of Pharmacy, Pharmacology Unit, University of Camerino, Camerino, Italy*

DUSTIN J. STAIRS • *Department of Psychological Science, Creighton University, Omaha, NE, USA*

ANDRIES VAN SCHUERBEEK • *Department of Pharmaceutical Sciences, Center for Neurosciences, Vrije Universiteit Brussel, Brussels, Belgium*

VINCENT VAN WAES • *Laboratoire de Recherches Intégratives en Neurosciences et Psychologie Cognitive, Université Bourgogne Franche-Comté, Besançon, France*

CONCEPCIÓN VINADER-CAEROLS • *Department of Psychobiology, Universitat de València, Valencia, Spain*

COURTNEY WILKINSON • *Psychology Department, Behavioral and Cognitive Neuroscience Program, University of Florida, Gainesville, FL, USA; Center for Addiction Research and Education (CARE), University of Florida, Gainesville, FL, USA*

RYANN WILSON • *Department of Psychiatry, University of Florida, Gainesville, FL, USA*

# Part I

# Individual Variables of Vulnerability to Addiction

# Chapter 1

# Influence of Sex on the Effects of Nicotine and Other Drugs of Abuse on Intracranial Self-Stimulation

Ranjithkumar Chellian, Ryann Wilson, Azin Behnood-Rod, and Adriaan W. Bruijnzeel

## Abstract

Drug abuse is one of the leading causes of loss of productivity, disease, and early death in the world. People typically start experimenting with drugs of abuse during adolescence, and about 20% of people who use drugs develop a substance abuse disorder. In general, drug use is more common in men than in women. However, when women use drugs, they are just as likely as men to develop a substance use disorder. Drugs of abuse induce euphoria, and the rewarding properties of drugs play an essential role in the early stages of drug use. After the development of dependence, cessation of drug use leads to somatic and affective withdrawal signs, which contributes to the maintenance of drug use. The intracranial self-stimulation (ICSS) method can be used to measure the acute rewarding properties of drugs and the dysphoria associated with drug withdrawal. One of the main advantages of the ICSS method is that it provides a quantitative measure of the rewarding aspects of drug use and the aversive aspects of withdrawal. The ICSS method can be used to identify sex differences in the rewarding and aversive aspects of drug use.

Key words Drugs of abuse, Animal models, Sex differences, Reward, Withdrawal, ICSS, Somatic signs

## 1 Introduction

Drug use is one of the most common causes of loss of productivity, disease (e.g., HIV, hepatitis C, lung cancer), and premature death [1]. Worldwide there are almost 300 million adults who use drugs of abuse, and about 10% of them suffer from a drug use disorder [1]. People often start using drugs because their friends use drugs, because they are curious about the effects of drugs, or to improve their mood [2, 3]. The types of drugs that are being used and the prevalence of drug use varies by country [4, 5]. For example, 2% of the students in Greece and 53% of the students in Scotland use cannabis [5]. The prevalence of substance use is high in the USA and Eastern European countries and relatively low in Africa and the Middle East [4]. Substance abuse is more common among men

María A. Aguilar (ed.), *Methods for Preclinical Research in Addiction*, Neuromethods, vol. 174,
https://doi.org/10.1007/978-1-0716-1748-9_1, © Springer Science+Business Media, LLC, part of Springer Nature 2022

than women [6]. This is most likely due to the fact that historically men are more likely to drink alcohol and use illicit drugs [7]. Tobacco is one of the most widely used drugs in the world and smoking is the leading preventable cause of early death [8]. Smoking is much more common in men than in women.

This sex difference might be due to a combination of physiological, psychological, and behavioral factors [9]. The worldwide prevalence of daily smoking is 25% for men and 5% for women [10]. In 2016, there were 63 million Americans who used tobacco products (cigarettes, cigars, smokeless tobacco, and pipes) [11]. In Western countries, smoking rates are on the decline, and 80% of tobacco users now live in low and middle-income countries [12]. Although the use of combustible tobacco products is on the decline in the USA, there has been a very strong increase in the use of e-cigarettes [13]. It has been estimated that 10.5% of middle school students and 27.5% high school students currently use e-cigarettes [14]. Males are also more likely to use e-cigarettes than females [15, 16]. Adolescent e-cigarette users are becoming nicotine dependent, and therefore e-cigarette use might lead to a new generation of people with a nicotine addiction [17].

## 2  Intracranial Self-Stimulation Methods

Intracranial self-stimulation (ICSS) methods were developed in the late 1950s [18, 19]. ICSS is an operant procedure in which the behavioral response to electrical stimulation of the reward system is measured. Some ICSS methods vary the stimulation frequency (Hz), while others change the stimulation intensity ($\mu A$) [20, 21]. Furthermore, some ICSS methods measure the response rate, while others measure reinforcement thresholds in rate-independent procedures [22]. In one common method, the frequency of the stimulation is changed, and immediately following the stimulation the number of responses of the animal is measured [21]. Each response is followed by an electrical stimulus that is the same as the noncontingent stimulus that the rats received before the self-stimulation period. There is a strong positive relationship between the stimulation frequency and the response rate. When the stimulation frequency is high the rats have a high response rate, and when the stimulation frequency is low the response rate is also low [21]. Another common method is the discrete-trial ICSS procedure. In this rate-independent procedure, the current is systematically changed, and the reward threshold is determined that maintains electrical self-stimulation [20, 23]. A decrease in reward thresholds reflects enhanced reward function (i.e., euphoria) and an increase in thresholds is indicative of impaired reward function (i.e., dysphoria) [24, 25]. In this chapter, we will provide the methods for a discrete-trial ICSS procedure.

## 3    Sex Differences in the Rewarding Effects of Drugs

Nicotine addiction is characterized by compulsive smoking and relapse after periods of abstinence [26, 27]. Craving for nicotine and relapse can occur even after a very long abstinence period [28]. In humans, nicotine has reinforcing effects such as stress relief, enhanced cognition, and mild euphoria [26, 29]. During the initiation of smoking, smokers use nicotine to modulate arousal and to control mood states [30]. The rewarding effects of nicotine play a critical role in the initiation of nicotine use. There are sex differences in the acute rewarding effects of nicotine. In a recent study, we evaluated sex differences in the rewarding effects of nicotine in adolescent and adult male and female rats [31]. Nicotine lowered the brain reward thresholds of the adult and adolescent male and female rats. The nicotine-induced decrease in the reward thresholds was the same in the adult male and adult female rats. However, nicotine induced a greater decrease in the reward thresholds of the adolescent female rats than the adolescent male rats [31]. Most studies have used place conditioning procedures to compare the rewarding effects of nicotine in males and females. Some of these studies suggest that nicotine induces place preference in male and female rodents but that higher doses of nicotine are required in females [32, 33].

Sex differences in the rewarding effects of other drugs have been explored as well. Stratman and Craft investigated the rewarding effects of amphetamine, cocaine, and morphine in the ICSS procedure in male and female rats [34]. They found that amphetamine and cocaine lower the brain reward thresholds, and there were no sex differences in the rewarding effects of these drugs. Cocaine and amphetamine also decreased the response latencies, and there was no difference in the effects of these drugs on the response latencies. A decrease in the latencies indicates that these drugs shorten the response times. Morphine did not lower the brain reward thresholds in the males and the females. Morphine increased the response latencies and thus indicating that morphine has sedative effects. However, it should be noted that an ICSS study with mice suggests that morphine is more rewarding in females than males [35]. Sex differences in the effects of 3,4-methylenedioxymethamphetamine (MDMA) have been explored with a rate dependent ICSS procedure [36]. MDMA had similar effects in males and females and increased ICSS rates that were maintained by low brain-stimulation frequencies and decreased rates maintained by high brain-stimulation frequencies. The facilitation of low ICSS rates reflects the abuse potential of MDMA, and the decrease in high ICSS rates reflects the abuse-limiting effects of MDMA. There were no major sex differences, but MDMA produced a more prolonged ICSS facilitation and also a longer period of heightened dopamine release in the females. Alcohol also lowers the brain reward thresholds in rats, but these effects are only observed when the rats self-administer the alcohol [37, 38]. Sex differences in the effects of alcohol on the brain reward thresholds have not been explored.

## 4    Sex Differences in Drug Withdrawal

Cessation of nicotine intake leads to somatic withdrawal signs and dysphoria [26]. In humans, withdrawal from nicotine leads to somatic withdrawal symptoms such as bradycardia, insomnia, gastrointestinal discomfort, increased appetite, and weight gain [39]. Nicotine withdrawal can also induce negative affective symptoms such as irritability, depressed mood, restlessness, anxiety, stress, difficulty concentrating, and craving [39, 40]. The rewarding effects of nicotine play a role in the initiation of smoking and withdrawal plays a role in the maintenance of smoking [26, 41]. In rodents, cessation of nicotine administration or blockade of nicotinic acetylcholine receptors (nAChR) in nicotine-treated rats leads to somatic withdrawal signs, including abdominal constrictions, facial fasciculation, increased eye blinks, and ptosis [42–45]. Although somatic withdrawal signs are easy to measure, they do not reflect the motivational state of the animal. With the ICSS method, it is possible to measure the rewarding effects of nicotine and the dysphoria associated with the cessation of nicotine administration [20]. The ICSS method has been used to investigate reward function in adult and adolescent male and female rats [34, 46, 47]. Acute administration of nicotine lowers brain reward thresholds, which reflects the reward enhancing effects of nicotine [48–50]. Spontaneous and precipitated nicotine withdrawal elevates brain reward thresholds demonstrating diminished reward function [25, 42]. This increase in brain reward thresholds has also been observed during withdrawal from other drugs of abuse such as cocaine, amphetamine, opiates, and alcohol [24, 42, 51–55]. Thus, brain reward thresholds, as assessed with the ICSS procedure in combination with physical withdrawal symptoms, provides a way to assess nicotine dependence and withdrawal in animal models. In a recent study, we investigated the effects of precipitated and spontaneous nicotine withdrawal on brain reward thresholds in male and female rats [56]. A low dose of mecamylamine elevated the brain reward thresholds of the nicotine-treated male rats but not those of the nicotine-treated female rats. Furthermore, mecamylamine induced more somatic withdrawal signs in the male rats that were treated with nicotine than in the female rats treated with nicotine. Removal of the minipumps led to a similar increase in brain reward thresholds in the males and the females treated with nicotine. This finding suggests that low doses of a nicotinic receptor antagonist induce a greater reward deficit in males than females, but there is no sex difference in spontaneous withdrawal.

Several studies have used the ICSS procedure to study alcohol withdrawal in male rats. These studies showed that chronic exposure to alcohol vapor or an alcohol liquid diet leads to dependence and elevated ICSS thresholds upon cessation of alcohol administration [55, 57]. We are not aware of any studies that used the ICSS procedure to investigate sex differences in the deficit in reward function associated with alcohol withdrawal. There is, however, evidence that alcohol withdrawal is more severe in male than female rats. Male rats display more anxiety-like behavior after cessation of alcohol administration and are more likely to experience seizures compared to female rats [58–60]. Rats withdrawing from opioids display severe somatic withdrawal signs, anxiety-like behavior, and a dysphoria-like state. Several studies have shown that male rats have elevated brain reward thresholds in the ICSS procedure [24, 54, 61]. The ICSS procedure has not been used to determine if there are sex differences in opioid withdrawal. However, male rats display more somatic withdrawal signs than females, and male rats also display somatic withdrawal signs for a more prolonged period [62].

## 5 Materials

### 5.1 Electrode Implantations

1. Stainless-steel bipolar electrodes (model MS303/2 P1 Technologies, Roanoke, VA, USA).

2. Wire cutter (Lindstrom Precision Tools, model 8142, ultra-flush, small head cutter).

3. Isoflurane–oxygen (1–3% isoflurane).

4. Small trimmer (Oster MiniMax Trimmer) or regular trimmer (Oster A5 clipper with size 40 blade).

5. Aseptic surgery scrub prep—Betadine Surgical Scrub (7.5% povidone–iodine).

6. Isopropyl alcohol (70%).

7. Model 940 Kopf stereotaxic frame with digital display console (David Kopf Instruments, Tujunga, CA; Fig. 1a).

8. Surgical instruments: curette, surgery scissors, and forceps (Roboz Surgical Instrument Company, Gaithersburg, MD, USA; Fig. 1b).

9. Disposable scalpels (size 10).

10. Cotton swabs and gauze pads.

11. High-temp surgical cautery pen (Bovie Medical Corp, Clearwater, FL, USA).

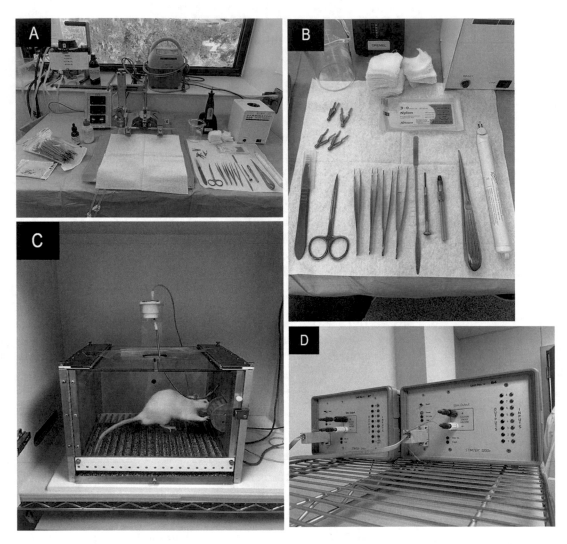

**Fig. 1** Intracranial self-stimulation surgeries and operant responding. The figures depict the surgery set up with stereotaxic frame (**a**), surgery tools (**b**), operant chamber (**c**), and ICSS stimulators (**d**)

12. Stainless-steel jeweler screws (shaft length 1.6 mm, shaft diameter 1.57 mm).

13. Microdrill with 1.07 mm (diameter) drill bit.

14. Dremel cordless drill with engraving cutter (Dremel 106, 1/8″ shank).

15. Dental Cement (Pharmaceutical grade Dental Acrylic Cement) and Teets Denture Material (methyl methacrylate liquid, CO-Oral-Ite Dental Mfg. Diamond Springs, CA, USA).

16. Spatula and dappen dishes.

17. Carprofen (5 mg/mL, once a day for 2 days) solution for pain treatment.

**5.2 Intracranial Self-Stimulation Procedure**

1. Operant conditioning chambers and sound-attenuating chambers (Med Associates, Georgia, VT; Fig. 1c).

2. Wheel manipulandum installed into the wall of the operant conditioning chamber.

3. Stimulators (Stimtek, Acton, MA; Fig. 1d).

**5.3 Osmotic Minipump Implantations**

1. Osmotic minipumps (Durect corporation, Cupertino, CA).

2. Isoflurane–oxygen (1–3% isoflurane).

3. #40 surgical clipper blade.

4. Aseptic surgery scrub prep—Betadine Surgical Scrub (7.5% povidone–iodine) and Becton Dickinson (BD) isopropyl alcohol swabs (70% isopropyl alcohol).

5. Surgical instruments: Scalpel, blunt tissue forceps, suture.

6. Carprofen.

**5.4 Somatic Withdrawal Signs**

1. Plexiglas observation chamber ($25 \times 25 \times 45$ cm; $L \times W \times H$).

2. Timer.

3. Checklist labeled with animal ID and abbreviations defined.

# 6 Methods

**6.1 Electrode Implantations**

1. Adolescent (postnatal day [P25] or older) or adult male and female rats (200–250 g) can be implanted with stainless-steel bipolar electrodes 11 mm in length by using stereotaxic procedures to target the posterior lateral hypothalamus. Rats are initially anesthetized with an isoflurane–oxygen vapor mixture and the depth of anesthesia determined by toe pinch. Anesthesia is maintained on isoflurane throughout the procedure.

2. Start a new surgery day with sterile instruments. All metal instruments, electrodes, and skull screws can be autoclaved. Between surgeries, clean the instruments with isopropyl alcohol and resterilize the instruments with a tabletop glass bead sterilizer.

3. Once anesthetized, the rats are prepared for aseptic surgery by removing the hair from the dorsal surface of the cranium with a narrow blade trimmer. The exposed area is then properly cleaned with Betadine solution followed by 70% isopropyl alcohol. Repeat cleaning procedure three times.

4. The rats' body temperature is maintained throughout the procedure with a "circulating water" heating pad that is automatically controlled and maintained and at 35 °C, approximately 2.5 °C below the body temperature of the rat. A nonmedicated, lubricating ophthalmic ointment is placed in the eyes.

5. The rat is then placed and secured in a model 940 Kopf stereo-taxic frame with the incisor bar set 5.0 mm above the interaural line (*see* **Note 1**). Coordinates for the stereotaxic procedures are obtained from the Paxinos and Watson rat brain atlas [63]. Stereotaxic coordinates for the electrode placements targeting the medial forebrain bundle are −0.5 mm anterior–posterior (AP), ±1.7 mm medial–lateral (ML) to bregma, and −8.3 mm dorsal–ventral (DV) from dura. The same coordinates can be used for adult male and female rats. Electrodes can also be implanted into adolescent rats (Postnatal Day 25), but then the coordinates need to be adjusted (−0.5 mm AP, ±1.48 mm ML, −8.3 DV).

6. A small incision (1 cm) is made in the skin overlying the skull and Bregma is exposed for subsequent determination of the implantation site. Retract the skin to allow adequate exposure of the bone surface.

7. With a curette gently scrape the skull to remove any connective tissue from the exposed skull. Wipe the skull with a sterile cotton swab to ensure that the surface is dry. Before proceeding with the rest of the surgery, stop bleeding from the skull or surrounding tissues with a cautery pen.

8. Use a sterile metal cutter (ultraflush cutter) to cut the electrode at 11 mm and use a sterile scalpel blade to gently separate the two coated wires of the electrode (*see* **Note 2**). The electrodes can also be purchased precut with the tips separated.

9. Secure the electrode to the stereotaxic apparatus with the electrode holder (*see* **Note 3**).

10. Determine the coordinates of bregma (intersection of coronal and sagittal suture). Once at Bregma move to the proper anterior–posterior and medial–lateral coordinates for electrode placement. Mark gently with a surgical pen or cautery pen.

11. Mark four more areas for placement of anchor screws (*see* **Note 4**).

12. Drill five holes (0.5 mm in diameter) through the cranial surface with a microdrill. The drill bit should only penetrate the thickness of the bone (*see* **Note 5**). If there is bleeding, use sterile cotton swabs to dry the surface of the skull.

13. Thread sterile microscrews into the four holes designated for the anchor screws leaving a thread or two of space between the head of the screw and the skull (*see* **Note 6**).

14. Lower the electrode to the dorsal–ventral coordinates targeting the posterior lateral hypothalamus.

15. Carefully apply dental cement around the electrode and screws with a spatula. Allow the cement to seep between the head of the screw and the skull (*see* **Note 7**).

16. Mold the dental cement to form a mushroom shaped cap. Smooth all the edges of the cap (*see* **Note 8**).

17. Allow the cap to harden. If the skin closes tightly around the cap, then the surgery is completed. However, if there is excessive space between the skin and the cap then add sutures anterior and/or posterior of the cap (sterile nonabsorbable monofilament 3-0 suture).

18. Following completion of surgery, apply Betadine ointment (10% povidone-iodine) to the incision area.

19. Administer Carprofen (5 mg/kg, sc) every 24 h for 2 days.

20. Allow a week for recovery before initiating the ICSS training.

**6.2 Intracranial Self-Stimulation Procedure**

1. A lead is connected to the electrode, and the rat is placed in an operant conditioning chamber. The operant conditioning chamber is place in a sound-attenuating chamber to prevent that noise distracts the animals during ICSS training and testing.

2. The lead will be connected to a stimulator, which generates electrical stimuli of different intensity levels (100 Hz of 0.1 ms rectangular cathodal pulses, 500 ms train duration, variable current intensity with levels ranging between 50 and 250 μA).

3. A computer program controls the stimulator and records the responses of the subject.

4. There are three stages of training for this operant procedure, simple fixed-ratio (FR) schedule of reinforcement, discrete-trial current-threshold procedure and modification of the psychophysical method of limits.

5. During the initial phase, rats are trained to respond on a FR1 schedule of reinforcement to turn a wheel manipulandum (5 × 7 cm; $W \times H$) embedded in the wall of the experimental chamber.

6. Each quarter turn of the wheel results in a delivery of a 0.5 s train of 0.1 ms cathodal square-wave pulses at a frequency of 100 Hz.

7. The training schedule can be repeated up to two times per day. However, training should be restricted to less than an hour per day. Successful acquisition of responding for stimulation on this FR1 schedule is defined as 100 reinforcements within 10 min.

8. Upon completion of the initial training stage, rats begin the second stage of training on a discrete-trial current-threshold procedure. This stage is divided into three phases in which the inter-trial interval and delay periods induced by time-out are

gradually increased until the animal acquires behavior during each part.

9. For all phases, each trial begins with the delivery of a noncontingent electrical stimulus, followed by a 7.5 s response window during which the animal can respond to receive a second contingent stimulus that is identical to the initial noncontingent stimulus.

10. A response during this 7.5 s response window is labeled as a positive response, while the lack of a response is labeled as a negative response. During the 2 s period immediately after a positive response, additional responses have no consequences.

11. During the first phase of training, there is a 3 s interval between trials, and responding during this interval results in a 1 s penalty delay. Rats should complete a minimum of 200 trials per session (*see* **Note 9**).

12. When subjects receive positive responses in 90% of the trials in 3 consecutive sessions, the subject can be moved to the second phase of training for this stage. During this phase, there is a 5 s interval between trials, and responding during this interval results in a 2 s penalty delay. Rats should complete a minimum of 200 trials per session (*see* **Note 10**).

13. When subjects receive positive responses in 90% of the trials in 3 consecutive sessions, the subject can be moved to the third phase of training for this stage. During this phase, there is a 10 s interval between trials, and responding during this interval results in a 5 s penalty delay. Rats should complete a minimum of 200 trials per session. During this phase, sessions are typically 30–40 min.

14. The third and final stage of training determines brain reward thresholds by using a modification of the psychophysical method of limits. The parameters used in this method are the same as those for the experimental phase.

15. During this phase, test sessions consist of four alternating series of descending and ascending current intensities starting with a descending series.

16. At the start of each trail, the subject receives a noncontingent sinusoidal electrical stimulus of 250 ms duration and 60 Hz frequency.

17. Subjects are required to rotate the wheel manipulandum one-quarter of a rotation to receive a contingent stimulus identical to the previously delivered noncontingent stimulus during a 7.5 s response window. This is defined as a positive response.

18. During the 2 s period immediately after a positive response, additional responses have no scheduled consequences.

19. If the subject does not respond within the 7.5 s, this is defined as a negative response.

20. Following a positive or negative response, there is an intertrial interval that averages 10 s (7.5–12.5 s). Following the intertrial interval the trial was terminated.

21. Any responding during the intertrial interval resulted in a 10 s delay before the start of the next trial.

22. Blocks of three trials are presented to the subject at a given stimulation intensity, and the intensity is altered systematically between blocks of trials by 5 µA steps.

23. The initial stimulus intensity is set 40 µA above the baseline current-threshold for each animal. Each test session typically lasts 30–40 min and provides two dependent variables for behavioral assessment: brain reward thresholds and response latencies.

24. The current threshold for a descending series was defined as the midpoint between stimulation intensities that supported responding (i.e., positive responses on at least two of the three trials) and current intensities that failed to support responding.

25. The threshold for an ascending series was defined as the midpoint between stimulation intensities that did not support responding and current intensities that supported responding for two consecutive blocks of trials.

26. Four threshold estimates were recorded, and the mean of these values was taken as the final threshold.

27. The time interval between the beginning of the noncontingent stimulus and a positive response was recorded as the response latency. The response latency for each test session was defined as the mean response latency on all trials during which a positive response occurred.

28. The rats are tested daily (30 min per session) and the total duration of the experiment is approximately 2 months. A computer will record the positive responses of the rats.

29. Rats should be tested until the threshold current is stable and there is less than 20% variability.

30. When the brain reward thresholds are stable, the rats can be prepared with osmotic minipumps to induce nicotine dependence.

### 6.3 Osmotic Minipump Implantations

1. Rats are implanted with osmotic minipumps filled with either saline or nicotine salt dissolved in saline subcutaneously utilizing aseptic surgical techniques.

2. This surgery is mostly done after the implantation of the electrodes, usually after the rats have undergone ICSS training and are stable on the ICSS procedure.

3. Osmotic minipumps are prepared immediately before implantation. The pumps should be filled with sterile saline for the control groups. Handle the pumps only with sterile gloves in order to maintain sterility.

4. The nicotine concentration must be adjusted to compensate for differences in body weight to ensure delivery of 9 mg/kg of nicotine salt per day (3.16 mg/kg/day nicotine base).

5. Inject sterile saline or the nicotine solution into the reservoir of the pump and then insert the flow moderator.

6. Rats are anesthetized with an isoflurane–oxygen vapor mixture (1–3% isoflurane) and depth of anesthesia determined by toe pinch. Anesthesia is maintained on isoflurane (1–2%) throughout the procedure.

7. Once anesthetized, the rats are prepared for aseptic surgery by removing the hair from one side flank with a #40 surgical clipper blade. The exposed area is then cleaned with Betadine solution followed by 70% isopropyl alcohol. Repeat cleaning procedure three times.

8. Subject's body temperature is maintained throughout the procedure with a "circulating water" heating pad that is automatically controlled and maintained and at 35 °C, approximately 2.5 °C below the body temperature of the rat. A nonmedicated, lubricating ophthalmic ointment is placed in the eyes.

9. Make a 1.5 cm skin incision posterior of the ribcage.

10. Make a subcutaneous pocket by separating the muscle and subcutaneous layer with sterile blunt scissors.

11. Place the tip of the scissors between the muscle and subcutaneous layer. Apply gentle pressure and gently open and close the scissors to excavate the pocket (*see* **Note 11**).

12. Remove the scissors in the open position to prevent the tearing of tissue. Repeat until the pocket is large enough to accommodate the pump (*see* **Note 12**).

13. Insert the minipump in the pocket.

14. Close the incision with staples (stapler, staple remover, and staples; Jorgensen Labs, Loveland Co 80538, USA) (*see* **Note 13**).

15. Apply Betadine ointment (10% povidone-iodine) to the incision area.

16. Administer Carprofen (5 mg/kg, sc) every 24 h for 2 days.

17. The duration of this surgery is approximately 10 min per rat.

**6.4 Somatic Withdrawal Signs**

1. Rats are habituated to the Plexiglas observation chamber (25 × 25 × 45 cm; $L \times W \times H$) for 5 min per day on 2 consecutive days prior to testing.

2. Rats are injected with 2 mg/kg of the nicotinic receptor antagonist mecamylamine to induce withdrawal. The injections should be administered 10 min before observing the somatic signs.

3. Rats are placed in the observation chamber and observed for 10 min.

4. The following somatic signs should be recorded based on the checklist of nicotine abstinence signs.
   (a) Body shakes.
   (b) Cheek tremors.
   (c) Escape attempts.
   (d) Gasps.
   (e) Genital licks.
   (f) Head shakes.
   (g) Ptosis (*see* **Note 14**).
   (h) Teeth chattering.
   (i) Writhes.
   (j) Yawns.

5. The total number of somatic signs is defined as the sum of the individual occurrences.

6. For the final statistical analyses, the signs should be divided into the following categories.
   (a) Abdominal constrictions—includes gasps and writhes.
   (b) Shakes—includes head shakes and body shakes.
   (c) Facial fasciculations—includes cheek tremors, teeth chattering, ptosis, yawns.
   (d) Other—include signs that may occasionally occur such as escape attempts and genital licks.

## 7 Notes

1. Proper placement of the incisor bar is important for correct placement of the electrode.

2. The space between the electrode ends should just be wide enough to allow the scalpel blade to slip between the two ends. Do not fray the ends of the electrode by manipulating the ends too vigorously. Do not bend the electrode.

3. Make sure the electrode is straight in all angles. Gently straighten the electrode with sterile forceps.

4. To ensure a secure anchor, the screws should be placed on different skull plates. The screw placements must be far enough from the electrode placement to allow for proper placement of the top of the electrode.

5. The dorsal–ventral coordinates are measured from the level of dura. When creating the hole for the electrode, be sure not to damage dura to accurately measure this coordinate.

6. These screws are necessary to help anchor the electrode and only penetrate the thickness of the bone (approximately 1 mm). Do not thread screws flush to the skull. Leave space between the head of the screw and the skull (approximately a thread or two) to allow the dental cement and acrylic to seep between the screw and the skull. This will form the anchor necessary to hold the electrode in place.

7. The dental cement and acrylic should be mixed just prior to application.

8. Do not allow the cement to seep under the skin. Use the spatula to control the flow of the cement. The cement should not be too thin to manipulate but must not be too thick to form a seal around the electrode and screws. Do not leave any sharp edges on the head cap as these can irritate the skin.

9. Training can be repeated up to three times per day to maximize training efforts. However, training sessions should not exceed 1 h. The current may need to be increased or decreased if the rat is not responding.

10. These sessions are slightly longer but can be repeated up to two times per day to maximize training efforts.

11. Be careful not to puncture the muscle. This will cause damage to the muscle and will prolong recovery time.

12. The pocket should be large enough to allow positioning of the pump away from the site of the incision. If the pump rests on the incision site, it will erode through the skin through the site of the incision.

13. Closing the incision with staples is faster than closing the incision with sutures and the rats are less likely to remove staples than sutures.

14. Ptosis should be counted once per minute if present continuously.

## 8 Conclusion

The ICSS method is an outstanding method to investigate sex differences in the rewarding effects of drugs of abuse and drug withdrawal. The brain reward thresholds provide insight into the effects of drugs on the brain reward system. The response latencies indicate if a drug has stimulating or sedative effects. The ICSS method is particularly well suited to determine the rewarding and aversive properties of psychostimulants. The sedative properties of alcohol and opioids may mask the rewarding properties of these drugs.

## References

1. UNODC (2018) World drug report 2018. United Nations Publication, Sales No. E.18. XI.9

2. Mounts NS, Steinberg L (1995) An ecological analysis of peer influence on adolescent grade point average and drug use. Dev Psychol 31 (6):915

3. Abelson J, Treloar C, Crawford J, Kippax S, Van Beek I, Howard J (2006) Some characteristics of early-onset injection drug users prior to and at the time of their first injection. Addiction 101(4):548–555

4. Ritchie M, Roser M (2019) Drug use. OurWorldInData.org

5. Smart RG, Ogborne AC (2000) Drug use and drinking among students in 36 countries. Addict Behav 25(3):455–460

6. Becker JB, Hu M (2008) Sex differences in drug abuse. Front Neuroendocrinol 29 (1):36–47

7. Becker JB, McClellan ML, Reed BG (2017) Sex differences, gender and addiction. J Neurosci Res 95(1–2):136–147

8. Gowing LR, Ali RL, Allsop S, Marsden J, Turf EE, West R, Witton J (2015) Global statistics on addictive behaviours: 2014 status report. Addiction 110(6):904–919

9. Sieminska A, Jassem E (2014) The many faces of tobacco use among women. Med Sci Monit 20:153

10. Reitsma MB, Fullman N, Ng M, Salama JS, Abajobir A, Abate KH, Abbafati C, Abera SF, Abraham B, Abyu GY (2017) Smoking prevalence and attributable disease burden in 195 countries and territories, 1990–2015: a systematic analysis from the Global Burden of Disease Study 2015. Lancet 389 (10082):1885–1906

11. Thalheimer W, Cook S (2002) How to calculate effect sizes from published research: a simplified methodology. Work-Learning Research, p 1

12. WHO (2015) WHO report on the global tobacco epidemic, 2015: raising taxes on tobacco. WHO, Geneva

13. Gentzke AS, Creamer M, Cullen KA, Ambrose BK, Willis G, Jamal A, King BA (2019) Vital signs: tobacco product use among middle and high school students—United States, 2011–2018. Morb Mortal Wkly Rep 68 (6):157

14. Cullen KA, Gentzke AS, Sawdey MD, Chang JT, Anic GM, Wang TW, Creamer MR, Jamal A, Ambrose BK, King BA (2019) e-Cigarette use among youth in the United States, 2019. JAMA 322:2095. https://doi.org/10.1001/jama.2019.18387

15. Leventhal AM, Strong DR, Kirkpatrick MG, Unger JB, Sussman S, Riggs NR, Stone MD, Khoddam R, Samet JM, Audrain-McGovern J (2015) Association of electronic cigarette use with initiation of combustible tobacco product smoking in early adolescence. JAMA 314 (7):700–707

16. Hamilton HA, Ferrence R, Boak A, Schwartz R, Mann RE, O'Connor S, Adlaf EM (2014) Ever use of nicotine and nonnicotine electronic cigarettes among high school students in Ontario, Canada. Nicotine Tob Res 17(10):1212–1218

17. Morean ME, Krishnan-Sarin S, O'Malley SS (2018) Assessing nicotine dependence in adolescent e-cigarette users: the 4-item Patient-Reported Outcomes Measurement Information System (PROMIS) Nicotine Dependence Item Bank for electronic cigarettes. Drug Alcohol Depend 188:60–63

18. Olds J, Milner P (1954) Positive reinforcement produced by electrical stimulation of septal area and other regions of rat brain. J Comp Physiol Psychol 47(6):419–427

19. Olds J (1958) Self-stimulation of the brain: its use to study local effects of hunger, sex, and drugs. Science 127(3294):315–324

20. Der-Avakian A, Markou A (2012) The neurobiology of anhedonia and other reward-related deficits. Trends Neurosci 35(1):68–77

21. Carlezon WA Jr, Chartoff EH (2007) Intracranial self-stimulation (ICSS) in rodents to study the neurobiology of motivation. Nat Protoc 2 (11):2987–2995. https://doi.org/10.1038/nprot.2007.441

22. Marcus R, Kornetsky C (1974) Negative and positive intracranial reinforcement tresholds: effects of morphine. Psychopharmacologia 38 (1):1–13

23. Stein L, Ray OS (1959) Self-regulation of brain-stimulating current intensity in the rat. Science 130(3375):570–572

24. Bruijnzeel AW, Lewis B, Bajpai LK, Morey TE, Dennis DM, Gold M (2006) Severe deficit in brain reward function associated with fentanyl withdrawal in rats. Biol Psychiatry 59 (5):477–480

25. Igari M, Alexander JC, Ji Y, Qi X, Papke RL, Bruijnzeel AW (2013) Varenicline and cytisine diminish the dysphoric-like state associated with spontaneous nicotine withdrawal in rats. Neuropsychopharmacology 39:455–465

26. Bruijnzeel AW (2012) Tobacco addiction and the dysregulation of brain stress systems. Neurosci Biobehav Rev 36:1418–1441

27. Bruijnzeel AW, Gold MS (2005) The role of corticotropin-releasing factor-like peptides in cannabis, nicotine, and alcohol dependence. Brain Res Rev 49(3):505–528

28. Foll BL, Goldberg SR (2009) Effects of nicotine in experimental animals and humans: an update on addictive properties. In: Henningfield JE, London ED, Pogun S (eds) Nicotine psychopharmacology. Springer, Berlin, pp 335–367. https://doi.org/10.1007/978-3-540-69248-5_12

29. Heishman SJ, Kleykamp BA, Singleton EG (2010) Meta-analysis of the acute effects of nicotine and smoking on human performance. Psychopharmacology 210(4):453–469

30. Benowitz NL (2009) Pharmacology of nicotine: addiction, smoking-induced disease, and therapeutics. Annu Rev Pharmacol Toxicol 49:57–71

31. Xue S, Behnood-Rod A, Wilson R, Wilks I, Tan S, Bruijnzeel AW (2018) Rewarding effects of nicotine in adolescent and adult male and female rats as measured using intracranial self-stimulation. Nicotine Tob Res 22:172–179. https://doi.org/10.1093/ntr/nty249

32. Lee AM, Calarco CA, McKee SA, Mineur YS, Picciotto MR (2019) Variability in nicotine conditioned place preference and stress-induced reinstatement in mice: effects of sex, initial chamber preference, and guanfacine. Genes Brain Behav 19:e12601

33. Lenoir M, Starosciak AK, Ledon J, Booth C, Zakharova E, Wade D, Vignoli B, Izenwasser S (2015) Sex differences in conditioned nicotine reward are age-specific. Pharmacol Biochem Behav 132:56–62. https://doi.org/10.1016/j.pbb.2015.02.019

34. Stratmann JA, Craft RM (1997) Intracranial self-stimulation in female and male rats: no sex differences using a rate-independent procedure. Drug Alcohol Depend 46(1):31–40

35. Legakis LP, Negus SS (2018) Repeated morphine produces sensitization to reward and tolerance to antiallodynia in male and female rats with chemotherapy-induced neuropathy. J Pharmacol Exp Ther 365(1):9–19

36. Lazenka M, Suyama J, Bauer C, Banks M, Negus S (2017) Sex differences in abuse-related neurochemical and behavioral effects of 3, 4-methylenedioxymethamphetamine (MDMA) in rats. Pharmacol Biochem Behav 152:52–60

37. Bain GT, Kornetsky C (1989) Ethanol oral self-administration and rewarding brain stimulation. Alcohol 6(6):499–503

38. Moolten M, Kornetsky C (1990) Oral self-administration of ethanol and not experimenter-administered ethanol facilitates rewarding electrical brain stimulation. Alcohol 7(3):221–225

39. Hughes JR, Gust SW, Skoog K, Keenan RM, Fenwick JW (1991) Symptoms of tobacco withdrawal. A replication and extension. Arch Gen Psychiatry 48(1):52–59

40. Hughes JR, Hatsukami D (1986) Signs and symptoms of tobacco withdrawal. Arch Gen Psychiatry 43(3):289–294

41. Wesnes K, Warburton DM (1983) Smoking, nicotine and human performance. Pharmacol Ther 21(2):189–208

42. Epping-Jordan MP, Watkins SS, Koob GF, Markou A (1998) Dramatic decreases in brain reward function during nicotine withdrawal. Nature 393(6680):76–79

43. Hildebrand B, Nomikos G, Bondjers C, Nisell M, Svensson T (1997) Behavioral manifestations of the nicotine abstinence syndrome

in the rat: peripheral versus central mechanisms. Psychopharmacology 129(4):348–356

44. Malin DH, Lake JR, Newlin-Maultsby P, Roberts LK, Lanier JG, Carter VA, Cunningham JS, Wilson OB (1992) Rodent model of nicotine abstinence syndrome. Pharmacol Biochem Behav 43(3):779–784

45. Malin DH, Lake JR, Carter VA, Cunningham JS, Hebert KM, Conrad DL, Wilson OB (1994) The nicotinic antagonist mecamylamine precipitates nicotine abstinence syndrome in the rat. Psychopharmacology 115 (1–2):180–184

46. O'Dell LE, Bruijnzeel AW, Smith RT, Parsons LH, Merves ML, Goldberger BA, Richardson HN, Koob GF, Markou A (2006) Diminished nicotine withdrawal in adolescent rats: implications for vulnerability to addiction. Psychopharmacology 186(4):612–619

47. Bruijnzeel AW, Markou A (2004) Adaptations in cholinergic transmission in the ventral tegmental area associated with the affective signs of nicotine withdrawal in rats. Neuropharmacology 47(4):572–579

48. Bespalov A, Lebedev A, Panchenko G, Zvartau E (1999) Effects of abused drugs on thresholds and breaking points of intracranial self-stimulation in rats. Eur Neuropsychopharmacol 9(5):377–383

49. Nakahara D (2004) Influence of nicotine on brain reward systems: study of intracranial self-stimulation. Ann N Y Acad Sci 1025 (1):489–490

50. Paterson NE (2009) The neuropharmacological substrates of nicotine reward: reinforcing versus reinforcement-enhancing effects of nicotine. Behav Pharmacol 20(3):211–225

51. Leith NJ, Barrett RJ (1976) Amphetamine and the reward system: evidence for tolerance and post-drug depression. Psychopharmacologia 46(1):19–25

52. Lin D, Koob GF, Markou A (1999) Differential effects of withdrawal from chronic amphetamine or fluoxetine administration on brain stimulation reward in the rat--interactions between the two drugs. Psychopharmacology 145(3):283–294

53. Markou A, Koob GF (1991) Postcocaine anhedonia. An animal model of cocaine withdrawal. Neuropsychopharmacology 4(1):17–26

54. Schulteis G, Markou A, Gold LH, Stinus L, Koob GF (1994) Relative sensitivity to naloxone of multiple indices of opiate withdrawal: a quantitative dose-response analysis. J Pharmacol Exp Ther 271(3):1391–1398

55. Schulteis G, Markou A, Cole M, Koob GF (1995) Decreased brain reward produced by ethanol withdrawal. Proc Natl Acad Sci U S A 92(13):5880–5884

56. Tan S, Xue S, Behnood-Rod A, Chellian R, Wilson R, Knight P, Panunzio S, Lyons H, Febo M, Bruijnzeel AW (2019) Sex differences in the reward deficit and somatic signs associated with precipitated nicotine withdrawal in rats. Neuropharmacology 160:107756

57. Rylkova D, Shah HP, Small E, Bruijnzeel AW (2009) Deficit in brain reward function and acute and protracted anxiety-like behavior after discontinuation of a chronic alcohol liquid diet in rats. Psychopharmacology 203 (3):629–640

58. Henricks AM, Berger AL, Lugo JM, Baxter-Potter LN, Bieniasz KV, Petrie G, Sticht MA, Hill MN, McLaughlin RJ (2017) Sex-and hormone-dependent alterations in alcohol withdrawal-induced anxiety and corticolimbic endocannabinoid signaling. Neuropharmacology 124:121–133

59. Varlinskaya EI, Spear LP (2004) Acute ethanol withdrawal (hangover) and social behavior in adolescent and adult male and female Sprague-Dawley rats. Alcohol Clin Exp Res 28 (1):40–50

60. Alele PE, Devaud LL (2007) Sex differences in steroid modulation of ethanol withdrawal in male and female rats. J Pharmacol Exp Ther 320(1):427–436

61. Liu J, Pan H, Gold MS, Derendorf H, Bruijnzeel AW (2008) Effects of fentanyl dose and exposure duration on the affective and somatic signs of fentanyl withdrawal in rats. Neuropharmacology 55(5):812–818

62. Cicero TJ, Nock B, Meyer ER (2002) Gender-linked differences in the expression of physical dependence in the rat. Pharmacol Biochem Behav 72(3):691–697

63. Paxinos G, Watson C (1998) The rat brain in stereotaxic coordinates, vol 4. Academic Press, San Diego, CA

# Chapter 2

# Animal Models of Adolescent Binge Drinking

Santiago Monleón, Rosa Redolat, Aránzazu Duque, Patricia Mesa-Gresa, and Concepción Vinader-Caerols

## Abstract

The main alcohol consumption pattern in adolescence is binge drinking (BD), characterized by intermittent consumption of large quantities of alcohol in short periods of time. BD has serious biomedical consequences and it is a prominent risk factor for later development of alcohol use disorders. Animal models can be extremely valuable for studying the neurobehavioral mechanisms and consequences of BD. This chapter focuses on the three main animal models of adolescent BD: the voluntary consumption paradigms "drinking in the dark" and "two-bottle choice," and the nonvoluntary BD method "chronic intermittent ethanol administration." A battery of behavioral tests is also described, with especial interest in evaluating the effects of alcohol BD on different kinds of memory. The use of these animal models of BD in combination with behavioral tasks will contribute to progress our understanding of the neurobehavioral consequences of BD, as well as to the potential development of prevention and treatment programs.

**Key words** Animal model, Adolescence, Binge drinking, Alcohol, Rats, Mice

## 1 Introduction

Alcohol is one of the most widely consumed psychoactive substances in the world, especially among adolescents and young adults [1–3]. Alcohol exposure in adolescence can lead to adverse cognitive sequelae that can persist into adulthood, as neuromaturation continues throughout this period [4–6] and brain is subjected to numerous structural and functional changes [7–9].

The main alcohol consumption pattern in adolescence is binge drinking (BD) which has been defined by the National Institute on Alcohol Abuse and Alcoholism as a consumption that raises the blood alcohol concentration (BAC) to 0.8 g/L or more [10]. This pattern, as it is shown in Fig. 1 (adapted from ref. 11), is characterized by heavy episodic drinking behavior involving the consumption of large quantities of alcohol in a short period (about 2 h), followed by a period of abstinence, with a variability between 1 week and 1 month.

María A. Aguilar (ed.), *Methods for Preclinical Research in Addiction*, Neuromethods, vol. 174,
https://doi.org/10.1007/978-1-0716-1748-9_2, © Springer Science+Business Media, LLC, part of Springer Nature 2022

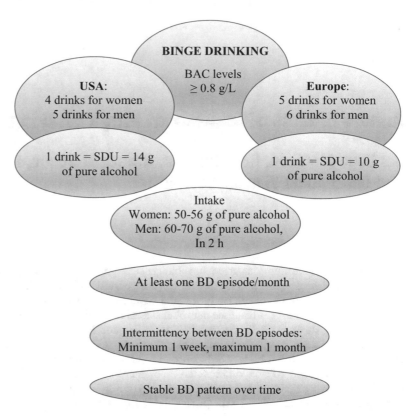

**Fig. 1** Binge drinking pattern criteria in humans. Number of drinks (1 drink = SDU: Standard Drink Unit) in USA and Europe for Binge Drinking's blood alcohol concentration (BAC) levels. Quantity: intake of 50–56 g of pure alcohol in women and 60–70 g in men, in 2 h. Frequency: at least one BD episode per month. Intermittency: abstinence between BD episodes over time (minimum 1 week, maximum 1 month). (Adapted from ref. 11)

It is noteworthy that young people who indulge in BD are frequently at a higher risk of developing an alcohol use disorder later in life (e.g., refs. 12–14). Thus, BD has become an important public health problem nowadays (e.g., ref. 15) and there is a critical need to improve our understanding of the neurobehavioral effects of this alcohol consumption pattern.

Although clinical research has provided important insights regarding the risks and consequences of BD, human studies cannot meticulously examine the effects of BD due to ethical limitations. Therefore, animal models (especially in rodents) can be extremely valuable in our understanding of the neurobehavioral mechanisms and consequences of BD [16, 17]. To date, many animal models have been developed for studying alcoholism in general. Traditionally, preclinical research in this field has used models of chronic ethanol drinking, but much less attention has been focused on binge ethanol drinking [18]. Recently, Jeanblanc et al. [17] have established the criteria for an animal model of BD, which include

(1) voluntary alcohol consumption, (2) the amount of intake required to reach the BAC or show signs of intoxication, (3) the speed of consumption, (4) the duration and intermittency of drinking patterns, (5) the development of brain and cognitive impairment as a result of drinking, and (6) the use of a model showing the interindividual differences of drinkers.

Among the preclinical models reproducing aspects of the BD alcohol consumption pattern, some of them have focused on voluntary binge-like ethanol consumption, such as the "drinking in the dark" (DID) and the "two-bottle choice" (TBC) paradigms (*see* ref. 16 for a review of voluntary BD models). An interesting review regarding animal models of BD has recently been published suggesting that these models based on voluntary intake could be more useful from a translational perspective [17]. These animal models could aid in the identification of the neurobiological and motivational factors related to impulsive alcohol drinking, especially in adolescents [17].

It is unlikely that a single rodent model can be developed to fully characterize ethanol abuse disorders and dependence. Therefore, much of the current work in the field relies on partial models that examine in detail specific facets of these disorders. For example, the two-bottle choice paradigm has been successfully used to examine ethanol preference, while other self-administration and relapse models have been used to investigate mechanisms contributing to the motivation to seek and consume ethanol. But these models do not accurately address the defining characteristics of binge ethanol drinking. Additionally, these self-administration methods usually require extensive training, the use of food and/or water deprivation, or sucrose fading techniques to elicit ethanol consumption. Also, rodents in these models do not always consume enough ethanol (or enough ethanol over a short enough period of time) to generate behavioral and/or pharmacological intoxication [18]. Therefore, the "chronic intermittent ethanol administration" (CIEA), a nonvoluntary BD model—without these handicaps—is complementary to those BD models and can be very useful for studying other aspects of BD consumption. The CIEA procedure seems to be a low-cost, simple, accessible behavioral protocol that efficiently induces BACs corresponding to BD.

Thus, the present chapter focuses on these three main animal models of adolescent binge drinking: the voluntary BD paradigms DID and TBC, and the nonvoluntary BD method CIEA.

On the other hand, animal models of ethanol administration have shown detrimental effects of alcohol on several behavioral tasks (e.g., refs. 4, 19–27). Nevertheless, much work remains to better understand the neurobehavioral mechanisms of BD and its consequences [28–29]. For this goal, a combination of an animal model of BD with diverse behavioral assessment seems to be useful. In this chapter, along with the main animal models of BD, a battery of behavioral tests will be described, with especial interest in evaluating the effects of alcohol BD on different kinds of memory. The use of animal models of BD in combination with behavioral tasks will contribute to progress our understanding of the neurobehavioral consequences of BD, as well as to the potential development of prevention and treatment programs.

## 2 Materials

### 2.1 Animals

Depending on the purpose and the experimental requirements of the study, rats and mice are commonly used from different suppliers, such as *Charles River*, *Harlan*, and *Janvier*. Other local and university sources of rodents may also be suitable. It is recommended to use the same strain of animals and the same supplier for the experiments of a series (*see* **Note 1**).

Adolescent, or periadolescent, animals are recommended for studying the neurobehavioral effects of BD (Fig. 2), as this consumption pattern is prevalent among adolescents and young adults in humans. Thus, differential effects of ethanol on memory have been reported in adolescent and adult rats [30]. Parallel developmental ages between rats and humans, as well as between mice and humans, have been estimated using behavioral and neurobiological milestones. Such equivalences are shown in Table 1 (Rats–Humans; adapted from ref. 31) and Table 2 (Mice–Humans, adapted from ref. 32).

It is important to consider the individual vulnerability of the animal model (rats or mice, strain, sex) used in the experiments. For example, Wistar rats show characteristics that make them interesting for this type of studies, since they seem more motivated to take alcohol when they are exposed to a discontinued self-administration model and they are not sensitive to the unpleasant consequences of consumption [33]. Different studies have

**Fig. 2** Adolescent mice are frequently used for evaluating neurobehavioral effects of BD and long-term effects can be observed in adult mice

**Table 1**
**Estimated parallel ages between the rat, associated developmental stage, and the human equivalent (adapted from ref. 31)**

| Rat ages (PNDs = postnatal days) | | | | | | | | |
|---|---|---|---|---|---|---|---|---|
| PNDs 1–7 | PNDs 8–21 | PNDs 21 | PNDs 22–27 | PNDs 28–42 | PNDs 43–60 | PNDs 61–75 | PNDs 76–90 | PNDs 90– |
| Human ages | | | | | | | | |
| Neonate | Prejuvenile | Weaning | Juvenile | Adolescent | Periadolescent | Early young | Young adult | Adult |
| −3 to 0 months | 0–6 years | 6 years | 7–12 years | 13–18 years | 18–21 years | 21–24 years | 25–28 years | 28 years |

**Table 2**
**Estimated parallel ages between the mouse, associated developmental stage, and the human equivalent (adapted from ref. 32)**

| Mouse ages (PND = postnatal day) | PND 0–28 | PND 42 (6 weeks) | PND 70 (10 weeks) | PND 450 (15 months) | PND 540 (18 months) |
|---|---|---|---|---|---|
| Human ages | Weaning | Puberty | Adulthood | Reproductive senescence | Old age |
| | 0–6 months | 11.5 years | 20 years | 51 years | 65 years |

established that inbred rodents can bring greater benefits to BD research, such as Long Evans, Fischer and Wistar rats or C57BL/6 mice [17]. In fact, many studies have been conducted with these strains of rodents [34]. Although male subjects are typically used, studies including females—along with males—may be of particular relevance (*see* **Note 2**).

Animals must be single housed (at least 1 week prior to initiation of the alcohol consumption period) in the DID and TBC paradigms, while they are usually group housed in the CIEA model. Subjects are maintained in a temperature-controlled room (21 ± 2 °C) under a reversed light–dark cycle (*see* **Note 3**).

*2.2 Ethanol Solutions*    Ethanol (95%) diluted with physiological saline (0.9% NaCl) to a final concentration of 20% ethanol v/v (e.g., 210.5 mL 95% ethanol to make 1 L) is commonly used in this kind of studies (*see* **Note 4**). In the chronic intermittent ethanol administration paradigm, this solution is intraperitoneally administered in a volume of 1 mL/kg (0.001 mL/g) body weight for rats and 10 mL/kg (0.01 mL/g) for mice. The doses 3–4 g/kg of ethanol are usually selected in order to model the alcohol BD pattern (*see* **Notes 5** and **6**).

### 2.3  Apparatus

#### 2.3.1  Drinking in the Dark (DID)

Following Thiele et al. [35], the DID paradigm is performed using the following materials: standard mouse cages made of clear polycarbonate, with stainless-steel wire bar lid and food hopper; water bottles with sipper tube; constructed ball-bearing sipper tubes; a balance/scale; and binder clips (Fig. 3). Each sipper tube is made from a 10 mL serological pipet, a ball-bearing sipper, a section of heat shrink tubing, and a silicone stopper (*see* ref. 35 for details on the construction process and components of a ball-bearing sipper tube).

#### 2.3.2  Two-Bottle Choice (TBC)

The TBC paradigm can be performed using identical materials to the basic DID procedure, except that one additional ball-bearing sipper tube per cage will be needed. Some modifications of the cage lid may be adopted to accommodate the extra bottle [35]. Standard cage water bottles can be also used (Fig. 4).

#### 2.3.3  Chronic Intermittent Ethanol Administration (CIEA)

The only material needed in the CIEA paradigm is a syringe with needle for administrating ethanol (Fig. 5).

In order to evaluate the effects of alcohol BD on different kinds of memory, as well as to control some potential confounding variables (*see* **Note** 7), a battery of behavioral tests is used in our laboratory (e.g., ref. 36). The common tasks for mice included in this battery are described below. The three first tasks are used in order to evaluate several kinds of memory (recognition memory, emotional memory, and spatial memory) and the rest of the tasks are included as complementary tests in order to evaluate several confounding variables (locomotor activity, anxiety, and analgesia) (*see* **Note 8**).

#### 2.3.4  Recognition Memory: Novel Object Recognition (NOR)

The NOR test consists of an open field arena (example measures: height 35 cm, width 30 cm, length 60 cm) made of translucent Plexiglas, with two identical novel objects (Training phase) and a different one (Test phase) (Fig. 6).

#### 2.3.5  Emotional Memory: Inhibitory Avoidance (IA)

The IA apparatus for mice (Ugo Basile, Comerio-Varese, Italy) consists of a cage made of Perspex sheets and divided into two compartments (both 15 cm high × 9.5 cm wide × 16.5 cm long). The chambers are separated widthwise by a flat-box partition with an automatically operated sliding door at floor level. The floor is made of stainless-steel bars of 0.7 mm in diameter and situated 8 mm apart. The starting compartment is white and continuously illuminated by a light fixture fastened to the cage lid (24 V, 10 W, light intensity of 290 lux at floor level, measured with the Panlux Electronic2 photometer, manufactured by GOSSEN, Nürnberg, Germany), whereas the "shock" compartment comprises black Perspex panels and is kept in darkness at all times. This IA apparatus is placed within an isolation box (Fig. 7).

**Fig. 3** Drinking in the dark: aerial and side view of the mouse cage set up for the basic DID procedure. (Source: ref. 35)

*2.3.6   Spatial Memory: Morris Water Maze (MWM)*

The MWM for mice (Cibertec S.A., Madrid, Spain) consists of a circular pool made of black Plexiglas (1 m diameter and 30 cm high). The maze is filled with water to a depth of 15 cm and maintained at $24 \pm 1$ °C. A small platform ($6 \times 6$ cm) is submerged 1 cm below the surface of the water in the target quadrant (Fig. 8). Several extramaze cues, including laboratory equipment and posters, are available around the pool.

*2.3.7   Complementary Tests*

1. *Actimeter.* Locomotor activity can be tested by an actimeter. A version for mice (ACTIMET from Cibertec S.A., Madrid, Spain) consists of an infrared photocell system where the locomotor activity of the animal is measured. The photocell line is

**Fig. 4** Two bottle choice: front view of the basic TBC cage

**Fig. 5** Chronic-Intermittent Ethanol Administration: Ethanol intraperitoneal injection to mice

located 2.5 cm above the floor. Each photocell box (8.5 × 17 × 35 cm) has 16 photocells located along its long side (Fig. 9). The animals' behavior is continuously recorded and accumulated every minute for 5 min.

2. *Elevated Plus-Maze*: Anxiety and activity can be tested by an Elevated Plus-Maze (EPM). An EPM for mice (Cibertec S.A., Madrid, Spain) consists of two open arms ($30 \times 5$ cm² each) and two closed arms ($30 \times 15 \times 5$ cm³ each) which all fed into a common central square ($5 \times 5$ cm²). The maze is made of Plexiglas (black floor and walls) and is elevated 45 cm above the floor level (Fig. 10).

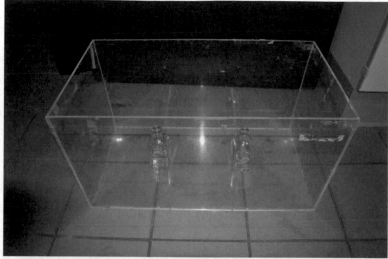

**Fig. 6** Novel object recognition test: front view of the NOR arena for training and test phases

3. *Hot Plate*: Pain sensitivity can be tested by a hot plate. A hot plate apparatus for mice (Mod. Socrel DS37, Ugo Basile, Varese, Italy) consists of a metal plate ($25 \times 25$ cm$^2$) located above a thermoregulator and a plastic cylinder (height 18 cm and diameter 19 cm) made of Plexiglas (Fig. 11).

## 3   Methods

### 3.1   Drinking in the Dark (DID)

The general DID procedure implemented by Rhodes et al. [37] in singly housed C57BL/6J mice (or other strains under consideration) involves replacing the water bottle with a bottle containing

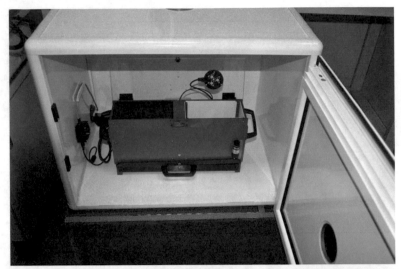

**Fig. 7** Inhibitory avoidance: IA apparatus (a step-through version of inhibitory avoidance conditioning for mice, placed inside an isolation box)

an ethanol solution (20%) and maintaining the mice on a 12-h light/dark cycle with limited access to ethanol over 4 consecutive days. This is a simple method to facilitate ethanol drinking in genetically predisposed mice. In this way, the water bottle is replaced with 20% ethanol for 2 or 4 h in the mouse's home cage, starting 3 h after lights shut off, and this procedure is repeated for 4 consecutive days. Mice are allowed access to the ethanol-filled ball-bearing DID sipper tubes for 2 h during Days 1–3; while on Day 4 of the sipper tube is generally left in place for 4 h (*see* **Note 9**). This animal model of binge-like ethanol drinking involves a procedure that promotes high levels of ethanol drinking and pharmacologically relevant blood ethanol concentrations in ethanol-

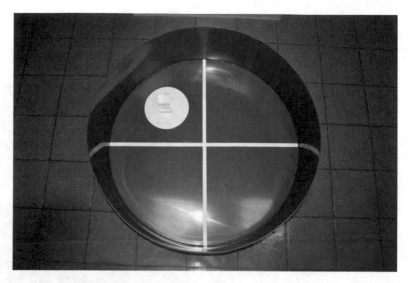

**Fig. 8** Morris water maze: aerial view of the MWM apparatus

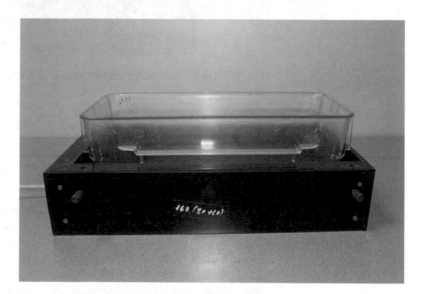

**Fig. 9** Actimeter: infrared photocell version of actimeter. This apparatus registers the spontaneous locomotor activity of the animals by means of an infrared photocell system

preferring strains of mice [38]. The most important factors for achieving binge-like levels of ethanol drinking and high BACs are: (a) the use of a line of mice with high ethanol preference (e.g., C57BL/6J mice or F1 hybrid cross between C57BL/6J and FVB/NJ), (b) limited access (2–4 h) to an ethanol solution in place of water, and (c) initiation of ethanol access which begins 3 h into the dark cycle. With this protocol, C57BL/6J mice

**Fig. 10** Elevated plus-maze: EPM apparatus for mice. This test measures anxiety and activity

typically exhibit binge-like drinking patterns that are associated with BACs greater than 100 mg/dl [35].

*3.2 Two Bottle Choice (TBC)*

In the TBC model, animals are usually first presented with two bottles of water, later replacing one water solution bottle by another containing increasing percentages of alcohol [39]. Long-term exposure to intermittent alcohol access in this paradigm induces binge-drinking, potentiated alcohol preference and high blood alcohol concentrations [40]. As in the DID protocol, animals must be singly housed. Three hours into the dark cycle, the home cage water bottles are removed and two new bottles (or 2 ball-bearing sipper tubes), one filled with ethanol solution and the other with fresh water, are inserted into the wire cage (*see* **Notes 10** and **11**). Animals are allowed access to these bottles for 2 h, at which point the amount of fluid consumed is read and recorded. On Day 4 of the procedure the bottles/sipper tubes are generally left in place for 4 h before fluid volumes are read and recorded [35] (*see* **Notes 12–15**).

*3.3 Chronic Intermittent Ethanol Administration (CIEA)*

Essentially, CIEA consists of the administration in rats or mice of intraperitoneal injections of ethanol (usual doses: 3–4 g/kg) for several consecutive days (usually 2 or 3 days) each week, in alternation with several days without injections, during several weeks (e.g.,

**Fig. 11** Hot plate: HP apparatus for mice. This test measures analgesia

refs. 21, 22, 41–45) (*see* **Notes 16–18**). A standard timeline of this protocol used in our laboratory [36, 46] is shown in Fig. 12.

Several versions of this model can be found in the literature, varying the injections and abstinence intervals. Thus, the administration pattern used by other authors consists on giving one dose of ethanol at 48 h intervals (i.e., every other day) over a 20 day period (e.g., ref. 41); one injection of ethanol every 72 h for 14 days (e.g., ref. 43); injections for 2 consecutive days, with gaps of 2 days without injections, during a 2-week period (e.g., ref. 44) or a 30 day period (e.g., ref. 45) (one of this variations is shown in Fig. 13).

As mentioned above, a battery of behavioral tests is used in our laboratory in order to evaluate the effects of alcohol BD on several kinds of memory and to control some potential confounding variables. The general procedure for each task of this battery is described below. Standard values of the different parameters are given.

| PND 23 | PND 24 | PND 25 | PND 26 | PND 27 | PND 28 | PND 29 | PND 30 | PND 31 | PND 32 | PND 33 | PND 34 | PND 35 | PND 36 |
|---|---|---|---|---|---|---|---|---|---|---|---|---|---|
|  |  |  |  |  |  |  | 💉 | 💉 | 💉 |  |  |  |  |

**Habituation Phase**                                                    **Treatment Phase**

| PND 37 | PND 38 | PND 39 | PND 40 | PND 41 | PND 42 | PND 43 | PND 44 | PND 45 | PND 46 | PND 47 | PND 48 | PND 49 | PND 50 |
|---|---|---|---|---|---|---|---|---|---|---|---|---|---|
| 💉 | 💉 | 💉 |  |  |  |  | 💉 | 💉 | 💉 |  |  |  |  |

**Treatment Phase**

| PND 51 | PND 52 | PND 53 | PND 54 |
|---|---|---|---|
| ACT | ACT | IA Test | HP |

**Behavioral Tests**

**Fig. 12** Standard experimental timeline of a CIEA protocol. Ethanol injections (e.g. 3 g/kg EtOH) are administered for 3 consecutive days each week for 3 weeks. *PND* post-natal day, *ACT* actimeter, *EPM* elevated plus maze, *IA Tr* inhibitory avoidance, training phase, *IA Test* inhibitory avoidance, test phase, *HP* hot plate

| PND 25 | PND 26 | PND 27 | PND 28 | PND 29 | PND 30 | PND 31 | PND 32 | PND 33 | PND 34 | PND 35 | PND 36 | PND 37 | PND 38 |
|---|---|---|---|---|---|---|---|---|---|---|---|---|---|
| 💉 | 💉 |  |  | 💉 | 💉 |  |  | 💉 | 💉 |  |  | 💉 | 💉 |

**Treatment**

| PND 39 | PND 40 | PND 41 | PND 42 | PND 43 | PND 44 | PND 45 | PND 46 | PND 47 | PND 48 | PND 49 | PND 50 | PND 51 | PND 52 |
|---|---|---|---|---|---|---|---|---|---|---|---|---|---|
| 💉 | 💉 |  |  | 💉 | 💉 |  |  | 💉 | 💉 |  |  | 💉 | 💉 |

**Treatment**

| PND 53 | PND 54 |
|---|---|
|  |  |

**Fig. 13** An alternative version of CIEA protocol: a 2-day-on, 2-day-off regimen during adolescence (PND 25–54) for a total of 16 exposures (e.g., ref. 45). *PND* post-natal day

### 3.4 Novel Object Recognition (NOR)

The NOR protocol usually consists of three phases.

- Habituation phase. Mice are habituated to exploring the open field for 5 min.

- Training phase. 1 day after Habituation, two identical novel objects are placed in the arena and mice are allowed to explore them for 10 min.

- Test phase. 1 h after Training, mice explore a novel object and a familiar one explored previously for 5 min. The location of the objects is counterbalanced across the trials in order to avoid preference for location.

All test sessions are recorded by a video camera and subsequently behaviorally analyzed by a blind researcher using appropriate software. An object is considered to have been explored when the head of the animal is 0.5 cm from the object or when it touches the object. In contrast, exploration is considered not to have taken place when an animal climbs onto an object or used it as a base to explore the environment. In order to measure recognition memory during the test phase, a discrimination index (total time spent exploring the novel object/total time devoted to exploration of both objects) is calculated for each group.

### 3.5 Inhibitory Avoidance (IA)

The IA (also called passive avoidance) task is one of the most common procedures for evaluating memory in animals (e.g., ref. 47), as this task is learned in a single trial, which facilitates the timing of drug administration. This is crucial in discriminating the effects of a drug on different memory processes, such as acquisition, consolidation or retrieval. The procedure consists of two phases:

- Training phase. After a period (e.g., 90 s) of adaptation to the light compartment, the door to the other compartment is opened for a maximum of 300 s. When the animal enters the dark compartment it receives an inescapable footshock of 0.3–0.5 mA that is delivered for 5 s.

- Test phase. Animals are placed once again in the light compartment of the apparatus and the procedure used in the training phase is repeated, but without the shock. The time taken to enter the dark compartment, defined as latency, is automatically measured in tenths of a second and recorded manually at the end of each phase. Crossing latencies longer than 300 s in the test phase resulted in the trial being terminated and a latency of 300 s recorded.

### 3.6 Morris Water Maze (MWM)

The MWM is employed in order to evaluate the effects on spatial learning. The typical procedure consists of two phases:

- Acquisition phase. Subjects perform 4 trials per day for 4 consecutive days. After an intertrial interval of 30 s the trial begins by placing the animal on the platform for 30 s. Mice are then placed in the water with their noses pointing toward the wall at one of the three starting points in a random manner. During this phase animals are allowed 60 s to find the hidden platform. If unable to do so, they are led to it by the experimenter. Animals are allowed to stay on the platform for 30 s, regardless of whether they had found it independently or after guidance. Starting positions are chosen at random from the three possible sites around the pool's perimeter, which are situated in each of the quadrants not occupied by the platform. The starting positions are determined so that two successive trials never begin from the same position.

- Retention phase (probe trial). On the fifth day the platform is removed and animals are allowed to swim for 100 s after starting in the opposite quadrant to that in which was the platform during acquisition. The probe trials are recorded by video camera. The measures obtained are escape latency (time to reach the submerged platform) during the acquisition trials and search time in each quadrant during the probe trial.

*3.7*
*Complementary Tests*

1. *Actimeter*: This infrared photocell system is used to evaluate the locomotor activity of the rodents. The animals' behavior is continuously recorded and accumulated every minute for a period of time (e.g., 30 min), obtaining standard units of activity.

2. *Elevated Plus-Maze (EPM)*: The EPM is used to measure anxiety and activity. The EPM protocol usually consists of a 5 min period which begins with the subject being placed in an open arm (facing the central square). All sessions are recorded by a standard video camera for subsequent analysis. The maze is cleaned after each subject. The number of entries into open and closed arms (arm entry is defined as all four paws entering an arm) is scored by a trained observer who is unaware of the treatment applied. This provides a measurement of anxiety, the percentage of open arm entries [(open/open + closed) × 100], and a measurement of activity (number of closed arm entries).

3. *Hot Plate*: The hot plate is used to assess nociceptive sensitivity. The metal plate is heated through a thermoregulator to a fixed temperature of 55 °C (the surface temperature is continuously monitored). Each mouse is placed on the hot plate inside a plastic cylinder to confine it to the heated surface. The latency to lift one or both hind paws is recorded in seconds and provides a nociceptive measurement (the lower the score, the higher the nociception). Animals that fail to lift their paws within 45 s are removed from the plate (to avoid thermal injury) and are assigned a response latency value of 45 s.

# 4 Notes

1. When performing a series of experiments, in order to make interexperiment comparisons and to draw broader conclusions, it is important not only to use the same strain of animals (a given rodent line may not provide the same effects as another line) but also to get these subjects from the same supplier: it has been reported that the behavioral performance of mice of the same strain can vary depending on the supplier [48].

2. It is especially relevant to include females in this kind of studies. As there is evidence showing that female adolescents are especially vulnerable to the neurotoxic effects of alcohol on

cognition (e.g., refs. 49, 50), researchers should include both male and female animals in their experiments in this field.

3. In the DID and TBC paradigms animals should be housed in the same room, with the same ambient temperature and 12-h light/dark cycle that they will ultimately be tested in the ethanol consumption. These paradigms take advantage of the nocturnal nature of mice, being assessed early during the dark phase of the light–dark cycle; this involves housing animals on a reverse 12-h light–dark cycle (i.e., with lights off in the morning, and lights coming back on again 12 h later in the late afternoon or early evening) so that assessment of drinking can be conducted during the normal workday for lab personnel. As these procedures are executed in a darkened colony room, a dim red light may be used to allow the experimenter to navigate the room [35].

4. A 20% ethanol solution is most commonly used in these models. Using concentrations of ethanol that are too weak (i.e., below 10%) with the DID and TBC procedures may prevent mice from consuming enough ethanol in the limited 2- to 4-h period of time to achieve binge-like BAC levels; while concentrations greater than 20% have an increasing tendency to leak [35].

5. In order to model the alcohol BD pattern, the doses 3–4 g/kg of ethanol are recommended in the CIEA model. Lower doses than 3 g/kg do not guarantee to reach the BACs corresponding to BD. This is more relevant for mice, which metabolize ethanol somewhat more quickly than rats and which require higher ethanol doses and BACs than rats to show equivalent behavioral responses [51].

6. The animal's health status must be checked after the alcohol administration. Given the hypothermia produced by high doses of ethanol, room temperatures below normal should be avoided [52].

7. In order to exclude that the observed effects of alcohol BD on cognition are not secondary to other behavioral effects of alcohol (e.g., motor effects), it is necessary to carry out complementary tests for controlling these potential confounding variables. Thus, when significant differences in these measures are observed between the treatment groups, a time period (several days) between the end of treatment and the start of the cognition tasks is necessary so that these differences disappear.

8. When a set of behavioral tests is run, the results of each task can be influenced by experience acquired in the previous tests of the battery. Therefore, it is recommended to random the tasks

order and to use a different order for different groups of animals.

9. Alternative periods of ethanol intake can be employed. For example, in cases where a pharmacological treatment is given prior to the alcohol intake period, it may be advantageous to assess intake in a shorter access period, particularly if the drug effect might be expected to wane before the end of a longer 4-h access period [35].

10. Animals may develop place/bottle preferences with the TBC procedure, regardless of the drinking solution in specific bottles. To avoid this potential confound, the position of the bottles must be swapped for each alcohol exposure session [39].

11. Because food may have been removed from the cage lid to accommodate the two bottles, the experimenter must decide whether food should be provided or remain absent during this period. If keeping food available is desired, a few food pellets can be placed in the bottom of the mouse cage [35].

12. In the DID and TBC paradigms it is advisable to quantify the loss of liquid due to evaporation. To estimate the amount of evaporation that may exist in the bottles, a bottle containing water and another with alcohol are placed in an empty cage each session [39].

13. Mice achieve lower BACs using the two-bottle variation relative to the one-bottle approach, limiting the potential utility of the TBC procedure in modeling binge-like ethanol drinking [35].

14. For the DID and TBC protocols several approaches have been used to produce a reliable demonstration of voluntary alcohol drinking. These include alcohol acclimatization (i.e., providing gradually increasing alcohol concentrations), taste adulteration (i.e., addition of sweeteners to the alcohol solution), and the use of prandial models, which take advantage of the postmeal drinking seen in rats [53].

15. In the DID and TBC paradigms, immediately upon withdrawal of bottles/sipper tubes, a blood sample is taken for later determination of blood ethanol content [35]. Although this is also convenient for CIEA, the dose of ethanol usually used in this model (3–4 g/kg) guarantees to a large extent reach BACs above the value established for BD. Thus, Nogales et al.—administering 3 g/kg of ethanol—found a mean BAC of 125 mg/dL, much higher than the BD criterion (80 mg/dL) [42].

16. In the CIEA model, it is recommended to administer the alcohol injections in a different room from the testing room,

in order to avoid any phenomenon of place conditioning which could affect the subject's performance.

17. Although rodent models of voluntary binge-like ethanol consumption show greater face validity, it must be noted that getting ethanol into animals does not require the face validity of the voluntary intake route to provide useful behavioral phenotypes associated with BD pattern. Thus, the CIEA model—in combination with a battery of behavioral tests—seems to be a good tool for studying the neurobehavioral consequences of BD.

18. The CIEA model, as well as some other BD paradigms, allows combining the alcohol injections/consumption with the administration of other drugs (either drugs of abuse or therapeutic drugs). This can be very useful for studying polyconsumption, a very common way of drug abuse; as well as for testing substances for potential prevention and treatment strategies.

## 5    Advantages and Limitations

A general advantage of using these animal models lies in the life cycle of rodents as the period of adolescence in these animals is quite short, and therefore the time for subjects to get used to binge drinking is reduced. Moreover, in relation to the described protocols, another positive point is that they are easy to be implemented by experimenters and they are also easy to be learned by animals. An inherent limitation of these paradigms is that it is difficult to use them to demonstrate the animal's motivation to obtain alcohol, as in operant models in which animals must perform a certain task to receive alcohol [53].

The DID and TBC paradigms are animal models of free consumption, which means that animals choose freely between drinking or avoiding alcohol without undergoing any injection and minimizing stress conditions to which animals are subjected in other animal models of BD [39]. On the other side, a major limitation associated with voluntary ethanol consumption procedures is that this consumption is typically weak, and often requires water-depriving the animals to incentivize drinking or initially pairing ethanol with a more salient reward [54]. Another common limitation is the risk of ethanol evaporation.

More specifically, the DID paradigm promotes pharmacologically meaningful BACs in an experimenter-defined limited access time frame. Perhaps one of the most salient advantages of the DID procedure is its high face validity in terms of a model of human binge drinking because mice exhibit binge-like drinking under conditions of short-term oral ethanol intake. This model easily allows assessing the effects of pharmacological compounds on binge-like ethanol intake, especially in cases where the actions of a drug are short-term. On the other hand, due to exclusive access to ethanol in a specific period of time, the DID procedure is related to several limitations: water is replaced with ethanol in the dark when high intake is observed in rodents; the alteration of normal eating patterns is frequent, which could lead to a nutritional deficit; and the lack of choice because animals have not access to an alternative source of liquid different from ethanol when the water bottle is removed from the animal's cage during this specific period of time, raising the possibility that the high levels of ethanol intake stems from the mice being forced to choose between ethanol consumption or thirst resulting from fluid deprivation [38]. Moreover, only C57BL/6J strain achieves high ethanol plasma concentrations, so the use of the DID protocol is more appropriate in this strain of mice [51, 55].

When it comes to CIEA procedure, a clear advantage is that the experimenter knows the exact dose which is being administrating to animals, together with the fact that the influence of taste can be avoid [53]. Moreover, the CIEA model, as well as some other BD paradigms, allows for combining the alcohol injections with the administration of other drugs (either drugs of abuse or therapeutic drugs). A critical limitation of this model is that it is a forced consumption paradigm, which could make animals motivation to drink ethanol questionable, even more than in the previously described protocols.

## 6   Future Research

Studies evaluating the impact of adolescent intermittent ethanol exposure on behavior and neurobiology in preclinical models could contribute to a better understanding of long-term consequences of binge drinking in humans [56]. Translational models both in rodents [57] and in rhesus monkeys [58] could aid to investigate developmental brain changes linked to alcohol abuse during adolescence. In prior studies intermittent exposure to ethanol during adolescence has been linked to reductions of adolescent hippocampal neurogenesis that could continue during adulthood even after ethanol cessation. Recent studies could contribute to a better understanding of the main mechanisms (epigenetic, neurodegenerative, neuroimmune, etc.) underlying to these changes observed on the developing brain following exposure to binge drinking [9, 59]. The preclinical imaging assessment of the effects of alcohol on the developing brain during adolescence could also aid to a better understanding of the neurobiological changes after intermittent alcohol exposure during diverse developmental periods [29, 60].

**Fig. 14** Environmental enrichment may have a protective and therapeutic role for ethanol binge-drinking in mice [64]

Recent studies can contribute to a better knowledge of the neural circuits related to compulsive drinking [61]. Taking into account this new research it would be necessary to further explore individual differences, both at behavioral and neurobiological levels, in compulsive drinking both in mice and in humans [62]. Furthermore, sex differences [56, 59], environmental factors [63, 64] epigenetic mechanisms [56], and individual vulnerability [33] should also be considered in future research because they could explain some of the differences obtained between laboratories [64]. These factors and the combination of multiple mechanisms could be at the base of the development of alcohol addiction [33]. For example, animal models could contribute to a better knowledge about factors that underlie gender differences in binge drinking, such as the influence of the age of heavy drinking onset and its interaction with sex [65]. This developmental approach to binge drinking can also contribute to the design of more effective strategies aimed to the prevention of alcohol use during adolescence and to the development of new interventions [66]. Recent experimental evidence obtained in animal models of environmental enrichment (Fig. 14) suggests that complex housing conditions may have protective and even therapeutic effects for ethanol binge drinking in rodents. Further research is needed to explore the neurobiological mechanisms related to the effects of environmental enrichment on ethanol binge drinking during adolescence.

There is a need for reducing alcohol consumption [67] and developing strategies to address the consequences of binge drinking during adolescence [5, 62]. As emphasized by the 'Neurobiology of Adolescent Drinking in Adulthood Consortium' (NADIA

Consortium), we need to explore the behavioral and neurobiological changes induced by alcohol intermittent exposure in order to promote strategies aimed to the prevention, treatment or even reversal of the behavioral changes induced by adolescent intermittent exposure to ethanol [5]. This reversal of AIE-induced behavioral changes in adults has been tested trough both pharmacological and nonpharmacological (i.e., exercise) interventions [56]. For example, there is preliminary evidence suggesting that due to brain plasticity and neuroadaptation mechanisms, discontinuation of binge drinking may lead to a recovery (both neural and cognitive). More research is needed to understand brain mechanisms related to this partial recovery and which domains are more implicated [5].

As suggested by Spear [68], we need to further explore if there are timing-specific effects of ethanol exposure at different periods during adolescence. Results obtained in studies performed with this developmental approach could contribute to identify the more vulnerable periods to the effects of binge drinking. Prior studies suggest that consequences of exposure to alcohol during early adolescence differ from those observed with later exposure since this last one more clearly impact cognitive tasks [68]. It has also been proposed the need of obtaining longitudinal data in order to evaluate main neurocognitive changes related to binge drinking and their relationship with neurobiological changes and brain activation [9]. These studies are also critical for designing prevention or intervention strategies aimed to minimize long-term consequences of binge drinking [68] and reducing the rate of alcohol use disorder [9]. It has been recently proposed that prevention and intervention strategies should take into account exposure age to binge drinking [68, 69].

## Acknowledgments

This work was partly supported by the PROMETEO-II/2015/020 from "Generalitat Valenciana."

## References

1. Hibell B, Guttormsson U, Ahlström S, Balakireva O, Bjarnason T, Kokkevi A, Kraus L (2007) Substance use among students in 35 European countries (The 2007 European School Survey Project on alcohol and other drugs, ESPAD, Report). The European Monitoring Center for Drugs and Drug Addiction (EMCDDA), Stockholm

2. Chavez PR, Nelson DE, Naimi TS, Brewer RD (2011) Impact of a new gender-specific definition for binge drinking on prevalence estimates for women. Am J Prev Med 40:468–471

3. Bagley SM, Levy S, Schoenberger SF (2019) Alcohol use disorders in adolescents. Pediatr Clin N Am 66:1063–1074

4. Schulteis G, Archer C, Tapert SF, Frank LR (2008) Intermittent binge alcohol exposure during periadolescent period induces spatial

working memory deficits in young adult rats. Alcohol 42:459–467

5. Lees B, Mewton L, Stapinski LA, Squeglia LM, Rae CD, Teesson M (2019) Neurobiological and cognitive profile of young binge drinkers: a systematic review and meta-analysis. Neuropsychol Rev 29:357–385

6. Thorpe HHA, Hamidullah S, Jenkins BW, Khokhar JY (2019) Adolescent neurodevelopment and substance use: receptor expression and behavioral consequences. Pharmacol Ther 206:107431

7. Dahl RE (2004) Adolescent brain development: a period of vulnerabilities and opportunities. Ann N Y Acad Sci 1021:1–22

8. Monti PM, Miranda R Jr, Nixon K, Sher KJ, Swartzwelder HS, Tapert SF, White A, Crews FT (2005) Adolescence: booze, brains, and behavior. Alcohol Clin Exp Res 29:207–220

9. Jones SA, Lueras JM, Nagel BJ (2018) Effects of binge drinking on the developing brain. Alcohol Res 39:87–96

10. National Institute on Alcohol Abuse and Alcoholism (2004) NIAAA Council approves definition of binge drinking. Department of Health and Human Services. NIAAA Newslett 3:3

11. Vinader-Caerols C, Monleón S (2019) Binge drinking and memory in adolescents and young adults. In: Palermo S (ed) Inhibitory control training - a multidisciplinary approach. InTech, Rijeka

12. DeWit DJ, Adlaf EM, Offord DR, Ogborne AC (2000) Age at first alcohol use: a risk factor for the development of alcohol disorders. Am J Psychiatry 157:745–750

13. Olsson CA, Romaniuk H, Salinger J, Staiger PK, Bonomo Y, Hulbert C, Patton GC (2016) Drinking patterns of adolescents who develop alcohol use disorders: results from the victorian adolescent health cohort study. BMJ Open 6: e010455

14. Chung T, Creswell KG, Bachrach R, Clark DB, Martin CS (2018) Adolescent binge drinking. Alcohol Res 39:5–15

15. Kuntsche E, Kuntsche S, Thrul J, Gmel G (2017) Binge drinking: health impact, prevalence, correlates and interventions. Psychol Health 32:976–1017

16. Fritz BM, Boehm SL II. (2016) Rodent models and mechanisms of voluntary binge-like ethanol consumption: examples, opportunities, and strategies for preclinical research. Prog Neuro-Psychopharmacol Biol Psychiatry 65:297–308

17. Jeanblanc J, Rolland B, Gierski F, Martinetti MP, Naassila M (2019) Animal models of binge drinking, current challenges to improve face validity. Neurosci Biobehav Rev 106:112–121

18. Sprow GM, Thiele TE (2012) The neurobiology of binge-like ethanol drinking: evidence from rodent models. Physiol Behav 106:325–331

19. Cippitelli A, Zook M, Bell L, Damadzic R, Eskay RL, Schwandt M, Heilig M (2010) Reversibility of object recognition but not spatial memory impairment following binge-like alcohol exposure in rats. Neurobiol Learn Mem 94:538–546

20. Maldonado-Devincci AM, Badanich KA, Kirstein CL (2010) Alcohol during adolescence selectively alters immediate and long-term behavior and neurochemistry. Alcohol 44:57–66

21. Sanchez-Roige S, Peña-Oliver Y, Ripley TL, Stephens DN (2014) Repeated ethanol exposure during early and late adolescence: double dissociation of effects on waiting and choice impulsivity. Alcohol Clin Exp Res 38:2579–2589

22. Lacaille H, Duterte-Boucher D, Liot D, Vaudry H, Naasila M, Vaudry D (2015) Comparison of the deleterious effects of binge drinking-like alcohol exposure in adolescent and adult mice. J Neurochem 132:629–641

23. Vetreno RP, Crews FT (2015) Binge ethanol exposure during adolescence leads to a persistent loss of neurogenesis in the dorsal and ventral hippocampus that is associated with impaired adult cognitive functioning. Front Neurosci 9:35

24. Beaudet G, Valable S, Bourgine J, Lelong-Boulouard V, Lanfumey L, Freret T, Boulouard M, Paizanis E (2016) Long-lasting effects of chronic intermittent alcohol exposure in adolescent mice on object recognition and hippocampal neuronal activity. Alcohol Clin Exp Res 40:2591–2603

25. Marco EM, Peñasco S, Hernández MD, Gil A, Borcel E, Moya M, Giné E, López-Moreno JA, Guerri C, López-Gallardo M, Rodríguez de Fonseca F (2017) Long-term effects of intermittent adolescent alcohol exposure in male and female rats. Front Behav Neurosci 11:233

26. Contreras A, Morales L, del Olmo N (2019) The intermittent administration of ethanol during the juvenile period produces changes in the expression of hippocampal genes and proteins and deterioration of spatial memory. Behav Brain Res 372:112033

27. Contreras A, Polín E, Miguéns M, Pérez-García C, Pérez V, Ruiz-Gayo M, Morales L, Del Olmo N (2019) Intermittent-excessive

and chronic-moderate ethanol intake during adolescence impair spatial learning, memory and cognitive flexibility in the adulthood. Neuroscience 418:205–217

28. Waszkiewicz N, Galińska-Skok B, Nestsiarovich A, Kułak-Bejda A, Wilczyńska K, Simonienko K, Kwiatkowski M, Konarzewska B (2018) Neurobiological effects of binge drinking help in its detection and differential diagnosis from alcohol dependence. Dis Markers 2018:5623683

29. Fritz M, Klawonn AM, Zahr NM (2019) Neuroimaging in alcohol use disorder: from mouse to man. J Neurosci Res

30. Markwiese BJ, Acheson SK, Levin ED, Wilson WA, Swartzwelder HS (1998) Differential effects of ethanol on memory in adolescent and adult rats. Alcohol Clin Exp Res 22:416–442

31. Bell RL, Rodd ZA, Engleman EA, Toalston JE, McBride WJ (2014) Scheduled access alcohol drinking by alcohol-preferring (P) and high alcohol-drinking (HAD) rats: modeling adolescent and adult binge-like drinking. Alcohol 48:225–234

32. Dutta S, Sengupta P (2016) Men and mice: relating their ages. Life Sci 152:244–248

33. Heilig M, Augier E, Pfarr S, Sommer WH (2019) Developing neuroscience-based treatments for alcohol addiction: a matter of choice? Transl Psychiatry 9:255

34. Ghosh Dastidar S, Warner JB, Warner DR, McClain CJ, Kirpich IA (2018) Rodent models of alcoholic liver disease: role of binge ethanol administration. Biomolecules 8:3

35. Thiele TE, Crabbe JC, Boehm SL II (2014) "Drinking in the Dark" (DID): a simple mouse model of binge-like alcohol intake. Curr Protoc Neurosci 68:9.49.1–9.49.12

36. Monleón S, Duque A, Vinader-Caerols C (2020) Emotional memory impairment produced by binge drinking in mice is counteracted by the anti-inflammatory indomethacin. Behav Brain Res 381:112457

37. Rhodes JS, Best K, Belknap JK, Finn DA, Crabbe JC (2005) Evaluation of a simple model of ethanol drinking to intoxication in C57BL/6J mice. Physiol Behav 84:53–63

38. Thiele TE, Navarro M (2014) "Drinking in the dark" (DID) procedures: a model of binge-like ethanol drinking in non-dependent mice. Alcohol 48:235–241

39. García Pardo MP, Roger Sánchez C, De la Rubia Orti JE, Aguilar Calpe MA (2017) Animal models of drug addiction. Adicciones 29:278–292

40. Carnicella S, Ron D, Barak S (2014) Intermittent ethanol access schedule in rats as a preclinical model of alcohol abuse. Alcohol 48:243–252

41. White AM, Bae JG, Truesdale MC, Ahmad S, Wilson WA, Swartzwelder HS (2002) Chronic-intermittent ethanol exposure during adolescence prevents normal developmental changes in sensitivity to ethanol-induced motor impairments. Alcohol Clin Exp Res 26:960–968

42. Nogales F, Rua RM, Ojeda ML, Murillo ML, Carreras O (2014) Oral or intraperitoneal binge drinking and oxidative balance in adolescent rats. Chem Res Toxicol 27:1926–1933

43. Sánchez P, Castro B, Torres JM, Ortega E (2014) Effects of different ethanol-administration regimes on mRNA and protein levels of steroid 5α-reductase isozymes in prefrontal cortex of adolescent male rats. Psychopharmacology 231:3273–3280

44. de Oliveira BMT, Telles TMBB, Lomba LA, Correia D, Zampronio AR (2017) Effects of binge-like ethanol exposure during adolescence on the hyperalgesia observed during sickness syndrome in rats. Pharmacol Biochem Behav 160:63–69

45. Broadwater MA, Lee SH, Yu Y, Zhu H, Crews FT, Robinson DL, Shih YI (2018) Adolescent alcohol exposure decreases frontostriatal resting-state functional connectivity in adulthood. Addict Biol 23:810–823

46. Monleón S, Duque A, Mesa-Gresa P, Redolat R, Vinader-Caerols C (2019) An animal model of alcohol binge drinking: chronic-intermittent ethanol administration in rodents. In: Kobeissy FH (ed) Psychiatric disorders: methods and protocols. Methods in molecular biology series, vol 2011, 2nd edn. Humana Press, Totawa, NJ

47. Gold PE (1986) The use of avoidance training in studies of modulation of memory storage. Behav Neural Biol 46:87–98

48. Parra A, Rama E, Vinader-Caerols C, Monleón S (2013) Inhibitory avoidance in CD1 mice: sex matters, as does the supplier. Behav Process 100:36–39

49. Alfonso-Loeches S, Pascual M, Guerri C (2013) Gender differences in alcohol-induced neurotoxicity and brain damage. Toxicology 311:27–34

50. Vinader-Caerols C, Talk A, Montañés A, Duque A, Monleón S (2017) Differential effects of alcohol on memory performance in adolescent men and women with a binge drinking history. Alcohol Alcohol 52:610–616

51. Crabbe JC, Harris RA, Koob GF (2011) Preclinical studies of alcohol binge drinking. Ann N Y Acad Sci 1216:24–40

52. Knapp DJ, Breese GR (2012) Models of chronic alcohol exposure and dependence. In: Kobeissy F (ed) Psychiatric disorders. Methods in molecular biology (methods and protocols), vol 829. Humana Press, New York, NY, pp 205–230

53. Tabakoff B, Hoffman PL (2000) Animal models in alcohol research. Alcohol Res Health 24:77–84

54. Kuhn BN, Kalivas PW, Bobadilla A-C (2019) Understanding addiction using animal models. Front Behav Neurosci 13:262

55. Crabbe JC, Metten P, Huang LC, Schlumbohm JP, Spence SE, Barkley-Levenson AM, Finn DA, Rhodes JS, Cameron AJ (2012) Ethanol withdrawal-associated drinking in the dark: common and discrete genetic contributions. Add Gen 1:3–11

56. Crews FT, Robinson DL, Chandler LJ, Ehlers CL, Mulholland PJ, Pandey SC, Rodd ZA, Spear LP, Swartzwelder HS, Vetreno RP (2019) Mechanisms of persistent neurobiological changes following adolescent alcohol exposure: NADIA consortium findings. Alcohol Clin Exp Res 43:1806–1822

57. Holgate JY, Shariff M, Mu EW, Bartlett S (2017) A rat drinking in the dark model for studying ethanol and sucrose consumption. Front Behav Neurosci 11:29

58. Shnitko TA, Liu Z, Wang X, Grant KA, Kroenke CD (2019) Chronic alcohol drinking slows brain development in adolescent and young adult nonhuman primates. eNeuro 6: ENEURO.0044-19.2019

59. Macht V, Crews FT, Vetreno RP (2020) Neuroimmune and epigenetic mechanisms underlying persistent loss of hippocampal neurogenesis following adolescent intermittent ethanol exposure. Curr Opin Pharmacol 50:9–16

60. Courtney KE, Li I, Tapert SF (2019) The effect of alcohol use on neuroimaging correlates of cognitive and emotional processing in human adolescence. Neuropsychology 33:781–794

61. Siciliano CA, Noamany H, Chang CJ, Brown AR, Chen X, Leible D, Lee JJ, Wang J, Vernon AN, Vander Weele CM, Kimchi EY, Heiman M, Tye KM (2019) A cortical-brainstem circuit predicts and governs compulsive alcohol drinking. Science 366:008–1012

62. Nixon K, Mangieri RA (2019) Compelled to drink: why some cannot stop. Science 366:947–948

63. Kim J, Park A (2018) A systematic review: candidate gene and environment interaction on alcohol use and misuse among adolescents and young adults. Am J Addict

64. Rodríguez-Ortega E, de la Fuente L, de Amo E, Cubero I (2018) Environmental enrichment during adolescence acts as a protective and therapeutic tool for ethanol binge-drinking, anxiety-like, novelty seeking and compulsive-like behaviors in c57bl/6j mice during adulthood. Front Behav Neurosci 12:177

65. Wilsnack RW, Wilsnack SC, Gmel G, Kantor LW (2018) Gender Differences in Binge Drinking. Alcohol Res 39:57–76

66. Quadir SG, Guzelian E, Palmer MA, Martin DL, Kim J, Szumlinski KK (2019) Complex interactions between the subject factors of biological sex and prior histories of binge-drinking and unpredictable stress influence behavioral sensitivity to alcohol and alcohol intake. Physiol Behav 203:100–112

67. Wood AM, Kaptoge S, Butterworth AS, Willeit P, Warnakula S, Bolton T (2018) Risk thresholds for alcohol consumption: combined analysis of individual-participant data for 599,912 current drinkers in 83 prospective studies. Lancet 391:1513–1523

68. Spear LP (2015) Adolescent alcohol exposure: are there separable vulnerable periods within adolescence? Physiol Behav 148:122–130

69. Hiller-Sturmhöfel S, Spear LP (2018) Binge Drinking's Effects on the Developing Brain-Animal Models. Alcohol Res 39:77–86

# Chapter 3

# Prepulse Inhibition and Vulnerability to Cocaine Addiction

## M. Carmen Arenas, Sergio Pujante-Gil, and Carmen Manzanedo

## Abstract

Prepulse inhibition (PPI) of the startle reflex is the most common sensorimotor gating index, representing the brain's ability to filter out irrelevant stimuli and prevent information overload. The relationship between PPI impairment and pathology has been widely reported. A PPI deficit is considered an endophenotype of schizophrenia and it is observed in other disorders. Recently, we observed that animals with low PPI are more likely to display behaviors induced by cocaine, which can increase the risk of developing a cocaine use disorder. Therefore, we consider that a PPI deficit could represent a biomarker of vulnerability to the development of a Substance Use Disorder (SUD). The objective of this chapter is to describe PPI, its relation with cocaine addiction and to provide guidelines on how to evaluate it in rodents and humans in order to be able to identify subjects with a PPI deficit.

**Key words** Prepulse inhibition, Startle reflex, Cocaine, Vulnerability, Substance use disorder

## Abbreviations

BLA     Basolateral amygdala
CNS     Central nervous system
CSPP    Corticostriatopallidopontine
DA      Dopamine
dB      Decibels
DP      Dual pathology
ISI     Interstimuli interval
mPFC    Medial prefrontal cortex
NAcc    Nucleus accumbens
OFC     Orbitofrontal cortex
PPI     Prepulse inhibition
SR      Startle reflex
SUD     Substance use disorder
VTA     Ventral tegmental area

María A. Aguilar (ed.), *Methods for Preclinical Research in Addiction*, Neuromethods, vol. 174,
https://doi.org/10.1007/978-1-0716-1748-9_3, © Springer Science+Business Media, LLC, part of Springer Nature 2022

## 1   Introduction: Inhibition Prepulse of Startle Reflex

Prepulse inhibition (PPI) is defined as a measure of sensory–motor synchronization based on the startle reflex (SR) response. The PPI of the startle reflex is considered a model of preattentional inhibitory function and takes place when a high intensity stimulus is preceded by a less intense stimulus, causing the attenuation of the SR. It is an adaptive phenomenon for the subject, as it avoids an overload of stimulation. This phenomenon allows to regulate the sensory input, filtering out the irrelevant or distracting stimuli thereby avoiding the alteration of sensory information, which allows the selective and efficient processing of information of a relevant nature [1–4].

Numerous studies have shown that measuring PPI levels can be useful in the field of mental illnesses, in both research with humans [2, 3, 5] and animal models [4, 6, 7]. Deficits in PPI have been observed in people with different mental disorders in which the presence of abnormalities in sensory, cognitive and/or motor inhibition mechanisms is significant, with schizophrenia being the most characteristic [2, 8–10]. These disorders present common anatomical structures and neurobiology with PPI [3, 4]. In general, the results obtained in the past few years point to the fact that PPI can be a marker of vulnerability for schizophrenia [4, 10], as well as other mental disorders [3–5, 8, 9, 11]. This is why the PPI paradigm is being used to screen new drugs for treatments of such disorders [12]. Recently, it has been reported that the baseline PPI level of mice can predict their sensitivity to the effects of cocaine [13, 14], whereas a PPI deficit could represent a biomarker of vulnerability to the development of a substance use disorder (SUD).

### 1.1   Concept

PPI is based on the startle reflex, which is a spontaneous, unconditioned and rapid response, issued in a few milliseconds, before an unexpected, high-intensity stimulus caused by any sensory modality. It is characterized by the coordinated contraction of skeletal muscles, together with heart rate acceleration. This response overrides the conscious behavior that the individual is performing at that moment. However, the SR may be attenuated if a low intensity stimulus precedes a high intensity stimulus. In other words, although SR is an automatic reaction, its expression can be modulated by the previous presence of less intense stimuli. This phenomenon is known as PPI. Therefore, PPI is the reduction of SR due to the introduction of a weak sensory stimulus (known as prepulse) that precedes the introduction of a more intense sensory stimulus (known as pulse), which is the generator of the startle. Thus, the prepulse, which does not cause SR by itself, intervenes as an inhibitory barrier decreasing the response to the sensory stimulus or pulse [1–4, 9, 15, 16].

This phenomenon is considered a simple operational measurement of sensory information processing and sensory-motor synchronization, characterized by its ability to filter out the relevant information for the subject preventing an overload of stimuli [3, 9]. PPI is an automatic process, which is not linked to learning and which maintains certain stability over time [7, 17]. In general, acoustic stimuli are used to generate PPI, although it can be provoked by stimuli of any sensory modality (visual, cutaneous, etc.). It has also been observed to occur in a large number of species, which allows its use both in basic and clinical research [5]. In humans, the ocular component of the startle reflex is measured by the electromyographic record of the orbicularis oculi muscle [1, 9, 18]. In rodents, the skeletal motor response of the whole body to an auditory stimulus is evaluated [19]. Both evaluations have similar properties and reflect the activity of several neural circuits [20, 21]. At the neuroanatomical level, SR is generated in structures in the reticular formation of the brainstem responsible for integrating sensory signals [1–4, 15, 16]. However, PPI, as an inhibitory control/regulatory mechanism, involves cortical and subcortical structures [3, 9, 15, 16]. Therefore, this measure can be used to evaluate the function of complex structures [1–4].

PPI is based on mainly automatic preattentional mechanisms, and therefore it can be observed with intervals between short stimuli (interstimuli interval, ISI <60 ms), but it reaches its maximum amplitude with ISIs between 60 and 120 ms. When a larger interval is used (ISI 120–240 ms), it is possible to modulate PPI levels by instructing the subjects to pay attention to the stimuli. Thus, with higher ISIs, the subject can focus their attention on the prepulse and/or the pulse, to either acknowledge it or ignore it, thereby altering their PPI levels. In consequence, PPI, being an automatic process, depends on the experimental design used, as well as the instructions given, hence the importance in the unification of the methodology used [5, 17].

| 1.2 History | The term "prepulse inhibition" was coined by James R. Ison and Geoffrey R. Hammond in 1971. However, there are historical precedents related to the term prior to this date. In 1863, Ivan M. Sechenov first described the startle reflex, as a result of his studies on the neurophysiology of central nervous system (CNS) inhibition. Well into the twentieth century, Helen Peak demonstrated in 1939 the modulation of the inhibitory reflex in the auditory system. But it was not until 1965 when Howard S. Hoffman and John L. Searle observed that the startle reflex response diminished when preceded by a sound. In addition, they pointed out that this occurred when the sound appeared between 20 and 500 ms before the presentation of the startle stimulus [22]. Later on, in 1978, David L. Braff and his collaborators in Dr. Enoch Callaway's laboratory at University of California San |

Francisco demonstrated for the first time that schizophrenic patients presented a deficit in their PPI levels, leading to a large number of studies that have revealed the relationship between this paradigm and schizophrenia [9].

The first explanatory theories surrounding PPI are found in the sequential information processing model described by classical cognitivism [23]. But it is in the late 1970s, with the development of connectionist models, that the first attempts to explain PPI alterations in schizophrenic patients took place. Schizophrenia came to be understood as a "disconnection syndrome" that affected the connections between certain cortical and subcortical regions through the thalamus [24]; disconnections that are considered responsible for the deficits observed in the PPI levels of schizophrenic patients [8, 9].

Studies that use the PPI paradigm, especially in terms of translational research, have increased since the mid-1970s, due to two complementary factors. On the one hand, this neurobiological measure has proven to be useful to delve into the principles of reflex modulation as a sensory-motor filtering mechanism; and on the other, a PPI deficit or alteration has been associated with certain mental pathologies [3, 8, 9]. In recent years, this paradigm has started to be used in drug addiction. An alteration in PPI levels has been observed in users of different drugs, although very few studies have been carried out so far [3, 5].

### 1.3 Neurobiology of PPI

Three differentiated parts are distinguished in the PPI phenomenon: the structures that control the startle response, those that trigger the PPI and those that modulate it (see Fig. 1). The intervening structures present neuronal connections with a sequential and parallel action.

The circuit begins with an acoustic input into the cochlear nuclei (dorsal cochlear nucleus, ventral cochlear nucleus and cochlear nucleus root), which is part of the primary auditory pathway. When an acoustic stimulus exceeds a certain decibel level (dB) (>80 dB), the information passes to the ventrolateral tegmental nucleus and the pontine-caudal nucleus of the reticular formation. The latter is one of the most basal nuclei in the brainstem and has several projections that go directly to the motor neurons, which gives it a significant involvement in the primary circuit of the SR. Lastly, the motor neurons activate the motor response, which completes the SR circuit [1, 5].

The excitatory input that triggers the PPI in response to the auditory stimulus originates in the auditory circuit of the inferior colliculus in the mesencephalon, where the auditory information from the cochlear nuclei arrives. The inferior colliculus, in turn, activates the superior colliculus, whose projections are directed toward the peduncle-pontine tegmental nucleus that inhibits the pontine-caudal nucleus of the reticular formation. The inhibition of

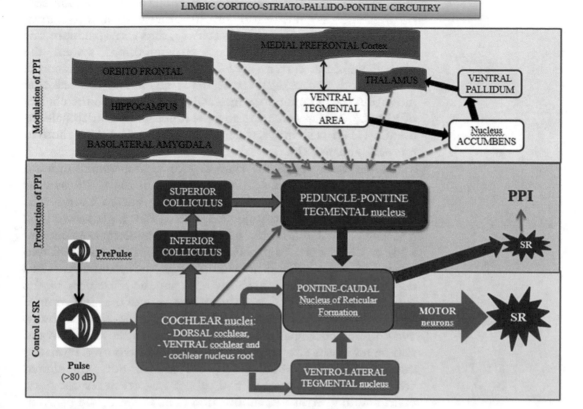

**Fig. 1** Neurobiology of prepulse inhibition (PPI) of startle reflex (SR) to acoustic stimulus. (Modified from Arenas et al. [78])

this nucleus causes a downregulation of the startle response, which allows for measuring the effect of PPI. The structures involved have different implications. The inferior colliculus seems to be important for the suppression of SR caused by an auditory stimulus, while the superior colliculus is considered crucial in the integration of information from different sensory modalities (visual and tactile), although both colliculi are involved in attenuating PPI, as demonstrated by studies on injuries [22, 25]. The tegmental peduncle-pontine nucleus also intervenes in the mediation of PPI, since its partial or total injury causes a PPI disruption while its stimulation produces a PPI increase [22, 26, 27]. The latero-dorsal tegmental nucleus, the pars reticulata of the substantia nigra [22] and the dorsolateral region of the periaqueductal gray matter [16] are also involved in this circuit.

The modulation of PPI is performed by limbic and cortical structures that have reciprocal connections. The structures involved in this modulation include the ventral tegmental area (VTA), the substantia nigra and the thalamus, all three of which receive afferences from the tegmental peduncle-pontine and laterodorsal tegmental nuclei. Efferences and connections of the VTA, substantia nigra and thalamus in turn involve other cortico-limbic areas, including the nucleus accumbens (NAcc), the ventral pallidum, the basolateral amygdala (BLA), the septohippocampal system, the medial prefrontal cortex (mPFC) and the orbitofrontal cortex (OFC). Both networks, the one that triggers and the one that modulates are in continuous interaction to determine the efficacy of PPI [16, 28]. In general terms, the limbic corticostriatopallidopontine (CSPP) circuitry is the main modulator of PPI [3], both in rats and humans [21].

It has been suggested that a poor neural communication between the limbic and cortical structures in the CSPP may be the basis of deficits in PPI. The NAcc is considered a primordial structure in the regulation and alteration of PPI, together with the adjacent cortical and limbic subregions. Hippocampal manipulation, or BLA, as well as dysfunctions in the mPFC and OFC also produce changes in PPI [3, 28]. In addition, the modulation of PPI depends on the interconnection between the structures of the entire CSPP circuit [3]. Thus, the involvement of the hippocampus in the disruption of PPI takes place through the NAcc pathway, while the hippocampal efferent connections to the mPFC would also be responsible for some PPI alterations, as reported by anatomical and functional studies [5]. On another note, the altered connection between cortical and subcortical structures can cause deficits in long-term PPI modulation. The lesion of the paraventricular nucleus of the thalamus impairs the PPI of the animals, since the connection with the mPFC, the cortex responsible for regulating the dopaminergic activity in the VTA, is interrupted. This causes the loss of regulation that the cortical glutamate exerts on the mesolimbic dopaminergic system [29]. In sum, different cortical and subcortical areas are involved in the regulation of PPI and the modulation through different routes of these structures can cause a disruption or reduction of PPI. The startle reflex circuit is uninhibited and the sensory-motor input is decompensated or destabilized. In such cases, a strong startle response would occur, even if the prepulse is present [1, 3, 5].

In terms of neurochemistry, dopamine (DA) is the reference neurotransmitter in both the production and modulation of PPI. The main changes in PPI levels are related to the disruptions in this neurotransmission system [3, 4, 6]. It is considered that a subject with impaired dopaminergic function is unable to filter out irrelevant information from the environment because their inhibitory functions regulated by this neurotransmitter are altered [30, 31]. It has been proposed that the SR in these subjects would remain high, even with the previous presentation of the prepulse, and consequently their PPI would be lower. The level of DA in the NAcc is important, since an increase in DA decreases PPI, as observed after the administration of dopamine agonists [32]. The level of DA in the prefrontal cortex modulates the dopaminergic signaling in the VTA; therefore, variations in the interrelation of these structures modify the availability of DA in NAcc, with deficits being observed in the PPI [29]. This variability, which may exist in the production and modulation of PPI, would result in differences in executive function and cognitive control between individuals with high or low PPI levels, as described above. The formation of strategies is more effective and the time of execution shorter in individuals with a high PPI [33, 34]. However, the dopaminergic system is not the only system involved. The dysregulation of PPI has been related to other neurotransmission systems, among which are the serotoninergic [35–37] and glutamatergic [4, 38] systems.

The role of dopaminergic receptors in the PPI has been studied by observing whether the blockade of DA receptors reduces or alters the decrease in PPI levels caused by dopamine agonists (amphetamine, apomorphine, cocaine or selective DA receptor agonists), using animals with and without genetic modification. The strategy is based on the premise that changes in dopaminergic sensitivity, the level of DA or the volume of DA receptors determine the PPI level.

Thus, it seems that in comparison to D1 receptors, the activation of D2 receptors is more effective to obtain a decrease in the PPI level when selective agonists are used [32, 39]. However, when antagonists are used, the blockade of the D1 receptors reverses the decrease in the PPI level induced by apomorphine, while the blockade of D2 receptors is not effective [40]. Doherty et al. [41] found that the blockade of the D1, D2 and D3 receptors prevents cocaine from decreasing the PPI level. Using mice with genetic modification for D1, D2, and D3 receptors, these authors observed that cocaine did not affect the PPI levels of animals without D1 receptors, but produced a decrease in the PPI level of mice without D2 receptors and a much greater decrease in animals without D3 receptors compared to their controls. In view of the results, the authors consider that it is difficult to specify the performance of each receptor separately, since the activation of the D1 receptors is necessary to enhance the D2 receptors' effect in the decrease of PPI levels; that is, there is a synergism between the receptors. Lastly, Yamashita and colleagues [42] report that mice without the dopamine transporter did not show a decrease in PPI induced by cocaine, highlighting again that it is the cocaine-induced increase in DA the one responsible for the decrease in PPI levels.

## 1.4 PPI in Psychiatric Disorders

A deficit in PPI is present in psychiatric pathologies that display a primordial dopaminergic alteration. Variations in DA levels and dopaminergic function, as well as the sensitivity of the receptors, have been shown to correlate with cognitive alteration. In clinical research, this translates into difficulties in the processing of information as a consequence of inadequate sensory inhibition, causing subjects to be unable to focus on the most relevant aspects of the environment [1, 3, 9]. Schizophrenia is the psychiatric disorder most closely related to a deficit in PPI [2], but this relation is also observed in other pathologies, such as neurocognitive and neurodevelopmental disorders, depression, anxiety, and hyperactivity, among others [2, 5, 9]. Even in pathologies with a marked genetic alteration such as Huntington's disease, 22q11 deletion syndrome and fragile X syndrome, a PPI deficit has also been observed [8]. In addition, lower levels of PPI are also present in relatives of schizophrenic and bipolar disorder patients, who are not affected by these disorders [2, 5]. Therefore, a PPI deficit is thought to constitute a neurophysiological endophenotype for schizophrenia and a genetically determined trait, as indicated by the differences observed in the PPI level, both in patients and their relatives [8, 9].

PPI is considered one of the most promising neurophysiological indices in Psychiatry translational research [4, 5, 7]. The benefits of its use have been mostly observed in the field of mental pathology, since PPI detects individual differences, is used in humans and rodents alike, and the neural substrates that regulate it are also involved in different pathologies. Although it should not be used as a clinical diagnostic tool, since a deficit in PPI is also found in healthy relatives of schizophrenic patients, this paradigm can help us understand the processes that contribute to presenting a deficit in the inhibitory behavioral mechanisms. It is important to emphasize that the current working hypothesis is not that the clinical symptoms are produced by a PPI deficit, but that the genetics and neurobiology of these two processes (the clinical symptoms and the PPI deficit) overlap. Therefore, it is considered particularly useful as a behavioral measure, which, combined with other tests and in an experimental context, can describe dysregulations observed in clinical conditions [5, 9].

The relationship of PPI and dopamine also raises its use in the field of addiction, and although studies are scarce so far, there is evidence pointing to its usefulness in SUDs. It has been demonstrated that drug consumption alters the mesolimbic dopaminergic system [43], which is also the system responsible for PPI [3, 6], which is the reason why, in this case, differences in the production of PPI according to drug use are expected. Kumari et al. [44] evaluated the change in PPI before and after the administration of a low dose of amphetamine (system activation) and observed that PPI does not vary in nonsmoking subjects, while it decreases in smokers. The study indicates that the PPI phenomenon is sensitive to dopaminergic manipulation depending on previous experience with drugs, pointing in this case to the fact that the subgroup of smokers presents dopaminergic sensitization in such a way that, with the same stimulus, a greater increase in the release of DA takes place. It has even been suggested that there may be an inheritable genetic component related to drug use, since a lower PPI has been observed in the children of alcoholic parents [45]. Thus, the neuroplasticity caused by chronic drug use will be reflected in the PPI levels.

Additionally, a research area of great interest is the relationship between the presence of mental disorders and compulsive drug use. Reports indicate that up to 70% of people who have been diagnosed with a SUD present an added psychiatric disorder, while 45–50% of patients diagnosed with a mental disorder have a history of a SUD [46, 47]. This concurrence of disorders, called dual pathology (DP), entails a greater severity in the symptoms of each disorder with a worse prognosis, and a greater functional disability for the individual. In specific populations, such as people in prison or in poverty, women or the elderly, the presence of DP is greater than in the rest of the population, and runs with more morbidity and mortality [48], becoming a serious public health concern.

PPI can be very useful to understand the etiology of DP too. Morales-Muñoz et al. [49] conclude that cannabis users who have difficulties in inhibition of the startle reflex may have a higher risk of developing cannabis-induced psychotic disorder, and a greater PPI has been shown to predict a longer successful quitting from tobacco smoking in posttraumatic stress disorder subjects [50].

## 1.5 Effects of Cocaine on PPI

Cocaine is the most commonly used illegal stimulant drug in Europe and in North America, being primarily consumed by young adults [51, 52]. Consumers of this substance constitute the third largest group that opts for outpatient treatment and polydrug use and mental disorder is characteristic in them. Cocaine users often also report the use of cannabis and alcohol as secondary drugs, and many opioid users also report secondary use of cocaine [53]. Psychiatric comorbidity is common in substance abuse in general [46, 54], and among cocaine users in particular [55]. As a consequence, the prognosis of addiction worsens [47] with all the clinical and social implications that this entails [48]. Among the most common psychopathologies found in cocaine users are cocaine psychosis, anxiety, mood, impulse control and personality disorders (borderline, antisocial, histrionic, narcissistic, passive-aggressive and paranoid the most frequent), traumatic disorders for residual attention deficit, dissociative states, cognitive alterations, and sexual dysfunctions [55]. Studies report the presence of DP in 50–75% of cocaine users, which means that these mostly young people meet the diagnostic criteria of another mental disorder together with the compulsive use of cocaine [46, 54, 55]. The direction of this relationship is yet unknown, that is, whether drug use precipitates the development of a mental disorder or whether vulnerability to psychiatric disorders triggers consumption. However, it does seem that an early contact with drugs increases the probability of developing DP [54, 55].

The main mechanism of action of cocaine lies in the mesolimbic and mesocortical dopaminergic pathways. Cocaine blocks the reuptake pump, that is, the DA transporter, and, as a result, increases both the amount of DA and the time of exposure of DA in the NAcc and in the rest of the structures receiving the projections of the VTA and the substantia nigra, such as the mPFC, the caudate and putamen nuclei, the hippocampus, and the amygdala [43]. In these structures, the increase in DA activates the dopaminergic pre- and postsynaptic receptors for a long time, which means that dopaminergic signaling is altered from the first consumptions, and sequential, progressive, and long-term changes are also produced depending on the subsequent consumption [43, 56].

It has been observed that, with chronic administration, the synthesis of DA decreases, and with it, its basal level, release, and turnover. The availability of transporters also decreases, along with D2/D3 receptors [57]. In addition, the number of D3 receptors is gradually increased, as is the distribution and interrelation between the D1 and D2 receptors, which present postsynaptic hypersensitivity [43, 56]. Through the activation of D1 receptors, cocaine induces molecular changes by activating the second messenger pathway, such as the cAMP. The cAMP pathway stimulates the production of intracellular proteins, promoting the expression of early-acting genes, which in regions such as the basal ganglia and the NAcc are responsible for the development of rigidity in the behavioral response, habit formation, and compulsive behavior which are part of the SUD [43, 56]. Furthermore, modifications in the expression and signaling of these receptors are linked to the appearance of mental disorders in individuals who have not used cocaine [58].

Nevertheless, although the actions of cocaine on the dopaminergic system are considered its main mechanism of action, cocaine also increases the levels of norepinephrine and, to a lesser extent, serotonin, since it blocks the transporters of these two neurotransmitters, which are very similar to the DA transporter, preventing its presynaptic reuptake. In addition, the glutamatergic system is of particular importance in the neuroplasticity induced by cocaine. The administration of cocaine produces hyperexcitability in the DA neurons through the NMDA and AMPA receptors located in the dopaminergic terminals of the VTA [43]. This action on the neuronal firing in the VTA alters the communication with the NAcc and is the beginning of the subsequent synaptic alterations in the structures connecting to the VTA and to the NAcc, such as the mPFC and the amygdala, among others [59]. The dopaminergic regulation from the VTA to the NAcc and the modulation from the mPFC to the NAcc are critical in the development of a SUD as well as in schizophrenia [30].

Thus, chronic use of cocaine sequentially produces permanent changes in the structure and cellular functioning of the brain, both at cortical and subcortical levels. A smaller volume of gray matter in the cortex, amygdala, and the striatum has been observed in cocaine addicts [56], along with an abnormal pattern of activity and cerebral metabolism in cortical areas (prefrontal, parietal, and temporal) and in subcortical structures such as the striate [60]. In addition, the presence of deficits in cognitive functions (lower score in attention and memory tests) and executive functions (lower performance with respect to flexibility and inhibition processes) is common in cocaine dependent subjects [61]. The cognitive and executive alterations are related to frontal lobe dysfunctions, and the neuropsychiatric alterations, which also observed in cocaine addicts, are related to the deterioration of the dopaminergic function. Alterations in the nigrostriatal system are related to the appearance of motor and extrapyramidal symptoms, as observed in Parkinson disorder or Tourette syndrome; and in the mesolimbic system, they are related to the appearance of delirium, psychosis and altered consciousness, as observed in dementia or schizophrenia [56].

The action of cocaine on the dopaminergic system [57] and the neuronal alterations caused by its consumption are well-known [60]; and it is known that this system of neurotransmission is crucial in the regulation of PPI [3, 4, 6]. Thus, PPI levels have been observed to be increased in both recreational and cocaine-dependent users with respect to the control group [62]. However, when their recent consumption is controlled through a urinalysis, PPI is shown to be lower in consumers who test positive in urine than in consumers who test negative. The latter seems to indicate that the subject's PPI level decreases under the acute effects of cocaine, and a "rebound" increase in PPI is observed during abstinence. In fact, abstinent cocaine users showed a higher PPI level compared to controls [63]. These changes in PPI are consistent with the neuroadaptations caused by the continued consumption of cocaine, specifically with a low DA level in the striatum, which tends to recover with abstinence [43, 57, 64]. This regulation has also been observed in alcohol-dependent patients. All had a lower PPI level than the controls, but the PPI improved (i.e., increased) after the detoxification treatment, although it did not reach the levels of inhibition observed in the subjects without dependence [18].

Studies conducted with animal models corroborate this conclusion. The acute administration of high doses of cocaine decreases the PPI levels in rats [65–67] and mice [41, 42]. The acute administration of amphetamine, another psychostimulant similar to cocaine, also causes a decrease in PPI levels in humans [20, 68, 69], rats [69, 70], and mice [71, 72]. Similarly, other drugs that increase the release of DA through indirect mechanisms, such as the antagonism of NMDA receptors or the activation of

5HT2A receptors, also cause the decrease in PPI levels. Thus, phencyclidine and ketamine [4, 71], as well as hallucinogenic drugs with serotoninergic profile (LSD, psilocybin, DMT, MDMA), decrease PPI levels in humans, rats and mice [35], because they also indirectly increase the dopaminergic function in the NAcc [30, 73, 74]. It is well known that this alteration in the mesolimbic dopaminergic pathway causes, to a greater or lesser extent, a potent addiction [75] and possible psychotic episodes observed in both humans and animal models [76].

Studies with animal models have allowed us to expand and clarify these results. Zhang and colleagues [32] demonstrated that the observed PPI levels depend on the existing level of DA in the NAcc. They found that the PPI level only decreases when a strong increase in DA happens in this region using a high dose of amphetamine, but that it does not change when the increase in DA is lower using a minimal dose of cocaine. In agreement with this study, Broderik and Rosenbaum [65] observed that the dose of 20 mg/kg of cocaine produced a greater decrease in PPI levels than that of 10 mg/kg in both male and female rats. However, taking into account that the continuous administration of cocaine causes a downregulation of dopaminergic function [43], an upregulation of PPI levels should be observed after a continuous consumption of the drug, similarly to that observed in abstinence, when normal dopaminergic activity is recovered. Three of the studies conducted with rats have proven this hypothesis [66, 67, 77]. Thus, Martínez et al. [67] observed that the PPI levels of rats with a subcutaneous cocaine implant for 5 days were lower than those of the controls on the third day of the implant, returning to similar levels 10 days after removing the implants. In the same line, Adams et al. [77] evaluated the effect of chronic cocaine administration on PPI levels with two administration protocols, one short, lasting 2 weeks, and a longer one, lasting 8 weeks. They did not observe any differences in the PPI levels of the animals and the controls, neither at the 3rd nor the 14th day after finishing each treatment. Byrnes and Hammer [66] performed a chronic administration of cocaine to develop dopaminergic neuroadaptations. As expected, on the first day of treatment (acute administration of cocaine), the PPI levels of the rats decreased, but on the seventh day of treatment, an increase in PPI levels was observed. One week after finishing the chronic administration, they tested a new acute dose of cocaine and checking that PPI levels had decreased again. Therefore, in summary, it seems that the release of DA induced by the acute administration of cocaine decreases PPI levels, while the regulation of the dopaminergic activity that occurs during cocaine abstinence produces a progressive increase in PPI levels [78].

It could be suggested that, after a prolonged period of abstinence, PPI levels should normalize, as results with animal models seem to indicate [66, 67, 77]. However, the pattern of consumption of cocaine in humans is intermittent from the beginning, with longer or shorter periods of abstinence and multiple relapses. The neuroadaptations that develop during abstinence are manifested after regaining contact with cocaine. In rats with self-administration experience, cocaine caused a greater release of DA in the NAcc after 30 days of abstinence [79], while environmental stimuli caused a lower release of DA in the NAcc [80]. In cocaine users, the brain reward circuit response is also altered (in particular, in the form of a reduced reuptake of DA in the NAcc and VTA). This altered response is also observed in the rest of structures of the CSPP circuit, especially the connectivity between them. Thus, there is a decrease in functional connectivity between the thalamus and the mPFC [31] and between the mPFC and the NAcc [59]. These changes may be expressed with a different oscillation of PPI levels, and in the long term, could be related to the presence of a SUD or DP.

It is well established that an increase in DA decreases PPI levels, while a reduction in DA either increased or did not have any effect over the levels of PPI. This correspondence goes beyond a linear relationship between PPI levels and DA concentration [78]. In healthy subjects with a low PPI baseline, the PPI levels were increased with the administration of clozapine (a D4 receptor antagonist), which did not occur in subjects with a higher PPI baseline [81]. This baseline-dependent variation in PPI has also been observed with the administration of memantine (an antagonist of glutamatergic NMDA receptors). Memantine increased PPI levels in individuals considered to have a low PPI, while it either did not produce alterations or it tended to decrease PPI in those with a higher PPI [4]. Therefore, the PPI baseline of each individual seems to constitute an index indicative of the response that will occur after pharmacological manipulation. Something similar happens after the administration of a low dose of alcohol, where the variation produced in the PPI is determined by the subject's baseline at the beginning of the test. Alcohol decreased PPI levels in subjects who had low inhibition, but increased it in those with an already high PPI. The authors note that the PPI baseline may be indicative of the response to alcohol, since in their study, they did not observe other differences in response to alcohol (e.g., negative or positive affect and stimulant or sedative effect) [82]. Similar results have been obtained using amphetamine, with a PPI reduction only in women categorized as having a high PPI baseline [20].

Consistent with these results, studies with animal models have also shown that the PPI levels observed after the administration of a drug are dependent on the subject's PPI baseline, which in turn is related to the basal dopaminergic function of the animal [78]. Amphetamine decreases PPI levels in rats that have a higher PPI baseline [20]. Also, cocaine administration decreases the PPI levels only in rats categorized as having a high sensitivity to apomorphine (a dopamine agonist) [83]. Similarly, amphetamine also decreases PPI levels with very low doses in animals that are more sensitive to apomorphine, while higher doses are needed to obtain the same effect in less sensitive rats [84]. In addition to differences in the response to cocaine and amphetamine, the most sensitive rats to apomorphine present differences in their dopaminergic response to novelty, such as an increase in DA and in the tyrosine hydroxylase enzyme in the NAcc [85, 86]. Using rats with a preference for alcohol, results have been obtained pointing in the same direction. Although the administration of a high dose of amphetamine decreases PPI levels in all rats, the decrease in the levels observed is significantly greater in animals that show a higher preference for alcohol [87]. Thus, understanding that the dopaminergic alteration caused by cocaine consumption can be considered a primary factor in the alteration of PPI levels, the PPI present in each subject before cocaine consumption could determine the response to the drug and be indicative of the state of the dopaminergic function, as well as the neuroadaptations that cocaine could produce. Bearing the aforementioned in mind can be essential to understand the dopaminergic alteration caused by different factors, such as drug use [78].

To sum up this section, it should be noted that the PPI baseline should not per se be considered indicative of a merely dopaminergic alteration. Subjects considered as high risk to present psychosis show a reduction of PPI levels with respect to the controls, but the study did not observe significant correlation between the concentration of striatal DA and the PPI levels in each group [88]. The presence of dopaminergic hypersensitivity and an increase in DA in the mesolimbic system [30], which can be caused both by the use of drugs (hallucinogens and psychostimulants) [35] and by the combination of stress and genetic factors [30], has been related to lower PPI levels and the disconnection in the structures that regulate PPI levels, such as the thalamus and the prefrontal cortex [89].

## 1.6 PPI as a Biomarker in Cocaine Use Disorder

Recently, our laboratory has used PPI to discriminate subpopulations with different vulnerability to cocaine effects [13, 14]. We have reported that the baseline PPI level of mice can predict their sensitivity to the conditioned rewarding and motor effects of cocaine [13, 14]. Thus, male and female mice with a low PPI level presented a lower sensitivity to the conditioned reinforcing effects and motor effects of the drug than those with a high PPI. The differences in the cocaine response between the mice with a lower and higher PPI could be explained in part by the differences found in the levels of D1 and D2 dopamine receptors in the striatum of these animals. Mice with a lower PPI, especially females, showed higher levels of D2 receptor expression than those with a higher PPI [13]. High D2 receptor levels in the striatum have been related to a lack of pleasant effects of the psychostimulants [90], which could explain the lower sensitivity of low-PPI mice to the motor and reinforcing effects of cocaine [13, 14]. Other studies have also demonstrated the importance of baseline PPI level to determine the effects of DA agonists [20, 69, 91, 92]. Thus, amphetamine largely reduced PPI only in individuals with the highest basal levels of PPI, in both men [67] and women [20]. Furthermore, pergolide and amantadine, other DA agonists, reduced PPI in men with high baseline PPI levels, but not in subjects with low baseline PPI levels [91].

Nevertheless, the subjects with a low PPI level seem to display higher neural changes caused by cocaine consumption, which is evidenced when they are exposed to high doses of this drug. Thus, the low-PPI mice showed stronger associative cocaine effects with environmental cues when they were exposed to high doses of cocaine. Specifically, male mice did not extinguish the conditioned preference induced by cocaine and females reinstated the preference with much smaller doses than animals with a high PPI [13]. In addition, low-PPI mice pretreated with cocaine presented higher hyperactivity induced by cocaine than those pretreated with saline, while animals with a high PPI level were less affected by pretreatment with cocaine [14]. Therefore, mice with a low PPI level seem more vulnerable to develop behavioral sensitization induced by the drug when they were exposed to intermittent administration of cocaine. Likewise, low-PPI mice showed a stronger behavioral sensitization to the effects of amphetamine than the high-PPI ones, showing that males characterized by low basal PPI may be more susceptible to the development of dopamine sensitization too [92]. Behavioral sensitization, also called motor sensitization, is the increased motor activity induced by cocaine after drug intermittent administration [93]. It is known that intermittent access to a drug is more effective in producing neuroplasticity and addiction-like behaviors [94, 95]. For this reason, the increased motor activity induced by cocaine after drug intermittent administration (behavioral or motor sensitization [93]) is considered an indicator of neural changes caused by cocaine consumption and it may underlie drug craving and relapse to cocaine use [96]. Therefore, the fact that the subjects with low PPI develop a stronger motor sensitization suggests that they have a greater risk to display a SUD. Accordingly, PPI levels can be useful to discriminate between animals with a higher vulnerability to the effects of drug, at least for cocaine, being more likely to display behaviors induced by the drug, which can increase the risk of developing a cocaine use disorder. Consequently, we suggest that PPI can be considered a biomarker of vulnerability to the development of a cocaine use disorder.

In addition, as already mentioned, the PPI is a natural phenomenon that happens extensively and it is possible to successfully replicate it in the laboratory with various kinds of stimulus intensities and experimental designs. Therefore, in the following sections, we provide a detailed description of the PPI protocol used in both mice and humans in our laboratory.

## 2 Guidelines to Study the Prepulse Inhibition in Experimental Animals

PPI of acoustic startle response is observable across many species, but rodents are the experimental animals that have been more broadly studied. The startle response is reflexive and can be reliably elicited with a high acoustic stimulus. The startle response in rodents is measured to record the activation of skeletal muscles in the whole body. Next, we will explain the method used in our laboratory.

*2.1 Apparatus*

The equipment required to evaluate the PPI in rodents is a computer, a tube with a sensor, an accelerometer and a sound-proof chamber (mod startle response CERS, CIBERTEC, S.A, Madrid, Spain). The Plexiglas tube (28 × 15 × 17 cm) with a platform contains a sensor that has two small transparent doors on each side and it allow the animal to enter and exit (*see* Fig. 2a, b). The upper part is made of a flexible material (silicone) with holes that allow the animal to breathe. This upper part can be adjusted on each side depending on the size of the animal, allowing the animal to be restrained and comfortable, without leaving gaps. The lower part is a platform that contains the sensor at its base to gauge the force exerted onto the platform when the animal gets startled after a stimulus appears. The animal's startle response is transduced by an accelerometer (black base under the sensory box), recording the measure of the amplitude of the startle response as maximum value. The movements caused by the startle are transduced by an accelerometer and the signal is collected and digitized by a microcomputer, which is also used to adjust the sensor plate pressure in three positions: up to 80, 300, and 600 g allowing the apparatus to be used by a mouse or a rat (*see* Fig. 3). The unit is located in a soundproof chamber (90 × 55 × 60 cm) (*see* Fig. 4) constantly lit with a 10 W lamp and equipped with a loudspeaker located in the interior of the roof of the box that produces a constant sound as background noise (65 dB). Two 28 cm speakers located 15 cm to the two sides of the Plexiglas box produce acoustic stimuli. These speakers are connected to an amplifier, which in turn is connected to a noise generator that manages the startle stimulus and a second noise generator that produces the signal corresponding to the prepulse. The computer runs the software (*Monitor Respuesta Startle*, MRS3; CIBERTEC, S.A, Madrid, Spain) that controls the presentation of the acoustic stimulus, soundproofed camera lighting and data collection (*see* Figs. 3 and 4).

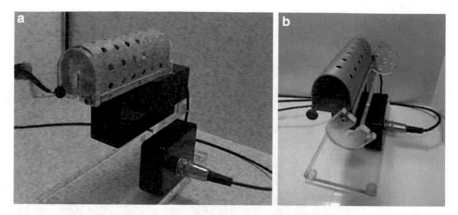

**Fig. 2** (**a**, **b**) The device is a Plexiglas tube with a sensor and two small transparent doors on each side that allow the animal to enter and exit. The upper part is made of a flexible and adjustable material with holes that allow the animal to breathe

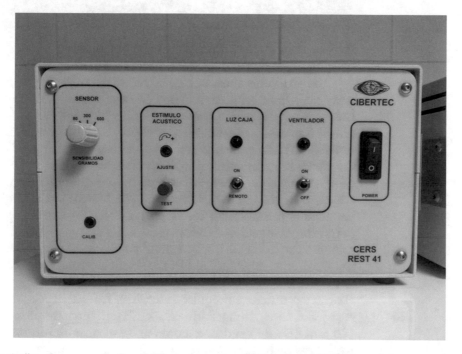

**Fig. 3** Controller of sensor unit. *Sensibilidad gramos*: sensitivity in gram; *Estímulo acústico*: acoustic stimulus; *Ajuste*: setting; *Luz caja*: light of box; *Remoto*: remote control; *Ventilador*: fan

**2.2 Stimuli Protocol Design**

There are many parameters and protocols of the pulse and prepulse stimuli used in the literature to evaluate the PPI in rodents. This is why, considering the literature, we carried out pilot experiments to obtain the most appropriate measure of PPI for our objective, a strong value of baseline PPI level in each animal, which was a good indicative of the state of its mesolimbic dopaminergic system. Next, we will explain the protocol employed by us.

**Fig. 4** Soundproof chamber where the sensor unit is placed

The procedure must be carried out in two phases, in 2 days; the first one is the Acclimation Phase. Mice are placed into the animal holder for 5 min with a constant 65-dB white noise as background noise, but no startle stimuli (*see* **Note 1**). The animal holder is cleaned after an animal is removed. The second phase, Day 2, is the block Prepulse Inhibition. This phase also starts with 5 min of 65-dB white noise, but it is followed by a program of stimuli as the white noise plays in the background (*see* **Note 2**). Although other parameters can be used, we employ a program consisting of two parts: the first one is a series of 50 trials of pulses of 120 dB to establish the base line. Hundred and twenty decibel is chosen as it is the maximum value reached in the startle response in pilot studies (*see* **Note 3**). In the second part, the prepulse inhibition is evaluated. Two different prepulse intensities (75 and 85 dB for 4 ms each) are used, along with two different interstimulus intervals (30 and 100 ms) and one single pulse at an intensity of 120 dB for 20 ms each. Thus, there are four types of prepulse–pulse trials: 75 dB/30 ms, 75 dB/100 ms, 85 dB/30 ms, and 85 dB/100 ms, all of them followed by a 120-dB pulse. To determine the value of the prepulse inhibition, we run the four types of prepulse–pulse trials, alongside single instances of the tones 75, 85, and 120 dB, 10 times each in pseudorandom order, totaling 70 trials with a 20-s interval between them (*see* **Note 4**). The total time for phase 2 is 45 min.

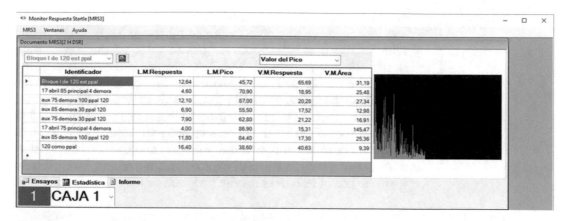

**Fig. 5** Image of a computer screen with the software

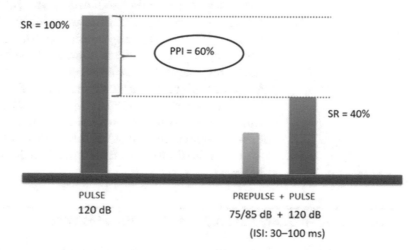

**Fig. 6** Diagram of prepulse inhibition (PPI) of startle reflex (SR) to acoustic stimulus and its parameters. *L. M. Respuesta*: mean latency of response; *L. M. Pico*: mean latency of top response; *V. M. Área*: average value of area of startle responses; *V. M. Respuesta*: average value of top of startle responses

**2.3 Software**

Figure 5 shows data recorded by the software. All trials carried out in the session are recorded in the Essays section, but the software in the Statistics section shows the average of the startle response of the animal in each block of trials: trials of 120-dB pulse, 75-dB pulse, and 85-dB pulse, trials of 75-dB prepulse and 120-dB pulse with an interstimulus interval (ISI) of 30 ms, 75-dB prepulse and 120-dB pulse with ISI of 100 ms, trials of 85-dB prepulse and 120-dB pulse with ISI of 30 ms, and trials of 85-dB prepulse and 120-dB pulse with ISI of 100 ms (*see* Fig. 6). These averages are shown in four different measures: Mean latency of response (mean time to display the startle response by the animal after the presentation of the pulse), mean latency of top response (mean time to display the highest startle response by the animal after the presentation of the pulse), average value of area of startle responses (average value of area of the curve of startle response by the animal), and average value of top of startle responses (mean value of the top of each startle response of animal). This last measure is used to obtain the PPI of the animal (*see* Fig. 5). Moreover, the screen can show us the diagram or compilation of all responses from which the computer gets the averages of each measure. Three commands at the top left on the screen allow for the necessary program adjustments, such as to introduce different pulses and prepulses and their administration sequences, as well as to save, export, and transform data, or help with the use of resources on the program.

### 2.4  Analysis of PPI

To obtain the measure of PPI, the startle response in the prepulse–pulse trials is averaged and divided by the mean value of the startle responses to the pulse and multiplied by 100. This shows the amount of remaining startle (in percent of the baseline startle) under different prepulse conditions for each animal. Thus, the percentage of PPI (%PPI) is obtained by subtracting the remaining startle response from 100% [%PPI = 100 − (100 × startle response for pulse with prepulse/startle response for pulse alone)].

The four prepulse–pulse types of PPI are used to divide the animals in a two-cluster analysis of K mean (cluster) in high or low PPI. Cluster centers must be determined separately by sex to allow for equal distribution of males and females in each PPI group (*see* **Note 5**).

## 3  Guidelines to Study the Prepulse Inhibition in Human

The PPI paradigm is commonly measured and calculated in humans using startle eyeblink electromyographic (EMG) studies. Thus, the most widely used methodological procedures for this paradigm will be the focus of the following sections. Establishing guidelines for human startle eyeblink is important to standardize the methodology and results across laboratories [97, 98]. This text focuses on the most commonly used and representative techniques and tools used to set up the PPI device for humans in our laboratory, according to the existing human startle eyeblink literature. The human startle response is a sensitive, noninvasive measure of activity of the CNS and is extensively used in research and clinical settings. It is mostly measured with the eyeblink reflex in the right orbicularis eye muscle, since this reflex is suspected to be stronger in the right side of the face [99], but both sides appear across the literature. The reflex is generated by acoustic stimuli often presented by headphones and in some cases by speakers. It must be taken into account that the resulting eyeblink has different characteristics than a spontaneous one, because it is evoked and thus has to be recorded and analyzed differently.

Therefore, the following section will be centered around how to create the stimuli protocol for its acoustic presentation with headphones, how to place the electrodes properly in the right orbicularis muscle to measure the eyeblink reflex elicited by these acoustic stimuli, and how to set up the hardware and software employed to use electromyography to establish the PPI percentage in humans in relation to addiction.

**3.1  Stimuli Protocol Design**

Existing literature is not very consistent on the parameters and values of the pulse and prepulse stimuli used. The sensorimotor gating, the CNS's ability to filter for relevant information and prevent overstimulation, is able to occur within a wide range of stimuli and intensities. For this, several conditions must be taken into account, some of them intrinsic to human and its attentional and preattentional function, which could also be influenced by several substances of abuse and the addictive behavior (*see* **Notes 6–10**).

White noise is the presented acoustic stimulus of choice, since it is more neutral than tones. It has been shown to evoke higher response magnitudes with higher probability and amplitude, and it is generated to contain frequencies in the 20 Hz to 20 kHz range [97].

Pulse intensity is allowed to exceed 100 dB if the duration is 40 ms. For example, a 115-dB Pulse for 40 ms is commonly employed. Another option could be a 100-dB Pulse with a duration of 50 ms. In this case, the loss of intensity is compensated with an increase in duration. Using background noise is highly recommended for several purposes: Firstly, it allows to create pulse and prepulse stimuli clearly distinct from background noise, with a basal reference (*see* **Note 11**). Furthermore, background noise enables the CNS to habituate to the white noise and masks the possible environmental noise out of the headphones. Seventy-decibel background noise is a good value that allows to create prepulses with a difference of 15 dB over it. This difference of 15 dB is recommended and well established by the literature [100] leading to 0.94 intraclass correlations in PPI measurements repeated at 1-month intervals for 3 months when this background noise to prepulse distinction is used. Accordingly, an 85-dB prepulse stimuli for 20 ms would be ideal. Several studies use different intensities of prepulse in the same test (e.g., 2, 4, 8 and 15 dB above background noise). This procedure is helpful when impairments in PPI caused by substances of abuse such as MDMA or cannabis are pursued [101]. If the purpose of the study is to establish the baseline PPI percentage as a predictor, 15 dB above background noise prepulse intensity should be enough to differentiate between subjects with high and low PPI (*see* **Note 12**).

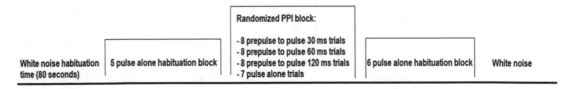

**Fig. 7** Stimuli protocol test

Pulse and prepulse stimuli should not appear suddenly, but rather after a period of only background noise (from 80 s up to 5 min) and one pulse alone block in order to habituate the subject to the white noise and startle reflex, respectively (*see* Fig. 7).

Various prepulse to pulse interstimulus intervals (ISI) are used in the same test. Prepulse to pulse intervals of 30–240 ms are typically utilized in human PPI experiments. Intertrial intervals (ITI) within the same block can vary from 8 to 30 s, all randomized (*see* Fig. 7). Maximal amplitude inhibition generally occurs within intervals of approximately 120 ms, and the opposite with 30 ms. A typical experimental design utilizes 30, 60, and 120 ms prepulse intervals [100]. Other studies utilize a 240 ms interval as well [50], but it must be taken into account that intervals of over 150 ms may not be considered preattentional because they reach consciousness.

In summary, a typical stimulus protocol configuration should include 70 dB background noise, 85 dB 20 ms prepulse, and 115 dB 40 ms pulse stimuli, with 30, 60, and 120 ms ISI between prepulse and pulse. Nevertheless, as mentioned above, others conditions and configurations have been successfully tested. In addition to this, it would be advisable to administer the test with background noise as habituation time, a pulse alone trial, PPI trial with randomized prepulse conditions (e.g., 30, 60, and 120 ms) and another pulse alone trial that allows to calculate habituation to the startle reflex as well (*see* Fig. 7).

*3.2 Hardware*

The electromyography device can be provided by several companies and brands. It should contain at least one module for the power supply (MP150 BIOPAC Systems. CIBERTEC, S.A, Madrid, Spain), one module for the acoustic stimuli presentation with a headphone jack and volume control (STM 100 BIOPAC Systems. CIBERTEC, S.A, Madrid, Spain), one module to create a stimulus analogic channel (UIM 100C BIOPAC Systems. CIBERTEC, S.A, Madrid, Spain), one EMG amplifier module (EMG 100C BIOPAC Systems. CIBERTEC, S.A, Madrid, Spain) with three Ag/AgCl electrodes plugged in, two active shielded electrodes for the orbicularis eye muscle and one isolated ground unshielded electrode for the forehead. The headphones module should have the volume control in the same position at all times in order to keep the created stimulus values constant (*see* **Note 13**).

It is crucial to have a sound level meter that is able to use a decibels scale (dB (A) SPL scale) to measure the stimulus intensity directly from the headphones. This matter will be further elaborated in the next section.

Additional disposable material, such as cleansing creams, conductive gels, adhesive collars and cotton cleaners may be needed, as well as a soundproof cabin to perform the test, ensuring the correct elimination of ambient noise.

## 3.3 Software

### 3.3.1 Filters and Channels

Implementing filters on the EMG device allows to record only the desired frequencies. An EMG device with an installed analogic band-stop filter is practical, but can be substituted by creating a digital filter of 50 or 60 Hz, operating as a wall. Beyond the band-stop filter, low-pass and high-pass filters should be added as well. It is common to apply a 500 Hz low-pass filter to avoid higher frequencies and a 1 Hz high-pass filter to avoid lower frequencies [97]. These options should be selected in the amplifier as well as in the software options, leading to an analog-to-digital conversion.

Once the signal frequencies have been filtered by the amplifier, it is possible to filter out the recorded EMG raw signal in the same way by adding digital filters (IIR-Filter) in the software (AcqKnowledge 5.0 BIOPAC Systems. CIBERTEC, S.A, Madrid, Spain). The commonly used values are 28 Hz low frequency and 500 Hz high frequency filters, in addition to the 50 Hz band-stop filter in case a digital one is required.

The integration of the EMG filtered signal is necessary in order to calculate appropriate values and visualize the evoked eyeblink signal correctly. There are two main methods to integrate the signal: The Root Mean Square and the Rectify. Selecting the Maximum (Max) value of an interval just under the stimulus provides the number with which to calculate the average for that kind of stimulus. Although two different methods may provide different Max values, the appearance of the signal should be equal and proportional. The typical number of samples recorded is 5, 20 or 30 samples. The higher the sample size, the higher the risk of losing or smoothing the Max value of the startle reflex (*see* **Note 14**).

Therefore, as shown in Fig. 8, a typical set up should have two analogical channels, one for the EMG raw signal (500 Hz low-pass and 1 Hz high-pass filtered) and another one for the stimuli protocol. In case the EMG device does not have an analogical band-stop filter, the commonly used digital channels would be a 50 Hz band-stop, 28 Hz low frequency and 500 Hz high frequency band-pass filters. The Root Mean Square or Rectify integration methods can be used indistinctly. Removing the baseline in the settings is not necessary when using the Root Mean Square with a properly functioning recording device (*see* **Note 15**), since the startle reflex signal does not have a specific value and it clearly stands out from the baseline signal.

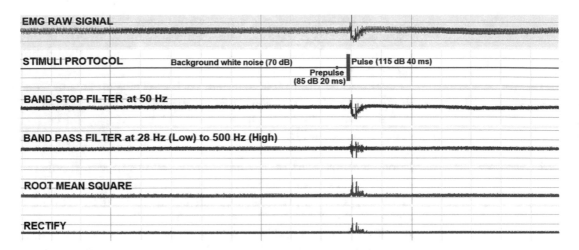

**Fig. 8** Software channels template

*3.3.2 Stimuli Protocol Creation*

Several methods and software applications can be used to create the acoustic stimuli. The main objective is to create a white noise stimulus that can be measured by a sound level meter directly from the headphones. For this, it is advisable to annotate the value given by the software for the stimulus (presumably in volts) to avoid repeating this step constantly.

The first step is to generate an empty recording with the duration of the test, and then transform it to white noise using the software settings. When an interval of white noise is selected, it is possible to obtain its maximum value in volts using software and decibels (dB) if measured in the headphones using a sound level meter. It is possible to manipulate the white noise sample at pleasure with software tools (addition, subtraction, multiplication and division), and it is necessary to do it until the desired value in dB is found. Then, it is advisable to annotate the maximum value of the sample in volts given by the software. Doing this prevents us from trying at random the next time. The values will be constant as long as the volume position in the analog device remains unchanged. The same operation must be carried out in order to find the values corresponding to the pulse and prepulse stimuli. When the 70-dB background noise software value is found, it is necessary to replicate it throughout the test as a baseline.

Having the relevant software values for pulse and prepulse at hand, it is possible to create a 115-dB pulse (40 ms) above the baseline background noise and then copy-paste it throughout the test in the desired positions. The same operation can be done with the 85-dB prepulses (20 ms). It must be taken into account that it is possible to copy and paste prepulses as well, but the prepulse to pulse interval (ISI) changes according to the trial (30, 60 and 120 ms). Once the stimuli protocol database is ready, it is possible to save it to be loaded later in the EMG template.

*3.3.3 Measurement*

PPI is commonly calculated in humans as $((A - B)/A) \times 100$, where $A$ = the startle only mean and $B$ = the prepulse mean. Larger numbers indicate greater PPI percentage. Only the mean from the PPI trial block and not from the habituation pulse alone trial blocks should be considered for PPI calculation.

Each mean must be calculated independently. Therefore, the proposed design includes a PPI percentage for 30, 60 and 120 ms. Hundred and twenty-millisecond trials usually have greater inhibition in healthy subjects, as the prepulse to pulse latency is closer to the threshold of consciousness. The opposite effect is found in 30-ms trials due to the short separation between prepulse and pulse.

Pulse alone trials in PPI block can be used to calculate the startle reflex magnitude, and pulse alone blocks to establish the habituation capability of the CNS. As such, we count on blocks 1 and 3 to establish habituation and block 2 to establish PPI and startle reflex magnitude (*see* Fig. 7). Thus, habituation is calculated as the percentage decrease in startle amplitude from pulse alone block 1 to pulse alone block 3. However, the sensitization response from pulse alone trial 1 to pulse alone trial 2 in block 1 is able to occur [102], giving rise to another measure allowed to include in the study, as well as another possible bias source to control. For this reason, habituation should be calculated from pulse alone trial 3 in block 1 through the rest of pulse alone trials in habituation blocks, and not from pulse alone trials 1 or 2.

Software programs may have a detector for each trial condition, searching for a specific time interval between prepulse and pulse in a specific channel (e.g., Root Mean Square channel). Then, all the values for that condition can be copied and exported for further average calculation. Though, each test design presents its different stimuli at the same point, and therefore the average of each condition can be calculated manually.

**3.4 Electrodes Placement**

A proper placement of the electrodes in the right orbicularis muscle is essential not only to obtain a good startle eyeblink measurement but also to not falsely classify a subject as a nonresponder. It is estimated that about 5–10% of people do not emit startle eyeblink in response to acoustic stimuli [97]. This physiological behavior of not blinking after an acoustic stimulus could be understood in evolutionary terms, with a small percentage of people generating a "freezing," instead of the classical "fight or flight" response to aversive stimuli. It is important to take into account that a subject that emits startle eyeblinks in the beginning but stops to do so after some time is not a nonresponder but a rapid-habituator in an adaptive and expected condition.

Good skin preparation is essential to eliminate potentially extraneous and biasing variables. Startle eyeblink is a very sensitive and delicate measure that usually generates values of 0–300 μV

[97, 98]. Considering that 1 V is a million microvolts, this part of the test should be performed following paramount and rigorous procedures. Firstly, the subject should be asked to clean the skin under their right eye and forehead with water and neutral pH soap and dry well. After this, the area is sometimes lightly scraped with an abrasive cream on a cotton, but it has to be done very carefully because the area is very close to the eye. It is also important not to leave behind any residue of abrasive cream or cotton, because this can interact with the signal recorded by the electrodes.

The first electrode to be placed on the prepared skin surface could be the negative one, then the positive one and lastly the one on the forehead. We must take into account that conductive gel use is necessary to transmit the muscular signal, as well as adhesives to paste the electrodes to the face. The most common are Ag/AgCL shielded electrodes with (19 mm OD, 4 mm ID) adhesive collars. The negative electrode must be placed just as close as possible to the lower eyelid, without touching it to avoid interference with normal blinking. It is possible to look at the pupil of the subject to locate the center of the eyelid. In order to facilitate the positioning, it is recommended to cut the electrode adhesive in the upper side, giving rise to an electrode well placed close to the lower eyelid without harming the normal blink pattern with the adhesive (*see* Fig. 9). The positive electrode must be placed about 1.5 cm next to the negative one and a little higher, following the natural orbicularis oculi curvature (*see* Figs. 10 and 11) and close to the lower eyelid on the side. The upper side of this electrode should be cut as well. It is also possible to cut the inner sides of both electrodes to avoid overlapping (*see* Fig. 9). In order to remove any possibility of the electrodes communicating because of gel excipient between them, this part of the face can be cleaned with a cotton swab. Electrodes work by detecting changes and differences between them, and a possible communication could eliminate the signal measurement, making it impossible to detect the generated startle eyeblink. The isolated ground electrode must be placed in the center of the forehead (*see* Figs. 10 and 11). Correct skin preparation and rigorous procedure can prevent signal noise, however it is possible to check it before the test with an Electrode Impedance Checker, connecting it to the electrodes. Impedance value should be less than 5 k$\Omega$. Further information about signal noise and how to reduce it can be found in **Note 16**. A cotton swab can be used after the test to properly clean the electrodes of conductive gel.

## 4    Notes

1. It is very important that the Acclimation Phase be carried out a day before the Test Phase. We should keep in mind that the apparatus records the startle response of the animal with an

**Fig. 9** Severed adhesive collars

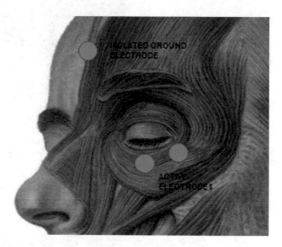

**Fig. 10** Right orbicularis eye muscle with electrodes

acoustic stimulus, therefore if the rodent is stressed from being in a new environment, its PPI level can be disrupted. This phase allows for the animal to get used to the apparatus and reduces its anxiety. During this phase, the rodent will calm down, stop exploring the environment and stop moving around.

2. The presentation of white noise is intended to point the animal's attention toward the acoustic sense; which has been verified to evoke a higher response, as we will explain in the human procedure. In pilot studies from our laboratory, we confirmed that a 65 dB white noise allow for the startle response of the animal with the main pulse.

3. To determine the main pulse, we must test the startle response to pulses of different intensities, beginning with 70 dB, increasing by 5 dB until 120 dB. Each pulse lasts 20 ms and occurs every 20 s. The selected main pulse must be the one that causes

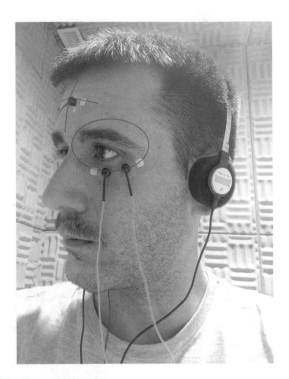

**Fig. 11** EMG set up with headphones

the highest startle response in the animal, following which the startle response is not higher even if pulse intensity increases. In pilot studies from our laboratory, we verified that OF1 mice reached a plateau in the startle response with the 110 dB pulse in males and with the 120 dB pulse in females.

4. The prepulses (75/85 dB) are introduced to verify that they are not acting as pulses and to confirm that only the 120 dB pulse is the main stimulus to induce the startle response in the animal. The selection of intensity of the prepulses was based on the literature [19].

5. Sex differences in PPI have been reported with women exhibiting lower PPI than men [103]; however, we have not observed significant sex differences in PPI in previous studies, maybe due to the use of several measures of PPI for obtaining of the average of PPI [13, 14]. However, we have observed sex differences in the SR in our studies; therefore, we consider it advisable to use a two-cluster analysis of K mean for the distribution between lower and higher PPI due to be determined separately by sex.

6. Affective and emotional states play a key role in modulating PPI [98]. Valence of perceived stimuli can be differentiated between pleasant and aversive, and the subject's arousal may contribute to the response given to these stimuli (*see* Fig. 12).

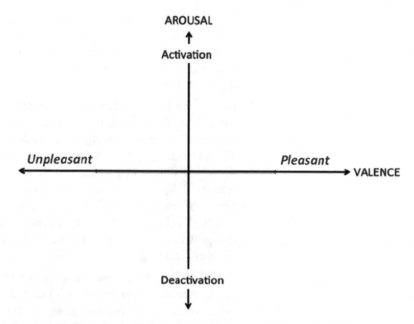

**Fig. 12** Valence and arousal axes. (Modified from Nesbitt et al. [98])

Aversive states lead to the tendency of perceiving the stimuli as more threatening, reducing the PPI. To equalize the emotional states of the subjects, it is important to explain the procedure in detail to relax them and reduce any possible anxiety caused by the novel context. It is also important to give each subject the same instructions (e.g., relax and keep their eyes open as they hear the sounds while looking at a fixed point). The subject's arousal index can be controlled by measuring the tension and heart rate using a tensiometer after the test, as well as the levels of induced affect and arousal using the Mood Grid Scale [104].

7. Several substances of abuse and drugs have been tested for their effects on PPI [100]. Stimulants such as cocaine, MDMA, nicotine or caffeine have been shown to increase the SR and diminish PPI, probably due to a state of induced hyper-arousal. In the case of nicotine, this effect can last up to 10 min [105]. Furthermore, caffeine doses equivalent to one or two cups of coffee (200 mg) have yielded opposite effects in low and high caffeine drinkers [106]. In low caffeine drinkers, caffeine withdrawal (14 h before test) led to decreased PPI and the absence of withdrawal led to PPI increased due to caffeine. The opposite pattern was evident in high caffeine drinkers. High caffeine drinkers led to increased PPI during withdrawal and decreased PPI in the absence of withdrawal. These results suggest that coffee and nicotine cigarette consumption in healthy subjects are potential sources of bias. Thus, for regular caffeine drinkers, it would be preferable to

drink coffee rather than experience withdrawal symptoms. Similarly, in subjects with high rates of nicotine consumption, such as schizophrenia patients, participants who smoke may lead to less biased effects than participants experiencing withdrawal anxiety [107].

8. According to Swerdlow, Hartman and Auerbach [108], sex is a variable to consider in order to not skew comparisons between men and women. Usually, men yield more robust and higher PPI percentages, while women with high progesterone levels in their mid-luteal phase show a decreased sensorimotor gating index [109]. Therefore, it is recommended to establish the phase of the menstrual cycle of female participants when PPI is tested, asking for their period duration mean and when their last period was. This enables us to cluster women according to their period phase.

9. A PPI deficit is observed in patients with various metal disorders, mainly schizophrenia, posttraumatic stress disorder, obsessive compulsive disorder, anxiety disorders and major depressive disorders [5]. Thus, asking for mental and physical diseases is advised. Additionally, being able to hear properly is a prerequisite of the test, because the most common form to present stimuli is acoustic.

10. Modality-specific directed attention may be considered as well. Some laboratories use a neutral video to maintain the subject's attention. However, enhanced visual attention could deteriorate acoustic attention, leading to decreased startle reflexes. Relaxing and maintaining the eyes open should be sufficient to perform the test.

11. It is highly recommended not to present prepulse and pulse stimuli directly, without using background white noise (e.g., 70 dB) as a baseline [100]. This makes it possible to have a basal reference, habituates the subject to the white noise, allows to calculate the dB difference between prepulse or pulse and background (e.g., 15 dB for prepulse) and masks the possible environmental noise.

12. The relationship between PPI and addiction can be either as a predictor of vulnerability [13, 50] or as a result of the consequences of substances of abuse [101]. If it is used as a predictor, the typical set up with one type of prepulse (e.g., 15 dB above background) might be enough. This allows to cluster the subjects (e.g., low or high) for further correlations with psychometric scales such as Alcohol, Smoking and Substances Involvement Screening Test (ASSIST) or Cannabis Abuse Screening Test (CAST). Contrarily, if PPI is used to measure the impairments caused by different substances of abuse, it is preferable to create a protocol with several kinds

of prepulses (e.g., 2, 4, 8, and 15 dB above background) since significant differences may appear in some conditions but not in others, as PPI is a natural phenomenon of the brain reproduced by acoustic stimuli in the laboratory.

13. It is important to keep the volume control in the headphones output module always in the same position. Stimuli values in Volts can be annotated when they are found, as well as the operations and procedure done for this purpose, in order to replicate the stimuli protocol if required. However, if the volume control position is changed, the values will not be valid anymore, as with the intensities in decibels presented to the subjects. The simplest method is to check that the volume control is always at maximum value.

14. The sample number in the integration methods may be a source of bias for the evoked blink due to its quick nature. The objective is to calculate the average for any kind of stimulus with the maximum value of each one, in short, the peak (*see* Fig. 8). If a high number of samples is chosen (e.g., 20 or 30), there is the risk of smoothing the peak too much, losing it partially in the integration channel.

15. When the Root Mean Square integration method is chosen, we have the option to remove the baseline or not. For the main PPI, it is important how large the recorded signal is, as well as its distinction from the baseline, taking care to minimize the recorded noise as much as possible (*see* **Note 16**). There is not a typical or "normal" value for PPI due to the differences in methodology and procedures across laboratories. The main factor is the inhibition percentage of each subject. It is possible to check the proper functioning of the EMG by recording an empty sample, without the electrodes plugged in. If the value is close to zero, there is no reason to remove the baseline in the Root Mean Square integration method. If the recorded signal is well above zero, it is possible to adjust it with a little screw on the EMG amplifier labeled "Zero."

16. Batteries and devices with electrical power may increase the noise recorded by the electrodes and bias the signal. Ideally, the cabin in which the test is done should be soundproof and have a Faraday cage. It is important to inform subjects not to carry with them their mobile phone during the EMG test. The cabin should include only the necessary laptops, because their batteries can contaminate the signal as well. The chargers should not be plugged in unless necessary. A partial solution could be to keep the EMG device as far away from any laptops, mobile phones, chargers, and plugs as possible. For this reason, a good cabin design is crucial. Placing the active electrodes too close between them can also increase the noise. Visualizing the signal

noise present in a selected interval is possible by performing a Fast Fourier Transform (FFT) analysis with the software, which provides through an algorithm a frequency analysis of the components of the selected sample interval, previously imperceptible to the human eye. If a 50 Hz band-stop filter is used as a wall, harmonics spikes at 100, 150 and 200 Hz in the FFT analysis may be getting through, which means that there is too much noise in the signal and it is necessary to reduce it.

## 5    Conclusion

In summary, the PPI of SR response to an acoustic stimulus is a sensorimotor gating index widely used to assay the preattentional inhibitory function of subjects. The PPI paradigm evaluates a very adaptive phenomenon involved in information selection processing, which is altered in some psychiatric disorders [3–5, 8, 9, 11]. The PPI test is widely used in animal models of diseases marked by a failure to inhibit irrelevant information in sensory, motor, or cognitive areas. Hence, a PPI deficit is observed in disorders with alterations in the cerebral dopaminergic system, such as the mesolimbic pathway; and it is considered a endophenotype of schizophrenia [4, 9, 10]. Recently, given the coincidence in neural structures regulating PPI and drug addiction [43, 78], we assayed the predictive capacity of PPI to identify the more vulnerable animals to the effects of cocaine [13, 14]. We have demonstrated that mice with a low PPI show more consequences after cocaine exposure, which suggests a higher risk to develop a SUD. The PPI paradigm identified individuals who were more likely to display behaviors induced by cocaine, which can increase the risk of developing a cocaine use disorder [13, 14]. Therefore, we consider that a PPI deficit could be a biomarker of vulnerability to display SUD with cocaine use.

The main advantages of the PPI paradigm aforementioned are: (1) it is a noninvasive test; (2) it provides an automatic measure; (3) it is not linked to learning; (4) it keeps certain stability over time; and (5) it occurs in a large number of species in a similar manner. These characteristics make the PPI test has a great translational value, allowing to research the risk factors of SUD and the vulnerability to develop cocaine addiction both in rodents and humans.

Considering all of the above, a PPI deficit indicates alterations in the mesolimbic dopaminergic system [3], and a subject with low PPI will be more vulnerable to the cerebral changes induced by cocaine use. Therefore, determining the PPI level of a subject allows us to know the individual degree of vulnerability to develop a SUD. However, we must never consider the PPI level as a direct marker of a pathology, as it is not a symptom. A PPI deficit does not predict a SUD, since some individuals at familiar risk of schizophrenia exhibit a PPI deficit, but they never develop the disorder.

Finally, identifying subpopulations with great vulnerability to cocaine effects can help us to study other risk factors such as stress. It is well known that stress is one of the main risk factors in the development of many psychiatric disorders such as schizophrenia [30], and substance use disorder [110], and this relation is mediated through the dopaminergic system. Thus, PPI could identify animals with a lower resilience, that is, those with a greater vulnerability to develop the behavioral consequences of social stress in relation to drugs. Therefore, obtaining psychophysiological markers that allow us to detect the most vulnerable subjects to develop a cocaine use disorder is a priority for further research aimed at preventing the consequences of using this drug.

## Acknowledgments

We wish to thank to Guillermo Chuliá for his editing of the manuscript and Dr Terry D. Blumenthal for sharing his technical experience in electromyography with us. This work was supported by the following research grants: Ministerio de Economía y Competitividad. Proyecto I+D+i PSI2015-69649-R; and UV-INV-AE19-1203315.

## References

1. Valls-Solé J (2004) Funciones y disfunciones de la reacción de sobresalto en el ser humano. Rev Neurol 39:946–955

2. García-Sánchez F, Martínez-Gras I, Rodríguez-Jiménez R, Rubio G (2011) Inhibición prepulso del reflejo de la respuesta de sobresalto en los trastornos neuropsiquiátricos. Rev Neurol 53(7):422–432

3. Swerdlow NR, Braff DL, Geyer MA (2016) Sensoriomotor gating of the startle reflex: what we said 25 years ago, what has happened since then, and what comes next. J Psychopharmacol 30(11):1072–1081

4. Swerdlow NR, Light GA (2016) Animal models of deficient sensorimotor gating in Schizophrenia: are they still relevant? Curr Top Behav Neurosci 28:305–325

5. Kohl S, Heekeren K, Klosterkötten J, Kuhn J (2013) Prepulse inhibition in psychiatric disorders - apart from schizophrenia. J Psychiatr Res 47:445–452

6. Vargas JP, Diaz E, Portavella M, López JC (2016) Animal models of maladaptive traits: disorders in sensorimotor gating and attentional quantifiable responses as possible endophenotypes. Front Psychol 19:7(206)

7. Miller EA, Kastner DB, Grzybowski MN, Dwinell MR, Geurts AM, Frank LM (2020) Robust and replicable measurement for prepulse inhibition of the acoustic startle response. Mol Psychiatry. https://doi.org/10.1038/s41380-020-0703-y

8. Braff DL, Greenwood TA, Swerdlow NR, Light GA, Schork NJ, Investigadores del Consortium on the Genetics of Schizophrenia (2008) Avances en la endotipificación de la esquizofrenia. World Psychiatry (Ed Esp) 6:11–18

9. Braff DL (2010) Prepulse inhibition of the startle reflex: a window on the brain in schizophrenia. Curr Top Behav Neurosci 4:349–371

10. Mena A, Ruiz-Salas JC, Puentes A, Dorado I, Ruiz-Veguilla M, De la Casa LG (2016) Reduced prepulse inhibition as a biomarker of schizophrenia. Front Behav Neurosci 10:202

11. Mao Z, Bo Q, Li W, Wang Z, Ma X, Wang C (2019) Prepulse inhibition in patients with

bipolar disorder: a systematic review and meta-analysis. BMC Psychiatry 19(1):282

12. Saletti PG, Tomaz C (2018) Cannabidiol effects on prepulse inhibition in nonhuman primates. Rev Neurosci 30(1):95–105

13. Arenas MC, Navarro-Francés CI, Montagud-Romero S, Miñarro J, Manzanedo C (2018) Baseline prepulse inhibition of the startle reflex predicts the sensitivity to the conditioned rewarding effects of cocaine in male and female mice. Psychopharmacology 235(9):2651–2663

14. Arenas MC, Blanco-Gandía MC, Miñarro J, Manzanedo C (2020) Prepulse inhibition of the startle reflex as a predictor of vulnerability to develop locomotor sensitization to cocaine. Front Behav Neurosci 13:296

15. Koch M, Schnitzler HU (1997) The acoustic startle response in rats—circuits mediating evocation, inhibition and potentiation. Behav Brain Res 89(1):35–49

16. Rohleder C, Wiedermann D, Neumaier B, Drzezga A, Timmermann L, Graf R, Leweke FM, Endepols H (2016) The functional networks of prepulse inhibition: neuronal connectivity analysis based on FDG-PET in awake and unrestrained rats. Front Behav Neurosci 10:148

17. Swerdlow NR, Bhakta SG, Rana BK, Kei J, Chou HH, Talledo JA (2017) Sensorimotor gating in healthy adults tested over a 15 year period. Biol Psychol 123:177–186

18. Marín-Mayor M, Jurado-Barba R, Martínez-Grass I, Ponce-Alfaro G, Rubio-Valladolid G (2014) La respuesta de sobresalto y la inhibición prepulso en los trastornos por uso de alcohol. Implicaciones para la práctica clínica. Clínica y Salud 25:147–155

19. Valsamis B, Schmid S (2011) Habituation and prepulse inhibition of acoustic startle in rodents. J Vis Exp 55:1–10

20. Talledo JA, Sutherland Owens AN, Schortinghuis T, Swerdlow NR (2009) Amphetamine effects on startle gating in normal women and female rats. Psychopharmacology 204(1):165–175

21. Kumari V, Antonova E, Zachariah E, Galea A, Aasen I, Ettinger U et al (2005) Structural brain correlates of prepulse inhibition of the acoustic startle response in healthy humans. NeuroImage 26(4):1052–1058

22. Fendt M, Li L, Yeomans JS (2001) Brain stem circuits mediating prepulse inhibition of the startle reflex. Psychopharmacology 156 (2–3):216–224

23. Fierro M (2011) El desarrollo conceptual de la ciencia cognitiva. Parte I. Rev Colomb Psiquiatr 40(3):519–533

24. Andreasen NC (1999) A unitary model of schizophrenia: Bleuler's "fragmented phrene" as schizencephaly. Arch Gen Psychiatry 56 (9):781–787

25. Ding Y, Xu N, Gao Y, Wu Z, Li L (2019) The role of the deeper layers of the superior colliculus in attentional modulations of prepulse inhibition. Behav Brain Res 364:106–113

26. Azzopardi E, Louttit AG, DeOliveira C, Laviolette SR, Schmid S (2018) The role of cholinergic midbrain neurons in startle and prepulse inhibition. J Neurosci 38 (41):8798–8808

27. Fulcher N, Azzopardi E, De Oliveira C et al (2020) Deciphering midbrain mechanisms underlying prepulse inhibition of startle. Prog Neurobiol 185:101734

28. Meng Q, Ding Y, Chen L, Li L (2020) The medial agranular cortex mediates attentional enhancement of prepulse inhibition of the startle reflex. Behav Brain Res 383:112511

29. Öz P, Kaya Yertutanol FD, Gözler T, Özçetin A, Uzbay IT (2017) Lesions of the paraventricular thalamic nucleus attenuates prepulse inhibition of the acoustic startle reflex. Neurosci Lett 642:31–36

30. Howes OD, McCutcheon R, Owen MJ, Murray RM (2017) The role of genes, stress, and dopamine in the development of schizophrenia. Biol Psychiatry 81(1):9–20

31. Zhang S, Hu S, Bednarski SR, Erdman E, Li CS (2014) Error-related functional connectivity of the thalamus in cocaine dependence. Neuroimage Clin 4:585–592

32. Zhang J, Forkstam C, Engel JA, Svensson L (2000) Role of dopamine in prepulse inhibition of acoustic startle. Psychopharmacology (Berlin) 149(2):181–188

33. Bitsios P, Giakoumaki SG, Theou K, Frangou S (2006) Increased prepulse inhibition of the acoustic startle response is associated with better strategy formation and execution times in healthy males. Neuropsychologia 44 (12):2494–2499

34. Giakoumaki SG, Bitsios P, Frangou S (2006) The level of prepulse inhibition in healthy individuals may index cortical modulation of early information processing. Brain Res 1078 (1):168–170

35. Halberstadt AL, Geyer MA (2018) Effect of hallucinogens on unconditioned behavior. Curr Top Behav Neurosci 36:159–199

36. Fendt M (1999) Enhancement of prepulse inhibition after blockade of GABA activity

within the superior colliculus. Brain Res 833 (1):81–85

37. Wang J, Song HR, Guo MN, Ma SF, Yun Q, Zhang WN (2020) Adult conditional knockout of PGC-1α in GABAergic neurons causes exaggerated startle reactivity, impaired short-term habituation and hyperactivity. Brain Res Bull 157:128–139

38. Bosch D, Schmid S (2008) Cholinergic mechanism underlying prepulse inhibition of the startle response in rats. Neuroscience 155 (1):326–335

39. Geyer MA, Krebs-Thomson K, Braff DL, Swerdlow NR (2001) Pharmacological studies of prepulse inhibition models of sensorimotor gating deficits in schizophrenia: a decade in review. Psychopharmacology (Berlin) 156(2–3):117–154

40. Ralph-Williams RJ, Lehmann-Masten V, Geyer MA (2003) Dopamine D1 rather than D2 receptor agonists disrupt prepulse inhibition of startle in mice. Neuropsychopharmacology 28(1):108–118

41. Doherty JM, Masten VL, Powell SB, Ralph RJ, Klamer D, Low MJ et al (2008) Contributions of dopamine D1, D2, and D3 receptor subtypes to the disruptive effects of cocaine on prepulse inhibition in mice. Neuropsychopharmacology 33(11):2648–2656

42. Yamashita M, Fukushima S, Shen HW, Hall FS, Uhl GR, Numachi Y et al (2006) Norepinephrine transporter blockade can normalize the prepulse inhibition deficits found in dopamine transporter knockout mice. Neuropsychopharmacology 31(10):2132–2139

43. Volkow ND, Morales M (2015) The brain on drugs: from reward to addiction. Cell 162 (4):712–725

44. Kumari V, Mulligan OF, Cotter PA, Poon L, Toone BK, Checkley SA et al (1998) Effects of single oral administrations of haloperidol and d-amphetamine on prepulse inhibition of the acoustic startle reflex in healthy male volunteers. Behav Pharmacol 9(7):567–576

45. Grillon C, Sinha R, Ameli R, O'Malley SS (2000) Effects of alcohol on baseline startle and prepulse inhibition in young men at risk for alcoholism and/or anxiety disorders. J Stud Alcohol 61(1):46–54

46. Arias F, Szerman N, Vega P, Mesías B, Basurte I, Morant C et al (2013) Abuso o dependencia a la cocaína y otros trastornos psiquiátricos. Estudio Madrid sobre la prevalencia de la patología dual. Revista de Psiquiatría y Salud Mental 6(3):121–128

47. Roncero C, Szerman N, Terán A, Pino C, Vázquez JM, Velasco E et al (2016) Professionals' perception on the management of patients with dual disorders. Pat Pref Adher 10:1855–1868

48. Szerman N, Marín-Navarrete R, Fernández-Mondragón J, Roncero C (2015) Patología dual en poblaciones especiales: una revisión narrativa. Revista Internacional de Investigación en Adicciones 1(1):50–67

49. Morales-Muñoz I, Martínez-Gras I, Ponce G, de la Cruz J, Lora D, Rodríguez-Jiménez R, Jurado-Barba R, Navarrete F, García-Gutierrez MS, Manzanares J, Rubio G (2017) Psychological symptomatology and impaired prepulse inhibition of the startle reflex are associated with cannabis-induced psychosis. J Psychopharmacol 31(8):1035–1045

50. Vrana SR, Calhoun PS, Dennis MF, Kirby AC, Beckham JC (2015) Acoustic startle and prepulse inhibition predict smoking lapse in posttraumatic stress disorder. J Psychopharmacol 29(10):1070–1076

51. European Monitoring Centre for Drugs and Drug Addiction (2018) Recent changes in Europe's cocaine market: results from an EMCDDA trendspotter study. Publications Office of the European Union, Luxembourg

52. European Monitoring Centre for Drugs and Drug Addiction (2019) European drug report 2019: trends and developments. Publications Office of the European Union, Luxembourg

53. OEDT (2016) Encuesta sobre Uso de Drogas en Enseñanzas Secundarias en España. ESTUDES 2014

54. Arias F, Szerman N, Vega P, Mesías B, Basurte I, Rentero D (2017) Trastorno bipolar y trastorno por uso de sustancias. Estudio Madrid sobre prevalencia de patología dual. Adicciones 29(3):186

55. González-Llona I, Tumuluru S, González-Torres MA, Gaviria M (2015) Cocaína: una revisión de la adicción y el tratamiento. Rev Asoc Esp Neuropsiq 35(127):555–571

56. Cadet JL, Bisagno V, Milroy CM (2014) Neuropathology of substance use disorders. Acta Neuropathol 127(1):91–107

57. Ashok AH, Mizuno Y, Volkow ND, Howes OD (2017) Association of stimulant use with dopaminergic alterations in users of cocaine, amphetamine, or methamphetamine: a systematic review and meta-analysis. JAMA Psychiatry 74(5):511–519

58. Ledonne A, Mercuri NB (2017) Current concepts on the physiopathological relevance of dopaminergic receptors. Front Cell Neurosci 11:27

59. Wang X, Liu L, Adams W, Li S, Zhang Q, Li B et al (2015) Cocaine exposure alters dopaminergic modulation of prefronto-accumbens transmission. Physiol Behav 145:112–117

60. Verdejo-García A, Pérez-García M, Sánchez-Barrera M, Rodriguez-Fernández A, Gómez-Río M (2007) Neuroimagen y drogodependencias: correlatos neuroanatómicos del consumo de cocaína, opiáceos, cannabis y éxtasis. Rev Neurol 44(7):432–439

61. Lorea I, Fernández-Montalvo J, Tirapu-Ustárroz J, Landa N, López-Goñi JJ (2010) Rendimiento neuropsicológico en la adicción a la cocaína: una revisión crítica. Rev Neurol 51:412–426

62. Preller KH, Ingold N, Hulka LM, Vonmoos M, Jenni D, Baumgartner MR et al (2013) Increased sensorimotor gating in recreational and dependent cocaine users is modulated by craving and attention-deficit/hyperactivity disorder symptoms. Biol Psychiatry 73(3):225–234

63. Efferen TR, Duncan EJ, Szilagyi S, Chakravorty S, Adams JU, Gonzenbach S, Angrist B, Butler PD, Rotrosen J (2000) Diminished acoustic startle in chronic cocaine users. Neuropsychopharmacology 22 (1):89–96

64. Volkow ND, Koob GF, McLellan AT (2016) Neurobiologic advances from the brain disease model of addiction. N Engl J Med 374 (4):363–371

65. Broderick PA, Rosenbaum T (2013) Sex-specific brain deficits in auditory processing in an animal model of cocaine-related schizophrenic disorders. Brain Sci 3 (2):504–520

66. Byrnes JJ, Hammer RP (2000) The disruptive effect of cocaine on prepulse inhibition is prevented by repeated administration in rats. Neuropsychopharmacology 22(5):551–554

67. Martínez ZA, Ellison GD, Geyer MA, Swerdlow NR (1999) Effects of sustained cocaine exposure on sensorimotor gating of startle in rats. Psychopharmacology (Berlin) 142:253–260

68. Hutchison KE, Swift R (1999) Effect of d-amphetamine on prepulse inhibition of the startle reflex in humans. Psychopharmacology 143:394–400

69. Swerdlow NR, Stephany N, Wasserman LC, Talledo J, Shoemaker J, Auerbach PP (2003) Amphetamine effects on prepulse inhibition across-species: replication and parametric extension. Neuropsychopharmacology 28:640–650

70. Zhang J, Engel JA, Söderpalm B, Svensson L (1998) Repeated administration of amphetamine induces sensitisation to its disruptive effect on prepulse inhibition in the rat. Psychopharmacology 135(4):401–406

71. Dulawa SC, Geyer MA (1996) Psychopharmacology of prepulse inhibition in mice. Chin J Physiol 39(3):139–146

72. Ralph RJ, Varty GB, Kelly MA, Wang YM, Caron MG, Rubinstein M, Grandy DK, Low MJ, Geyer MA (1992) The dopamine D2, but not D3 or D4, receptor subtype is essential for the disruption of prepulse inhibition produced by amphetamine in mice. J Neurosci 19(11):4627–4633

73. Yavas E, Young AM (2017) N-Methyl-d-aspartate modulation of nucleus accumbens dopamine release by metabotropic glutamate receptors: fast cyclic voltammetry studies in rat brain slices in vitro. ACS Chem Neurosci 8(2):320–328

74. Borroto-Escuela DO, Romero-Fernandez W, Narvaez M, Oflijan J, Agnati LF, Fuxe K (2014) Hallucinogenic 5-HT2AR agonists LSD and DOI enhance dopamine D2R protomer recognition and signaling of D2-5-HT2A heteroreceptor complexes. Biochem Biophys Res Commun 443(1):278–284

75. Dominici P, Kopec K, Manur R, Khalid A, Damiron K, Rowden A (2015) Phencyclidine intoxication case series study. J Med Toxicol 11(3):321–325

76. Ham S, Kim TK, Chung S, Im HI (2017) Drug abuse and psychosis: new insights into drug-induced psychosis. Exp Neurobiol 26 (1):11–24

77. Adams JU, Efferen TR, Duncan EJ, Rotrosen J (2001) Prepulse inhibition of the acoustic startle response in cocaine-withdrawn rats. Pharmacol Biochem Behav 68(4):753–759

78. Arenas MC, Caballero-Reinaldo C, Navarro-Frances CI, Manzanedo C (2017) Efecto de la cocaína sobre la inhibición por prepulso de la respuesta de sobresalto [Effects of cocaine on prepulse inhibition of the startle response]. Rev Neurol 65(11):507–519

79. Cameron CM, Wightman RM, Carelli RM (2016) One month of cocaine abstinence potentiates rapid dopamine signaling in the nucleus accumbens core. Neuropharmacology 111:223–230

80. Saddoris MP (2016) Terminal dopamine release kinetics in the accumbens core and shell are distinctly altered after withdrawal from cocaine self-administration. eNeuro 3 (5):ENEURO.0274-16.2016

81. Vollenweider FX, Barro M, Csomor PA, Feldon J (2006) Clozapine enhances prepulse inhibition in healthy humans with low but not with high prepulse inhibition levels. Biol Psychiatry 60(6):597–603

82. Hutchison KE, Rohsenow D, Monti P, Palfai T, Swift R (1997) Prepulse inhibition of the startle reflex: preliminary study of the effects of a low dose of alcohol in humans. Alcohol Clin Exp Res 21(7):1312–1319

83. van der Elst MC, Ellenbroek BA, Cools AR (2006) Cocaine strongly reduces prepulse inhibition in apomorphine-susceptible rats, but not in apomorphine-unsusceptible rats: regulation by dopamine D2 receptors. Behav Brain Res 175(2):392–398

84. van der Elst MC, Wunderink YS, Ellenbroek BA, Cools AR (2007) Differences in the cellular mechanism underlying the effects of amphetamine on prepulse inhibition in apomorphine-susceptible and apomorphine-unsusceptible rats. Psychopharmacology 190(1):93–102

85. van der Elst MC, Verheij MM, Roubos EW, Ellenbroek BA, Veening JG, Cools AR (2005) A single exposure to novelty differentially affects the accumbal dopaminergic system of apomorphine-susceptible and apomorphine-unsusceptible rats. Life Sci 76(12):1391–1406

86. van der Elst MC, Roubos EW, Ellenbroek BA, Veening JG, Cools AR (2005) Apomorphine-susceptible rats and apomorphine-unsusceptible rats differ in the tyrosine hydroxylase-immunoreactive network in the nucleus accumbens core and shell. Exp Brain Res 160(4):418–423

87. Bell RL, Rodd ZA, Hsu CC, Lumeng L, Murphy JM, McBride WJ (2003) Amphetamine-modified acoustic startle responding and prepulse inhibition in adult and adolescent alcohol-preferring and -nonpreferring rats. Pharmacol Biochem Behav 75(1):163–171

88. De Koning MB, Bloemen OJ, Van Duin ED, Booij J, Abel KM, De Haan L et al (2014) Pre-pulse inhibition and striatal dopamine in subjects at an ultra-high risk for psychosis. J Psychopharmacol 28(6):553–560

89. Falkai P, Rossner MJ, Schulze TG, Hasan A, Brzózka MM, Malchow B et al (2015) Kraepelin revisited: schizophrenia from degeneration to failed regeneration. Mol Psychiatry 20(6):671–676

90. Volkow ND, Wang GJ, Fowler JS, Logan J, Gatley SJ, Gifford A et al (1999) Prediction of reinforcing responses to psychostimulants in humans by brain dopamine D2 receptor levels. Am J Psychiatry 156:1440–1443

91. Bitsios P, Giakoumaki SG, Frangou S (2005) The effects of dopamine agonists on prepulse inhibition in healthy men depend on baseline PPI values. Psychopharmacology 182:144–152

92. Peleg-Raibstei D, Hauser J, Llano Lopez LH, Feldon J, Gargiulo PA, Yee BK (2013) Baseline prepulse inhibition expression predicts the propensity of developing sensitization to the motor stimulant effects of amphetamine in C57BL/6 mice. Psychopharmacology 225(2):341–352

93. Steketee JD, Kalivas PW (2011) Drug wanting: behavioral sensitization and relapse to drug-seeking behavior. Pharmacol Rev 63:348–365

94. Lüscher C, Malenka RC (2011) Drug-evoked synaptic plasticity in addiction: from molecular changes to circuit remodeling. Neuron 69:650–663

95. Kawa AB, Valenta AC, Kennedy RT, Robinson TE (2019) Incentive and dopamine sensitization produced by intermittent but not long access cocaine self-administration. Eur J Neurosci 50(4):2663–2682

96. Steketee JD (2005) Cortical mechanisms of cocaine sensitization. Crit Rev Neurobiol 17:69–86

97. Blumenthal TD, Cuthbert BN, Filion DL, Hackley S, Lipp OV, Van Boxtel A (2005) Committee report: guidelines for human startle eyeblink electromyographic studies. Psychophysiology 42(1):1–15

98. Nesbitt K, Blackmore K, Hookham G, Kay-Lambkin F, Walla P (2015) Using the startle eye-blink to measure affect in players. In: Serious games analytics. Springer, Cham, pp 401–434

99. Hager JC, Ekman P (1985) The assimetry of facial actions is inconsistent with models of hemispheric specialization. Psychophysiology 22:307–318

100. Braff DL, Geyer MA, Swerdlow NR (2001) Human studies of prepulse inhibition of startle: normal subjects, patient groups, and pharmacological studies. Psychopharmacology 156(2–3):234–258

101. Quednow BB, Kühn KU, Hoenig K, Maier W, Wagner M (2004) Prepulse inhibition and habituation of acoustic startle response in male MDMA ("Ecstasy") users, cannabis users and healthy controls. Neuropsychopharmacology 29(5):982–990

102. Meteran H, Vindbjerg E, Uldall SW, Glenthøj B, Carlsson J, Oranje B (2018) Startle habituation, sensory, and sensorimotor gating in trauma-affected refugees with

posttraumatic stress disorder. Psychol Med 49:581–589

103. Kumari V, Gray JA, Gupta P, Luscher S, Sharma T (2003) Sex differences in prepulse inhibition of the acoustic startle response. Personal Individ Differ 35(4):733–742

104. De la Casa LG, Mena A, Ruiz-Salas JC (2016) Effect of stress and attention on startle response and prepulse inhibition. Physiol Behav 165:179–186

105. Della CV, Höfer I, Weiner I, Feldon J (1998) The effects of smoking on acoustic prepulse inhibition in healthy men and women. Psychopharmacology 137:362–368

106. Swerdlow NR, Eastvold A, Gerbranda T, Uyan KM, Hartman P, Doan Q, Auerbach P (2000) Effects of caffeine on sensorimotor gating of the startle reflex in normal control subjects: impact of caffeine intake and withdrawal. Psychopharmacology 151 (4):368.378

107. Rubio G, López-Muñoz F, Jurado-Barba R, Martínez-Gras I, Rodríguez-Jiménez R, Espinosa R, Carlos LJ (2015) Stress induced by the socially evaluated cold-pressor test cause equivalent deficiencies of sensory gating in male subjects with schizophrenia and healthy controls. Psychiatry Res 228 (3):283.288

108. Swerdlow NR, Hartman PL, Auerbach PP (1997) Changes in sensorimotor inhibition across the menstrual cycle: implications for neuropsychiatric disorders. Biol Psychiatry 41(4):452–460

109. Kumari V, Konstantinou J, Papadopoulos A, Aasen I, Poon L, Halari R, Cleare AJ (2009) Evidence for a role of progesterone in menstrual cycle-related variability in prepulse inhibition in healthy young women. Neuropsychopharmacology 35(4):929–937

110. Ruisoto P, Contador I (2019) The role of stress in drug addiction. An integrative review. Physiol Behav 202:62–68

# Chapter 4

# Modulation of Effects of Alcohol, Cannabinoids, and Psychostimulants by Novelty-Seeking Trait

Claudia Calpe-López, M. Ángeles Martínez-Caballero, María Pilar García-Pardo, and María A. Aguilar

## Abstract

A main challenge for addiction research is to identify the factors that predispose individuals to drug addiction. Novelty- or sensation-seeking is a personality trait which partly explains individual differences in vulnerability to addiction. The hole board test is an animal model of novelty-seeking behavior used to study the influence of this behavioral trait in sensitivity to the rewarding effects of drugs. In our laboratory we have also used the hole board test to determine how the novelty-seeking phenotype modulates the long-lasting effects of adolescent exposure to psychostimulant drugs, alcohol, or cannabinoids on the subsequent rewarding properties of drugs of abuse in the conditioned place preference (CPP) paradigm. In this chapter we provide a detailed description of the hole board and CPP procedures and of different types of drug pretreatments. First, adolescent mice are classified as high novelty-seeking (HNS) or low novelty-seeking (LNS) according to the number of dips they perform in the hole board test. Next, the mice are exposed to repeated injections of cocaine, MDMA, alcohol or the cannabinoid agonist WIN 55212-2 during the adolescent period. Finally, the acquisition, extinction and reinstatement of drug-induced CPP is evaluated in HNS and LNS mice. Our results suggest that the high novelty-seeking endophenotype is a marker of enhanced susceptibility to the effects of environmental variables that increase the risk of drug addiction, such as adolescent drug exposure or stressful experiences. New knowledge concerning the contribution of novelty-seeking to the development of addiction is likely to contribute to the development of better preventive and treatment strategies for this disorder.

**Key words** Adolescence, Alcohol, Cannabinoids, Cocaine, Hole board, MDMA, Mice, Novelty-seeking

## 1 Introduction

### 1.1 The Novelty-Seeking Trait

A main challenge for addiction research is to identify the factors that predispose individuals to drug addiction in order to develop better preventive and treatment strategies [1]. Besides environmental factors, including exposure to the drug itself and stress, there are personality traits or vulnerable phenotypes (*see* Fig. 1) that may be present before the first experience of the drug and which explain

María A. Aguilar (ed.), *Methods for Preclinical Research in Addiction*, Neuromethods, vol. 174,
https://doi.org/10.1007/978-1-0716-1748-9_4, © Springer Science+Business Media, LLC, part of Springer Nature 2022

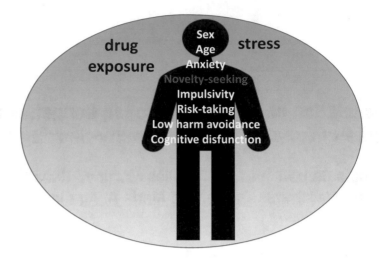

Environmental and individual factors associated with vulnerability to drug addiction

**Fig. 1** Environmental and individual factors associated with vulnerability to drug addiction

individual differences in sensitivity to the rewarding effects of drugs and vulnerability to develop addiction [2].

Endophenotypes are heritably determined quantitative traits associated with a disorder, and which connect genotype (predisposing genes) and phenotype (clinical symptoms of the disorder) [3]. Such traits are found in affected individuals, manifest themselves with and without the active illness/diagnosis, cosegregate according to the disorder within families, and are more prevalent among nonaffected family members than in the general population [4]. The high novelty-seeking profile, as well as cognitive dysfunction and anxious-impulsive personality traits [3], may be considered as an endophenotype for drug dependence.

Novelty- or sensation-seeking is a personality trait defined as a tendency to pursue novel and intense emotional sensations and experiences [5]. Individuals with a high novelty-seeking profile like taking risks and seeking new environments and situations to increase their level of arousal and intensify their experiences [6]. The recently deceased Marvin Zuckerman was the first author who defined the term of novelty- or sensation-seeking, and developed the most widely used sensation seeking scale (Zuckerman's Sensation Seeking Scale) [6]. Cloninger extended the work of Zuckerman and developed the Cloninger's Tridimensional Personality Questionnaire (TPQ), which also measured the novelty-seeking trait [7]. Other authors have extended our understanding of the novelty-seeking phenotype and its relationship with drug use disorders. Novelty-seeking is a multifaceted behavioral construct that includes personality characteristics usually displayed by individuals at a higher risk of developing drug addiction, such as thrill-seeking, novelty-preference, risk-taking, and harm avoidance [2, 6, 8–13]. There is a great deal of evidence of the relationship between the novelty-seeking trait and drugs of abuse. A high novelty-seeking profile predicts increased drug use during adolescence [2, 3, 8, 14–18]. For example, high-sensation seekers are more likely to experiment with tobacco, alcohol, marihuana and MDMA [19], display greater incentive motivation to self-administer amphetamine in a controlled laboratory study [20], and are more responsive to the positive subjective effects of this drug in various scales, such as the Profile of Mood States (POMS) and Visual Analogue Scale (VAS) [20, 21]. Furthermore, drug use itself enhances the novelty-seeking trait [22] and both novelty- and drug-seeking behavior are mediated by the mesolimbic dopaminergic system [23].

## 1.2 Adolescence and Novelty-Seeking

Adolescence is a critical developmental period characterized by hormonal changes and neuroplasticity, especially in the dopaminergic mesocorticolimbic areas, that contribute to the maturation of the central nervous system [24, 25]. In comparison to adults, adolescent subjects exhibit lower dopamine release in basal conditions and hypersensitivity of the dopaminergic striatal system to pharmacological or environmental stimuli (including drugs of abuse and exposure to novel environments) [26–30]. In addition, the prefrontal cortical system, related with planning, evaluation of consequences, decision-making, and inhibitory control of behavior, is still immature [31, 32]. Thus, the adolescent brain operates in a promotivational state due to a limited inhibitory capacity, poor regulatory control, and dopaminergic hyperactivity in the nucleus accumbens and amygdala [33].

Related with these changes in the dopaminergic systems [24, 34], adolescents display several personality characteristics that may explain why it is most likely that drug use begins during adolescence, such as impulsiveness, a natural drive to search for novel stimuli and sensations, willingness to take risks, low novelty- and risk-induced anxiety, and emotional liability [13, 31, 35, 36]. Furthermore, at this developmental stage the cognitive control and harm-avoidance systems are ineffective for limiting drug consumption [37, 38], and so the rewarding effects of addictive drugs predominate over their aversive effects [39, 40], thus augmenting the propensity to take drugs.

Although there is a positive correlation between early drug-taking and the development of drug dependence in adulthood [41–43], not all drug users progress to drug abuse, dependence, and addiction. Longitudinal studies point to an association between the novelty-seeking trait and the initiation of drug use [2, 44] and young adults with a high novelty-seeking profile show a greater sensitivity to the physiological and psychological effects of drugs in comparison with low novelty seekers [20, 21, 45–49]. In any case, the transition from voluntary to compulsive drug use that characterizes addiction depends on the combination of multiple factors, including the novelty-seeking trait.

### 1.3 Animal Models of Novelty-Seeking

Novelty-seeking is a complex behavior that involves the detection of changes in the environment (cognition), the natural tendency to approach and explore novel objects, and the individual's stress responsiveness [50]. Although rodents show an innate preference for novel objects over familiar ones [51], the behavior of an animal in a novel environment is conditioned by the interaction of several factors, including activity, motivation to explore and fear/anxiety [52, 53].

Different procedures allow the screening of novelty-seeking behavior in rats and mice, among which two of the most popular (and contrasting) are the inescapable vs. free choice novelty models. It is important to note that the procedure employed to measure novelty-seeking trait can influence the results obtained (*see* Table 1), particularly regarding the relationship between this endophenotype and vulnerability to drugs of abuse (*see* Table 2).

### 1.3.1 Inescapable Novelty Models

This approach, frequently known as novelty responding, is the most used to classify animals as high or low responders according to their locomotor activity in a new inescapable environment. In these procedures, animals are confined to the new place, without allowing them the freedom of choice to explore the novel object or environment. These models of novelty-seeking are artificial and have been criticized because it is unclear whether the locomotor reactivity displayed by rodents represents an escape or an exploratory behavior.

### Locomotor Reactivity in a Novel Open Field Environment

This test was the first animal model of individual differences in stress responsiveness to a novel inescapable environment [54]. Rodents are placed in a novel open field environment and the activity, distance traveled and rears are recorded over different periods of time (usually between 10 min and 2 h). Animals that exhibit high levels of novelty-induced locomotor reactivity are categorized as High Responders (HR) and those showing lower levels as Low Responders (LR).

A relationship between novelty-induced locomotor activity and drug-taking behaviors has been observed in the HR–LR model [55–57]. For example, HR rats exhibit high rates of behavioral sensitization to psychostimulant drugs and self-administer these drugs more readily than LR rats [54, 56]. In line with this, Davis et al. [58] suggested that "the HR–LR trait may tap into the broad dimension of behavioral disinhibition vs. behavioral control—a dimension that has been implicated in the vulnerability versus resilience to numerous psychiatric and addictive disorders."

**Table 1**
**Levels of novelty-seeking behavior in adolescent and young adult mice of both sexes in different paradigms of novelty-seeking**

| PARADIGM OF NOVELTY SEEKING | MICE | | | | Level of novelty seeking |
|---|---|---|---|---|---|
| | MALE | | FEMALE | | |
| Inescapable (locomotor activity) | ADOLESCENT | YOUNG ADULT | ADOLESCENT | YOUNG ADULT | low |
| NOVELTY-REACTIVITY (10 min in environment) | low | high | low | high | |
| NOVELTY-HABITUATION (1h in environment) | low | high | low | high | high |
| Free choice | | | | | |
| NOVEL OBJECT RECOGNITION TASK | high | low | high | low | |
| NOVEL ENVIRONMENT TEST | high | low | high | low | |
| HOLE BOARD | high | | | | |

**Activity in the Wheel Running Procedure**

In this model, rodents are classified as HR and LR according to their motor response to novelty on the first day they are placed on a running wheel [59]. It is important to note that the next day both groups showed a similar locomotor activity.

**Activity in an Exploration Box Test**

In this test, considered an indicator of inquisitive and inspective exploration, animals are placed in a box containing three unknown objects and one familiar food pellet (in the same location) on 5 consecutive days. According to the locomotor and rearing activity the animals display during exploration of the three unfamiliar objects, they are classified as high or low exploratory activity (HE or LE, respectively) [60, 61].

**1.3.2 Free-Choice Models**

The paradigms that evaluate preference for novel objects or environments in a free-choice procedure allow animals to choose or to reject novelty [2, 12, 62–64]. This is why they are currently considered a better measure of novelty-seeking and more suitable for categorizing the novelty-seeking trait.

**The Novel Environment Test**

This test is performed in a rectangular box divided by a partition into two compartments of equal size but different colors that are connected by an opening with a removable guillotine-type door located at floor level in the center of the partition. In order to establish familiarity with an environment, animals are placed in one of the compartments that they may explore freely with no access to the other compartment. In our laboratory, animals undergo a single habituation session of 15 min [65, 66], but some authors carry out more sessions [67]. After habituation, the guillotine-type door is raised, allowing free access to and exploration of both compartments for 20 min (test session). Usually the behavior of the animal is videotaped under red light and the number of transitions between the two compartments, the latency for

**Table 2**
**Acquisition of CPP induced by cocaine (1 mg/kg) according to the level of novelty-seeking shown by adolescent and young adult mice of both sexes in different paradigms of novelty-seeking**

| PARADIGM OF NOVELTY SEEKING | MICE | | | |
|---|---|---|---|---|
| | MALE | | FEMALE | |
| Inescapable (locomotor activity) | ADOLESCENT | YOUNG ADULT | ADOLESCENT | YOUNG ADULT |
| NOVELTY-REACTIVITY (10 min in environment) | **CPP** | No CPP | No CPP | No CPP |
| | No CPP | No CPP | No CPP | **CPP** |
| NOVELTY-HABITUATION (1h in environment) | No CPP | No CPP | No CPP | **CPP** |
| | No CPP | **CPP** | No CPP | No CPP |
| Free choice | | | | |
| NOVEL OBJECT RECOGNITION TASK | No CPP | No CPP | No CPP | No CPP |
| | No CPP | No CPP | No CPP | No CPP |
| NOVEL ENVIRONMENT TEST | No CPP | No CPP | No CPP | No CPP |
| | No CPP | **CPP** | No CPP | **CPP** |
| HOLE BOARD TEST | No CPP | No CPP | No CPP | **CPP** |
| | **CPP** | No CPP | **CPP** | No CPP |

Level of novelty seeking    **CPP** with cocaine

low    (1 mg/kg)

high

entering the novel environment and the percentage of time spent in the novel environment are scored. Depending on their percentage of novelty-preference animals are categorized as high or low novelty seekers (HNS or LNS). In fact, the novel environment test is the most popular method used to classify animals according to their high or low exploratory behavior.

**The Spontaneous Alternation Task**

In this paradigm, animals are placed in the stem of a T- or Y-maze [65, 68] that permits entry into one arm (trial 1). In some cases, animals may choose the arm (free trial procedure) and in other cases they are forced to enter one of the arms (forced trial procedure). In both cases, animals are confined to the arm for 30 s. In the test phase (trial 2) the animals are free to choose the same or the alternative arm. It is important to note that rodents show a higher preference for the novel arm in the forced trial procedure than in the free trial procedure. This appears to be the case because rodents have a bias for one side of the apparatus and choose freely this side on both occasions [68–70].

**The Novel Object Recognition Task**

Initially, this task was developed for evaluating directed exploration in rats [71], and was later adapted for mice, with minor modifications [51]. Besides measuring novelty-seeking, this test also involves recognition memory, detection, and processing of the novel object [72].

Animals are free to explore one or several different objects (usually between 2 and 8) in an open field or playground maze on consecutive days (during 3, 5, or 10 min) [66, 67, 71, 73,

74]. Once they have become habituated to these objects, one of them is replaced by a new object and the approach behavior of the animal to the novel vs. familiar object is assessed [75]. Animals respond by preferentially exploring the substituted objects over the nonsubstituted objects [74]. Typically scored behaviors are the latency to make contact with the novel object, the percentage of time spent exploring novel and familiar objects, and the number of approaches/explorations of the novel object. Object exploration is defined as intentional contact of the animal's nose or front paws with the novel object.

In our laboratory we used a procedure similar to that described by Frick and Gresack [74]. We performed a single trial novel object recognition in an open arena (24 × 24 cm) illuminated by a dim light (22 W, 1080 lm) with two small river stones (1.5 cm wide × 3 cm high) and a small color toy made of nontoxic plastic. These objects were fixed to the floor with Velcro tape in opposite corners of the open field (5 cm from the walls). A camera recorder fixed to the ceiling allowed the arena to be visualized. Each mouse completed a daily trial (lasting 10 min) on 4 successive days. In the habituation phase (trial 1), the mouse was placed in the center of the empty open field box and allowed to explore it freely. In the sample phase (trials 2 and 3), two of the stones were placed in opposite corners of the box and the mouse was allowed to explore them. In the test phase (trial 4), one of the stones was replaced with the small colored toy in order to assess novel object exploration.

**The Hole Board Test**

This procedure is briefly explained here because it will be described at greater length in the following sections. The apparatus consists of a square box with equidistant holes in the floor. Sensors inside the holes detect the number of head dips performed by the animals. The latency to the first head dip and number of dips are used to classify animals as high or low novelty seekers [65, 76, 77]. Furthermore, this test can also be used to estimate the emotional response of animals when facing an unfamiliar environment and to assess anxiety state [77–81]. According to the propensity of an animal to explore the apparatus (measured by the number of head dips) it is classified as high novelty seeker (HNS) or low novelty seeker (LNS) [65, 76]. Although this procedure shares the inescapable condition with the novelty response in the open field, the hole board test is considered a useful tool to evaluate novelty-preference in rodents [78, 82], as they may perform dipping behavior. In fact, some authors have verified that the results observed in the hole board are independent of locomotor activity [2, 81, 83].

**The Novel-Object Place Conditioning Paradigm**

In this model, first described by Bevins and Bardo [84], the conditioned rewarding effects of novelty are evaluated. The place conditioning apparatus consists of two well differentiated compartments (by means of distinctive stimuli such as the color of the walls or the texture of floor). During the Conditioning phase, animals are confined for 10–15 min to one compartment containing a new object (paired side) or to the other compartment, which is free of objects (unpaired side), for 5 or 8 consecutive days [84, 85]. On postconditioning (test) day, rodents are allowed to explore both compartments freely for 10 min without any object being present inside the apparatus. High novelty seekers spend more time in the novelty-paired compartment.

**Object Preference on the Hole Board**

This procedure combines the object preference and hole board tests. Two objects are placed in two separate holes of the hole board and the preference for novelty is calculated by the sum of the time engaged in head dipping in the holes that contain objects divided by the total time engaged in dipping [65].

### 1.4 Influence of Novelty-Seeking on Drug-Induced Reward

Empirical evidence indicates that the high novelty-seeking profile represents an increased risk of drug use in comparison to low novelty-seeking. However, novelty-seeking behavior does not always predict drug reward, probably due to variations in the procedure used to evaluate novelty-seeking (*see* Subheading 1.3) and/or the paradigms used to measure the rewarding effects of drugs of abuse. The most common animal models of reward are the drug self-administration and the conditioned place preference (CPP) paradigms.

Intravenous drug self-administration is a widely used animal model of human drug abuse and dependence, and allows the primary hedonic or reinforcing effects of drugs of abuse to be measured [86]. In a previous work we reviewed the main studies that have used this paradigm to evaluate the influence of the novelty-seeking phenotype on the reinforcing effects of psychostimulant drugs [87]. Most studies have used male adult rats and have evaluated the novelty-seeking trait by means of the locomotor response in an inescapable novel environment. Results obtained with this paradigm are controversial, since some studies have reported that HR rats self-administer more amphetamine and cocaine than LR rats [54, 57, 58, 62, 88–94] but not in others [64, 67, 90, 95–97]. Furthermore, HR and LR rats do not differ regarding the self-administration of methylphenidate [98] and MDMA [99]. Similar discrepant results have been obtained using free-choice novelty-seeking paradigms with HNS male adult rats self-administering more amphetamine, cocaine, and methylphenidate than their LNS counterparts in some studies [64, 67, 95, 97, 98], but not in others [62, 67]. In male adult mice, the novelty-seeking trait facilitates the self-administration of cocaine in inescapable and free-choice paradigms [100]. The influence of novelty-seeking on the reinforcing effects of alcohol has only been studied in two works, which also rendered divergent results. Bienkowski et al. [101] found that response to novelty in the open field and novel object test did not predict individual differences to oral operant ethanol self-administration. Conversely, Nadal et al. [102] reported a positive relationship between the level of activity in a novel environment and the acquisition of alcohol self-administration, although only under a FR3 schedule. No research to date has evaluated the effects of novelty-seeking on cannabinoid self-administration.

The conditioned place preference (CPP) induced by drugs of abuse has become a popular alternative to the drug self-administration paradigm for evaluating the sensitivity of individuals to the rewarding properties of these substances [103–106]. In the CPP paradigm, the incentive properties of a drug are assessed in drug-free animals according to the amount of time they spend in an environment previously paired with the drug's effects (see details of the procedure in Subheading 2.2.3). In a previous work we reviewed the main studies that have used this paradigm to evaluate the influence of the novelty-seeking phenotype on the reinforcing effects of psychostimulant drugs [87]. When assessing the locomotor response to novelty in an inescapable novelty-seeking paradigm, no differences were found between HR and LR rats in the CPP induced by cocaine or amphetamine [107–112], while HR rats were less sensitive to amphetamine CPP [113]. Similar negative results with respect to cocaine CPP (LR > HR, LR = HR) have been observed in mice [114–116], except in the case of young adult mice exposed to the new environment for only 10 min [114]. Conversely, most studies using free-choice novelty-seeking paradigms to classify animals have reported that HNS rats and mice acquired amphetamine and cocaine CPP more readily than LNS animals [66, 73, 111, 117]. Preference for novelty has also been related to a higher sensitivity to the rewarding effects of morphine [118–120] and nicotine [76], but not of alcohol [83].

Individual characteristics of experimental animals (species, strain, sex and age) and the paradigm of novelty-seeking used to categorize the animals as high or low novelty seekers are essential variables that can affect the results observed. As can be seen in Table 2, the differential sensitivity of HNS and LNS mice to the rewarding effects of 1 mg/kg of cocaine in the CPP paradigm has only been detected with some paradigms of novelty-seeking in function of the sex and age of mice. When we tested adolescents of both sexes, the higher sensitivity of HNS to cocaine CPP was detected only in the hole board paradigm. Similarly, when we evaluated adults of both sexes, the novel environment test was found to be useful to discriminate the higher sensitivity of HNS mice. Finally, the HR profile in an inescapable novel environment is also predictive of enhanced vulnerability to developing cocaine CPP in adult female mice.

**1.5 Novelty-Seeking and Adolescent Drug Exposure**

In a series of studies from our laboratory, we have evaluated how the novelty-seeking phenotype modulates the long-lasting effects of adolescent exposure to psychostimulant drugs, alcohol or cannabinoids on the subsequent rewarding properties of drugs of abuse in the CPP paradigm (*see* Table 3). Pretreatment with drugs of abuse during adolescence also induces other behavioral and biochemical effects that in some cases are influenced by the novelty-seeking trait (*see* Tables 4, 5, and 6).

*1.5.1 Cocaine*

In several studies performed in our laboratory we have observed that rodents that are more responsive to novelty (HR and HNS) develop CPP with a subthreshold dose of cocaine, which is ineffective in inducing CPP in low novelty seeker (LNS) and naïve mice [66, 114]. Moreover, we have seen that the level of novelty-seeking also influences the effects of cocaine exposure during adolescence on the subsequent vulnerability of mice to cocaine and MDMA [77] (*see* Table 3). Adolescent HNS mice exposed to cocaine binges are more sensitive to the conditioned rewarding effects of low doses of cocaine and MDMA in adulthood. Age is a critical factor in these effects, since exposure to the same schedule of cocaine binges in adult mice do not induce this subsequent enhanced vulnerability to the rewarding effects of cocaine and MDMA, irrespective of the novelty-seeking profile. Adolescent cocaine binges also produce subtle, long-term changes in the behavior of mice, particularly in those with an HNS profile, such as a decrease in exploratory behavior, increased locomotor reactivity and greater impulsivity-like behaviors [77] (*see* Tables 4, 5, and 6). These alterations could be associated with the increased vulnerability of HNS cocaine-treated mice to develop cocaine and MDMA CPP in adulthood.

*1.5.2 MDMA*

As commented on before, the novelty-seeking profile is not significantly correlated with either acquisition of MDMA self-administration or drug-seeking in rats [99]. However, we have observed that pretreatment with binges of cocaine during adolescence induces an increase in the conditioned rewarding effects of MDMA only in HNS mice [77] (*see* Table 3). Similarly, we have detected a higher sensitivity of HNS to the conditioned rewarding effects of low doses of cocaine and MDMA in mice exposed to binges of MDMA during adolescence [122] (*see* Table 3). In addition, only HNS mice show other behavioral consequences of adolescent exposure to MDMA binges, such as an increase in social and aggressive behaviors. HNS MDMA-treated mice engage in more social contacts, but also behave more aggressively, than LNS mice [122] (*see* Table 4). On the other hand, a long-lasting anxiolytic effect after exposure to a high dose of MDMA (20 mg/kg) was observed only in LNS mice (they spent more time in the open arms of the EPM than HNS mice). Thus, it can be deduced that LNS mice display less emotional reactivity than their HNS counterparts

**Table 3**

Mouse studies of the influence of novelty-seeking profile in the long-term effects of adolescent drug exposure on later sensitivity to drug reward (CPP)

| Animal | Sex | Age hole board PND | Pretreatment Drug | PND | Dose | CPP PND | Drug | Dose mg/kg | Sensitivity to drug reward | Reference | Conclusion |
|---|---|---|---|---|---|---|---|---|---|---|---|
| OF1 mice | Male | Adolescent, 31 | Saline | 33–34, 37–38, 41–42, 45–46 | | 67–74 | Cocaine | 1 | HNS = LNS, no CPP | [121] | Adolescent HNS exposed to alcohol are more vulnerable to reinstatement of cocaine CPP |
| | | | Alcohol binge | | 2.5 g/kg | 67–74 | Cocaine | 1 | HNS = LNS, CPP; reinstatement only in HNS | | |
| | | | | | 2.5 g/kg | 67–74 | Cocaine | 6 | HNS = LNS, CPP; reinstatement only in HNS | | |
| | | | | | 0 g/kg | 67–78 | MDMA | 1.25 | HNS = LNS, no CPP | | |
| | | | | | 2.5 g/kg | 67–78 | MDMA | 1.25 | HNS = LNS, CPP | | |
| | | | | | 2.5 g/kg | 67–78 | MDMA | 2.5 | HNS = LNS, CPP | | |
| OF1 mice | Male | Adolescent, 33 | Saline | 34–35, 36–38, 42–45 | | 67–74 | Cocaine | 1 | HNS = LNS, no CPP | [77] | Adolescent HNS exposed to cocaine are more sensitive to the rewarding effects of cocaine and MDMA CPP in adulthood. Cocaine exposure during adulthood do not increase the rewarding effects of cocaine |
| | | | | | | 67–78 | MDMA | 1.25 | HNS = LNS, no CPP | | |
| | | | Cocaine binge | | 5–10–15 mg/kg | 67–74 | Cocaine | 1 | CPP only in HNS | | |
| | | Young adults, 60 | | | | 67–78 | MDMA | 1.25 | CPP only in HNS | | |
| | | | Saline | 61–62, 63–65, 68–72 | | 93–100 | Cocaine | 1 | HNS = LNS, no CPP | | |
| | | | | | | 93–104 | MDMA | 1.25 | HNS = LNS, no CPP | | |
| | | | Cocaine binge | | 5–10–15 mg/kg | 93–100 | Cocaine | 1 | HNS = LNS, no CPP | | |
| | | | | | | 93–104 | MDMA | 1.25 | HNS = LNS, no CPP | | |
| OF1 mice | Male | Adolescent, 28 | Saline | 33, 34, 40, 41 | | 67–74 | Cocaine | 1 | HNS = LNS, no CPP | [122] | Adolescent HNS exposed to MDMA are more sensitive to the rewarding effects of cocaine and MDMA in adulthood |
| | | | | | | 67–78 | MDMA | 1.25 | HNS = LNS, no CPP | | |
| | | | MDMA binge | | 10 mg/kg | 67–74 | Cocaine | 1 | CPP only in HNS | | |
| | | | | | | 67–78 | MDMA | 1.25 | CPP only in HNS; reinstatement only in HNS | | |

(continued)

**Table 3**
(continued)

| Animal | Sex | Age hole board PND | Pretreatment PND | Pretreatment Drug | Pretreatment Dose | CPP PND | CPP Drug | Dose mg/kg | Sensitivity to drug reward | Reference | Conclusion |
|---|---|---|---|---|---|---|---|---|---|---|---|
| OF1 mice | Male | Adolescent, 26 | 26–30 | | | 30–41 | WIN | 0.05 | HNS = LNS, no CPP | [123] | Adolescent HNS are more sensitive to the rewarding effects of WIN 55212-2. Adolescent HNS exposed to this cannabinoid agonist are more vulnerable to the rewarding effects of cocaine (longer CPP and reinstatement) |
| | | | | | | 30–41 | WIN | 0.075 | CPP only in HNS | | |
| | | | | VEHICLE | 0 mg/kg | 34–45 | WIN | 0.05 | HNS = LNS, no CPP | | |
| | | | | WIN 55212-2 | 0.1 mg/kg | 34–45 | WIN | 0.05 | HNS = LNS, no CPP | | |
| | | | | SR 141716A | 1 mg/kg | 34–45 | WIN | 0.05 | HNS = LNS, no CPP | | |
| | | | | VEHICLE | 0 mg/kg | 34–41 | Cocaine | 1 | HNS = LNS, no CPP | | |
| | | | | WIN 55212-2 | 0.1 mg/kg | 34–41 | Cocaine | 1 | HNS = LNS, CPP; reinstatement only in HNS | | |
| | | | | SR 141716A | 1 mg/kg | 34–41 | Cocaine | 1 | HNS = LNS, no CPP | | |
| | | | | WIN 55212-2 | 0.1 mg/kg | 34–41 | Cocaine | 6 | HNS = LNS, CPP and reinstatement CPP duration: HNS > LNS | | |

**Table 4**

Mouse studies about the influence of the novelty-seeking profile in the long-term effects of adolescent drug exposure on later social behavior

| Animal | Sex | Age hole board PND | Pretreatment PND | Drug Binge | Dose | Social interaction test | Reference |
|---|---|---|---|---|---|---|---|
| OF1 mice | Male | Adolescent, 31 | 33–34, 37–38, 41–42, 45–46 | Alcohol | 0 g/kg<br>1.25 g/kg<br>2.5 g/kg | PND 71, HNS = LNS<br>PND 71, HNS = LNS<br>PND 71, HNS = LNS (HNS > HNS saline in non-social exploration) (HNS < HNS saline in threat and attack) | [121] |
| OF1 mice | Male | Adolescent, 33 | 34–35, 36–38, 42–45 | Cocaine | 0, 0, 0 mg/kg<br>5–15–25 mg/kg | PND 82–84, HNS = LNS<br>PND 82–84, HNS = LNS | [77] |
| OF1 mice | Male | Adolescent, 28 | 33, 34, 40, 41 | MDMA | 0 mg/kg<br>10 mg/kg<br><br>20 mg/kg | PND 75, HNS = LNS<br>PND 75, increased social investigation, threat and attack only in HNS<br>HNS > LNS in social investigation, threat and attack<br>PND 75, increased attack only in HNS<br>HNS > LNS in attack | [122] |

**Table 5**
**Mouse studies about the influence of the novelty-seeking profile in the long-term effects of adolescent drug exposure on later behavior in the elevated plus maze**

| Animal | Sex | Age Hole board PND | Pretreatment PND | Drug Binge | Dose | Elevated plus maze (EPM) | Reference |
|---|---|---|---|---|---|---|---|
| OF1 mice | Male | Adolescent, 31 | 33–34, 37–38, 41–42, 45–46 | Alcohol | 0 g/kg 1.25 g/kg 2.5 g/kg | PND 67, HNS = LNS PND 67, HNS = LNS (time in OA, %time OA: HNS > HNS saline) PND 67, HNS = LNS (time in OA, %time OA, HNS > HNS saline) (entries OA, HNS > HNS saline, LNS > LNS saline) | [121] |
| OF1 mice | Male | Adolescent, 33 | 34–35, 36–38, 42–45 | Cocaine | 0, 0, 0 mg/kg 5–15–25 mg/ kg | PND 67, HNS = LNS PND 67, HNS = LNS, cocaine pre-treatment reduced latency to enter OA only in LNS | [77] |
| OF1 mice | Male | Adolescent, 28 | 33, 34, 40, 41 | MDMA | 0 mg/kg 10 mg/kg 20 mg/kg | PND 64, HNS = LNS PND 64, increased time and %time in OA only in LNS PND 64, increased time and %time in OA only in LNS time and % time in OA, HNS < LNS | [122] |

**Table 6**

**Mouse studies about the influence of the novelty-seeking profile in the long-term effects of adolescent drug exposure on later behavioral and neurochemical measures**

| Animal Sex | Age | Pretreatment PND | Drug Binge | Dose | Other measures | Reference |
|---|---|---|---|---|---|---|
| OF1 mice | Male Adolescent, 31 | 33–34, 37–38, 41–42, 45–46 | Alcohol | 0 g/kg 1.25 g/kg 2.5 g/kg | Novel object. PND 70, HNS = LNS<br>Novel object. PND 70, HNS = LNS (%time exploring novel object, LNS < LNS saline)<br>Novel object. PND 70, latency to explore novel object: HNS < LNS (%time exploring novel object: LNS < LNS saline) (number of explorations: LNS < LNS saline) (latency to explore novel object: LNS > LNS saline) (number of explorations: HNS < HNS saline) | [121] |
| OF1 mice | Male Adolescent, 33 | 34–35, 36–38, 42–45 | Cocaine | 0, 0, 0 mg/kg 5–15–25 mg/ kg | Actimeter. PND 68–69, HNS = LNS<br>PND 68–69, HNS = LNS, cocaine pretreatment increased activity in both groups | [77] |
| OF1 mice | Male Adolescent, 28 | 33, 34, 40, 41 | MDMA | 0 mg/kg 10 mg/kg 20 mg/kg | Biogenic amines. PND 76, HNS < LNS striatal DOPAC<br>PND 76, increases striatal DA and DA turnover only in LNS<br>PND 76, HNS < LNS striatal DOPAC and DA turnover reduces striatal DA and increases DA turnover in LNS<br>HNS < LNS striatal DA turnover | [122] |

[122] (*see* Table 5). The long-term effect of MDMA binges on striatal DA is also modulated by the novelty-seeking trait, as MDMA exposure induces a significant decrease of striatal DA and an enhanced DA turnover in this structure only in LNS mice. Furthermore, irrespective of the MDMA treatment, we found that LNS exhibited lower levels of DOPAC, which may be related with the behavioral effects observed [122] (*see* Table 6). The changes in the behaviors evaluated in the study suggest that experience of MDMA during adolescence alters the way in which subjects interact with their environment in adulthood; thus, certain individuals will be more prone to suffering long-lasting effects. In particular, HNS become more sensitive to the rewarding effects of cocaine and MDMA; furthermore, they engage in more social interaction but are more aggressive, which could also increase drug use.

*1.5.3 Alcohol*

Several studies have revealed a relationship between ethanol use and the novelty-seeking trait. High sensation seekers show increased alcohol intake and experience more positive effects after its consumption [46]. Furthermore, the novelty-seeking trait predisposes individuals to develop alcohol-related problems, including alcoholism and enhanced vulnerability to relapse after periods of detoxification [124, 125].

In experimental animals, adolescent exposure to chronic alcohol increases the tendency of animals to engage in more exploratory or novelty-seeking behaviors [22], and exposure to alcohol binges during adolescence induces long-lasting behavioral consequences that are influenced by the novelty-seeking phenotype [121]. As shown in Table 3, although adolescent male mice exposed to a binge pattern of EtOH develop CPP with subthreshold doses of cocaine and MDMA regardless of their novelty-seeking profile, HNS animals have been found to require more extinction sessions for cocaine CPP to be extinguished than their LNS counterparts (8 sessions vs. 4 sessions). Furthermore, after extinction is achieved, only HNS mice show reinstatement of cocaine CPP by a priming dose of this drug [121]. Extinction measures the motivational properties of drugs, which are reflected by the persistence of drug-seeking behavior in the absence of the drug, while reinstatement of the extinguished preference is a reliable model of the craving and relapse that characterize drug addiction [106]. Thus, these results suggest that the HNS trait enhances motivation for cocaine and vulnerability to develop addiction in mice exposed to alcohol during adolescence.

In contrast with that observed with cocaine, the acquisition, extinction, and reinstatement of the CPP induced by MDMA is not affected by the novelty-seeking profile of mice exposed to alcohol during adolescence (*see* Table 3). Such results have been observed by Bird and Schenk [99], who reported that the level of novelty-seeking does not significantly correlate with MDMA self-administration in rats. The differential modulation by the novelty-seeking phenotype of the effects of adolescent EtOH exposure on cocaine and MDMA reward might be due to the unique pharmacology of MDMA, which, unlike other psychostimulant drugs, preferentially enhances synaptic 5-HT [126]. In fact, although serotonin mediates the response to novelty-seeking [127, 128], DA plays a more important role [23].

Other long-term effects of exposure to EtOH during adolescence are modulated by the novelty-seeking trait [121]. In particular, EtOH pretreatment is known to exert an anxiolytic effect only in HNS mice (*see* Table 4). Similarly, adolescent exposure to EtOH induces changes in aggressive behavior in HNS mice only, for instance, an increase in nonsocial exploration and a reduction in threat and attack (*see* Table 5). Finally, in a novel object recognition task, adolescent alcohol exposure was shown to increase the number of explorations of the novel object only in HNS; conversely, in LNS mice, alcohol pretreatment reduced the number of explorations and the time spent exploring the novel object, while it increased the latency to explore it (*see* Table 6).

*1.5.4  Cannabinoids*

Consumption of cannabis, the most used illegal drug, usually begins during adolescence.

Recently, the problematic use of cannabis has increased in adolescent individuals, a fact that can induce negative consequences, including enhanced vulnerability to develop dependence and/or higher propensity to later consumption of other drugs [129, 130] such as cocaine [131–133]. In humans, cannabinoid pre-exposure increases the severity of cocaine withdrawal symptoms and relapse to cocaine dependence [134]. Similarly, in adolescent rodents, exposure to cannabinoid agonists increases self-administration [135–140] and modifies the acquisition and reinstatement of the CPP induced by different drugs of abuse, such as morphine [141], MDMA [142] and cocaine [123].

The novelty-seeking phenotype influences the sensitivity of mice to the conditioned rewarding effects of the cannabinoid agonist WIN 55212-2. In particular, we have observed that HNS mice acquire CPP after conditioning with 0.075 mg/kg, a dose that is ineffective in inducing conditioned reward in LNS [123]. In the study in question, we also observed the influence of the novelty-seeking trait on the effects of adolescent cannabinoid exposure (*see* Table 3). In HNS and LNS pre-exposed to the CB1 agonist WIN 55212-2 no differences were observed between the two in the subsequent CPP induced by this drug. Furthermore, pretreatment with WIN 55212-2 increases the rewarding effects of a low dose of cocaine (1 mg/kg), irrespective of the novelty-seeking profile of the mice. However, it is important to note that HNS mice are more resistant to the extinction of CPP and more sensitive to reinstatement of CPP after extinction. In particular, HNS mice require twice as many extinction sessions as LNS mice to achieve extinction of a cocaine CPP. Furthermore, priming with 0.5 mg/kg and 3 mg/kg reinstates the CPP induced by 1 mg/kg and 6 mg/kg of cocaine, respectively, in HNS mice only. On the other hand, pretreatment with the cannabinoid antagonist rimonabant does not increase the sensitivity of mice to the conditioned rewarding effects of WIN 55212-2 or cocaine [123]. These results support the idea that not all subjects are equally vulnerable to the sensitization of the brain reward system induced by stimulation of the cannabinoid system during adolescence, but that those with an HNS profile will be particularly affected.

## 2   Evaluating How Behavior in the Hole Board Test Modulates the Effects of Adolescent Drug Exposure on Subsequent Drug-Induced CPP

### 2.1   Materials

#### 2.1.1   Subjects

Strain

With the exception of the studies discussed in the last section of the chapter, we always use mice of the OF1 strain.

The **Age** of animals is a relevant factor (*see* **Notes 1** and **2**); for example, novelty-seeking behavior is typically more pronounced in adolescent than in adult rodents (*see* **Note 1**). In addition, adolescent rodents show enhanced sensitivity to the rewarding properties of novelty [85] and psychostimulant drugs [28, 143–149], which is reflected by an enhanced vulnerability to developing CPP [39, 145].

Adolescent mice also experience less aversive effects of addictive drugs and reduced withdrawal symptoms (*see* [39]), and this is important to bear in mind when selecting the drug dose to evaluate the effects of novelty-seeking in drug-induced CPP (*see* **Note 2**).

To evaluate the novelty-seeking profile and to model adolescent drug exposure, we usually choose to employ mice of PND 26–45, which is considered a conservative age range during which animals of both genders and most breeds are expected to exhibit adolescent-typical neurobehavioral characteristics [36, 150].

**Sex** is another factor that must be considered. Although the estrous cycle does not affect the locomotor response to novelty-induced behavioral tests [58, 151], sex differences have been reported in the novel object recognition task [74, 151] (*see* Table 1). Furthermore, it is known that women show a faster onset of addiction and require treatment sooner than men [152–155]. Sex differences in the rewarding effects of drugs of abuse have also been observed in rodents [152, 154–156] (*see* **Note 3**).

To study the influence of the novelty-seeking profile on vulnerability to cocaine CPP we have used early–middle adolescent (PND

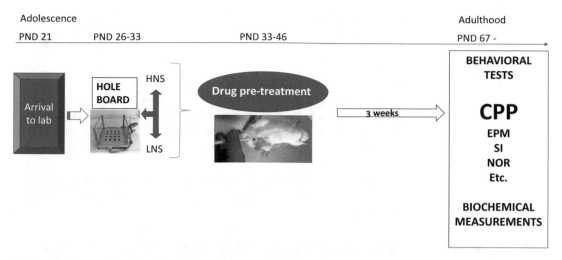

**Fig. 2** Timeline of studies about the influence of novelty-seeking profile in the long-term effects of adolescent drug exposure on behavioral tests (*CPP* conditioned place preference, *EPM* elevated plus maze, *SI* social interaction, *NOR* novel object recognition) and biochemical measures

28–35) and late adolescent (PND 49–56) mice of both sexes of the OF1 outbred strain [66, 114]. As can be seen in Table 2 the results obtained are in function of the paradigm of novelty-seeking used.

However, to evaluate the consequences of previous drug exposure for the subsequent rewarding effects of drugs of abuse according to novelty-seeking profile, we typically use male adolescent mice [77, 121–123]. In brief, the experimental steps are as follows (*see* Fig. 2 and the detailed schedules in Tables 3, 4, 5, and 6):

1. Mice are acquired commercially from the supplier Charles River (France).

2. Mice arrive at the laboratory on PND 21.

3. The hole board test is performed between PND 26 and PND 33 (early adolescence).

4. Typically, mice receive the drug treatment between PND 33 and PND 46 (middle adolescence). However, in the case of cannabinoids adolescent treatment is administered during early adolescence (PND 26–30).

5. Behavioral tests and biochemical measurements are performed in adulthood, usually from PND 67.

Conditions
in the Laboratory

When the animals arrive at the laboratory, they are housed in groups of four in plastic cages (28 × 28 × 14.5 cm) for 5–10 days prior to initiating experimental procedures. The conditions of the laboratory are as follows:

- Constant temperature (21 ± 2 °C),
- Relative humidity of 60%,

- Reversed light schedule (white lights on: 19.30–07.30 h),
- Food and water available ad libitum (except during the behavioral test).

Procedures involving mice and their care are always conducted in conformity with national, regional and local laws and regulations, which are in accordance with the European Communities Council Directives (2010/63/EU). The protocols are always approved the University of Valencia's Ethical Committee on Animal Experimentation.

Animals are handled briefly on the 2 days preceding initiation of the experimental procedures in order to reduce their stress levels (*see* **Note 4**).

2.1.2  *Drugs of Abuse*    **Alcohol**: *According to the National Institute on Alcohol Abuse and Alcoholism, an alcohol binge can be defined as a pattern of alcohol consumption that results in a blood alcohol concentration of 0.08-g% or higher, that is, the consumption of five (four in the case of females) or more drinks in the space of about 2 h. To imitate this pattern of binge consumption, we administer twice-daily injections (separated by a 4-h interval) of 1.25 or 2.5 g/kg ethanol (Merck, Madrid, Spain) several times for a week (for more details see Subheading 2.2.2). The control group is injected with repeated injections of physiological saline (NaCl 0.9%), also used to dissolve the ethanol, following the same schedule as the binge pattern.*

**Cocaine**: *(Laboratorios Alcaliber S.A., Madrid, Spain) is usually administered to induce CPP or as pretreatment during adolescence. This drug is diluted in physiological saline (0.1 mg/mL), which is administered to controls. The cocaine dose most used for CPP (1 mg/ kg) is based on previous studies [157] in which it was shown to be a subthreshold dose, that is, ineffective in inducing CPP in naïve mice [66, 157] (see **Note 5**). When administered during adolescence in order to evaluate its long-term effects, mice are frequently treated with cocaine binges. We usually employ one of two schemes of increasing doses from 5 to 10 mg/kg or from 5 to 25 mg/kg (for more details see Subheading 2.2.2).*

**MDMA**: *(3,4-methylenedioxy-metamphetamine hydrochloride; racemic mixture) is commercially acquired from Laboratorios Sigma-Aldrich (Spain) or provided by the Agencia Española del Medicamento, Ministerio de Sanidad, Política Social e Igualdad, Madrid, Spain. For CPP experiments we use doses of 1 or 2.5 mg/kg of MDMA dissolved in physiological saline, which are ineffective in naïve mice [158]. For pretreatment binges, we use 10 or 20 mg/kg of MDMA (for more details see Subheading 2.2.2).*

**WIN 55212-2**: *is a cannabinoid CB1 agonist used to study the effects of cannabinoid drugs such as cannabis. WIN 55212-2 is commercially acquired from Tocris, Biogen Científica, S.L. (Madrid,*

*Spain) and dissolved with saline and Tween 80 (Sigma-Aldrich, Madrid, Spain) at 0.01% (0.01 mL of Tween dissolved in 100 mL of saline) (see* **Note 6***). The doses administered in the CPP paradigm (0.05, 0.075 mg/kg) are selected on the basis of previous studies which have confirmed the subthreshold dose that does not induce CPP in naïve mice [141, 159]. For adolescent pretreatment, a higher dose of WIN 55212-2 (0.1 mg/kg) is used; mice receive one injection per day for 5 days (see more details in Subheading 2.2.2). Such cannabinoid exposure during adolescence is effective in increasing the CPP induced by MDMA in adulthood [142]. To evaluate the involvement of CB1 receptors in the effects of WIN 55212-2 (or other drugs of abuse), we use* **SR 141716A** *(rimonabant), commercially acquired from Sanofi Recherche (Montpellier, France), administered at a dose of 1 mg/kg, since this dose blocks CB1 receptors and does not act as an inverse agonist of these receptors [160].*

All injections are administered intraperitoneally, at a constant volume of 10 mL/kg (0.01 mL/g) (*see* **Note 7**).

| | |
|---|---|
| *2.1.3 Apparatus*<br><br>Hole Board | The hole board consists of a box (28 × 28 × 20.5 cm) with walls made of clear Plexiglas. In the floor of the box there are 16 equidistant holes with a diameter of 2.3 cm (*see* Fig. 2). Photocells below the surface of the holes detect the number of times the mouse performs a head dip (Med Associates, CIBERTEC, SA, Spain). A computerized system records the number of times a mouse explores a specific hole and the total frequency of dips (Activity Monitor v.7) (*see* Figs. 3 and 4). |
| Conditioned Place Preference Boxes | The apparatus consists of eight identical Plexiglas place conditioning boxes comprised of two equally sized compartments (30.7 × 31.5 × 34.5 cm) separated by a gray central area (13.8 × 31.5 × 34.5 cm). The compartments have different colored walls (black vs. white) and distinct floor textures (smooth in the black compartment and rough in the white one). Four infrared light beams in each compartment of the box and six in the central area allow the position of the animal and its crossings from one compartment to the other to be recorded (*see* Fig. 4). The equipment is controlled by an IBM PC computer using MONPRE 2Z software (CIBERTEC, SA, Spain) (*see* Fig. 4). |
| **2.2 Methods**<br><br>*2.2.1 Hole Board* | As commented on before, the hole board test evaluates the tendency of rodents to explore a new environment in a free-choice procedure. This test was developed in 1962–1964 by Boissier and Simon [78, 82], and is a simple and useful procedure to assess the response of an animal to an unfamiliar setting [81]. The exploratory behavior measured in this test is the number of head dips, which represents exploratory tendencies distinct from general locomotor |

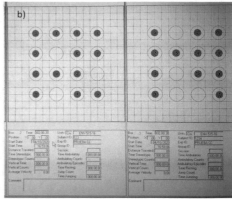

**Fig. 3** The hole board test. (**a**) Two hole board apparatuses connected to the interphase and the computer that controls the apparatus. (**b**) Register of the mouse's behavior (dips, distance traveled, velocity, etc.) on the computer screen. Holes visited are marked in yellow; the number inside the hole indicates the number of dips in the hole performed by the mouse

**Fig. 4** Experimental room for place conditioning. Each computer controls four CPP boxes. Each box has two different compartments (black and white) separated by a small central gray corridor (see more details in the text). A table in the middle of the room is used to administer injections immediately before placing the mouse in the corresponding compartment

activity. Number of head dips is a useful measure to study the relationship between novelty-seeking and drug abuse [83].

Adolescent animals perform the hole board test after a 5–10-day period of acclimatization to the laboratory, during the dark phase (9.00–12.00 h). The illumination in the experimental room is provided by four neon tubes fixed to the ceiling (light intensity of 110 lux at 50 cm above floor level) (*see* **Note 8**).

At the beginning of the test, the mouse is placed in one corner of the hole board and allowed to explore it freely for 10 min. The total number of heap dips and the latency to perform the first head dip is scored (*see* **Note 9**). The box is carefully cleaned with 70% alcohol at the end of each test.

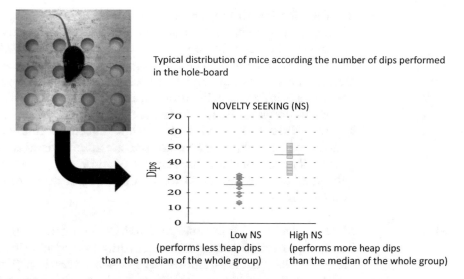

Fig. 5 Classification of mice as HNS and LNS according to the number of dips performed in the hole board

We use a median-split analysis to study the effects of novelty-seeking on the behavioral effects of different drugs of abuse [77, 114, 121–123]. Mice are defined as high novelty seekers (HNS) or low novelty seekers (LNS) according to whether the number of head dips they perform is higher or lower than the median of the group (HNS or LNS, respectively). We use the median to classify mice as HNS versus LNS, because it is a standard measure when a set of data has been arranged in order of magnitude and is less affected by outliers (*see* Fig. 5). As shown in Table 2, the hole board is the best test to predict which adolescent male mice will be more sensitive to the conditioned rewarding effects of cocaine in the adulthood (*see* **Note 10**). After the mice have been classified as HNS or LNS, they are randomly assigned to a drug treatment, ensuring that in each group of treatment approximately half the mice are HNS and the other half are LNS (*see* Fig. 5).

*2.2.2 Pharmacological Treatment*

After the adolescent mice have been classified as HNS or LNS (according to the number of dips performed in the hole board), both groups are treated with a drug of abuse, usually between PND 28–42, considered a conservative age range during which mice are expected to exhibit neurobehavioral characteristics typical of adolescence [150].

Cocaine Binges

To evaluate how the novelty-seeking trait influences the effects of adolescent cocaine exposure (PND 34–45) and the subsequent rewarding properties of cocaine and MDMA, a pretreatment of three injections per day is administered at 1-h intervals, with the dose being increased at different points of this 12-day period. From PND 34–35, mice receive 5 mg/kg; from PND 36 to PND

38, they receive 10 mg/kg; and from PND 41–45, they receive 15 mg/kg. For comparison, the same schedule of administration of cocaine can be administered to adult mice (PND 61 and 72). To evaluate how such adolescent cocaine binges alter behavior in the elevated plus maze (anxiety) and social interaction test (agonistic encounter with a cospecific mouse), the same schedule of cocaine administration is applied, but with higher doses (5, 15 and 25 mg/kg) (*see* **Note 11**). Physiological saline injections following the same schedule as that used for cocaine are administered to control groups. The long-term consequences of these binges on different behavioral tests are studied 3 weeks after the last drug administration (*see* Fig. 2).

**MDMA Binges**

Mice receive eight injections of MDMA (at doses of 0, 10, or 20 mg/kg) over a 2-week period according to the following schedule: twice-daily administrations (with a 4-h interval: at 9 am and 1 pm) on 2 consecutive days each week. In this way, mice are injected on PND 33, 34, 40, and 41. Behavioral tests are performed 3 weeks after exposure had finalized, when mice have entered adulthood. We use doses of 10 and 20 mg/kg to evaluate the long-term effects of binges on social behavior, anxiety and biochemical measures (dopamine, serotonin, metabolites and turnover), while only the dose of 10 mg/kg is used to evaluate the long-term effects of MDMA binges on drug-induced CPP.

**Ethanol Binges**

Mice are treated with EtOH (1.25 or 2.5 g/kg) or physiological saline on 2 consecutive days, at 48-h intervals, over a 14-day period to simulate a binge pattern such as that engaged in by human adolescents and young adults [161]. Each mouse receives 16 injections, according to the following schedule: twice daily injections (with a 4-h interval) on 2 consecutive days separated by an interval of 2 days during which no injections were administered. Animals are injected on PND 33, 34, 37, 38, 41, 42, 45, and 46. The control group receives the same schedule of injections, but of physiological saline. The long-term consequences of these binges on anxiety levels (elevated plus maze and social interaction test) and spontaneous motor activity are studied 3 weeks after the last administration of EtOH. To evaluate the consequences of adolescent ethanol binges on the subsequent rewarding effects of cocaine and MDMA we administer binges of 2.5 g/kg.

**Pretreatment with Cannabinoid Drugs**

Animals receive a daily injection of their respective treatment (vehicle, 0.1 mg/kg of WIN 55212-2, or 1 mg/kg of rimonabant) for 5 days (PND 26–30) and, after an interval of 3 days without any treatment, the CPP procedure is initiated [123].

2.2.3 *Conditioned Place Preference*

The CPP paradigm evaluates the positive and pleasant properties of stimuli (including the rewarding effects of addictive drugs) [103–106] in a rapid and simple way [162]. In this paradigm, contextual or environmental stimuli acquire secondary appetitive properties (conditioned rewarding effects) when paired with a primary reinforcer [104, 105]. Conditioned reward implies that animals attribute positive incentive value to the cues associated with the primary reinforcer (the drug of abuse), and thus perform free or voluntary responses to obtain access to said cues [163]. Under appropriate conditions, CPP can be sensitive to a wide range of substances, including opiates, psychostimulants, alcohol, and cannabinoids [162].

The procedure, unbiased in terms of initial spontaneous preference, is performed as described previously [164]. In the first phase, referred to as preconditioning (Pre-C), mice are allowed access to both compartments of the apparatus for 15 min (900 s) per day on 2 consecutive days. On Day 3, the time spent in each compartment over a 900-s period is recorded during the dark phase (between 10:00 and 14:00 h). We use a counterbalanced design to assign the mice of each group to the drug- and vehicle-paired compartment. After assigning the compartments, we perform an analysis of variance (ANOVA) with the data of the time spent in each compartment during the preconditioning phase. There must not be significant differences between the time spent in the compartment paired with the drug and that spent in the compartment paired with vehicle. This is an important requirement of the experimental procedure that avoids any preference bias prior to conditioning (*see* **Note 12**).

The second phase (conditioning) lasts 4 or 8 days, in function of the drug of abuse used to induce place conditioning. In the CPP induced by cocaine, the mice undergo 2 pairings per day, on 4 consecutive days. Animals receive an injection of physiological saline immediately before being confined to the vehicle-paired compartment for 30 min. After an interval of 4 h, they receive an injection of cocaine immediately before being confined to the drug-paired compartment for 30 min. The order of injections (cocaine or saline) is alternated every day. In the CPP induced by MDMA and the CB1 agonist, mice undergo only one pairing per day: animals conditioned with MDMA receive an injection of MDMA immediately before confinement in the drug-paired compartment for 30 min on Days 4, 6, 8, and 10, and receive physiological saline before confinement in the vehicle-paired compartment for 30 min on Days 5, 7, 9, and 11. Confinement is carried out in both cases by closing the guillotine door that separates the two compartments. The central area of the apparatus is never used during conditioning.

During the third phase, known as postconditioning (Post-C), the guillotine door separating the two compartments is removed (Day 8 in the case of cocaine, and Day 12 in the case of MDMA and

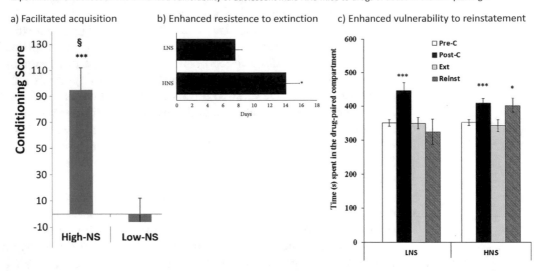

**Fig. 6** Experimental evidence of the enhanced vulnerability of adolescent male HNS mice to drugs of abuse in the CPP paradigm. (**a**) Facilitated acquisition. (**b**) Enhanced resistance to extinction. (**c**) Enhanced vulnerability to reinstatement

WIN 55212-2), and the time spent by the untreated mice in each compartment during a 900-s observation period is recorded. Post-C tests are performed in the dark phase between 10:00 and 14:00 h.

The difference in seconds between the time spent in the drug-paired compartment during the Post-C test versus the Pre-C phase is a measure of the degree of conditioning induced by the drug. If this difference is positive, then the drug has induced a preference for the drug-paired compartment, while the opposite indicates that an aversion has developed.

The rewarding effects of drugs of abuse are evaluated 3 weeks after the end of adolescent drug pretreatments. Usually, subthreshold doses of drugs are used in order to evaluate the sensitivity of HNS and LNS mice to the conditioned reinforcing effects of these drugs (*see* **Note 5**); we generally use cocaine (1 mg/kg), MDMA (1.25 mg/kg), and WIN 55212-2 (0.05 and 0.075 mg/kg). The doses selected to induce CPP are based on previous studies [142, 157–159]. As can be seen in Fig. 6 and Table 3 there is a body of experimental evidence for the enhanced vulnerability of adolescent male HNS mice to drugs of abuse in the CPP paradigm. HNS mice acquire drug-induced CPP more readily (Fig. 6a), and pretreatment with psychostimulant drugs during adolescence enhances the CPP induced by these drugs of abuse only in HNS mice (*see* Table 3).

In some experiments, the groups that acquire CPP undergo extinction sessions to evaluate the persistence of CPP and the subsequent vulnerability to drug-induced reinstatement of CPP. During each extinction session, the mouse is placed in the apparatus, without the guillotine doors separating the compartments, for 15 min. The extinction session is repeated every 72 h, until the time spent in the drug-paired compartment by each group is similar to that of Pre-C and differs from that of Post-C (Student's *t*-tests). In this way, all the animals in each group undergo the same number of extinction sessions, independently of their individual scores, as the criterion for extinction is a lack of significant differences with respect to Pre-C values. The extinction of CPP must always be confirmed in a subsequent session performed 24 h after the last extinction session. We have observed that HNS mice show enhanced resistance to the extinction of drug-induced CPP (Fig. 6b).

To evaluate vulnerability to drug-induced reinstatement, the effects of a priming dose of the drug used to induce CPP—typically half of the dose administered during conditioning (0.0375 mg/kg of WIN 55212-2, 0.5 or 3 mg/kg of cocaine)—are evaluated 24 h after confirmation of extinction. Reinstatement tests are the same as for Post-C (free ambulation for 15 min), except that mice are tested 15 min after administration of the drug employed during the conditioning phase. Reinstatement tests are always performed in the dark phase, between 10:00 and 14:00 h. We have observed that HNS mice exhibit enhanced vulnerability to the extinction of drug-induced CPP (Fig. 6c).

### 2.3 Statistical Analysis

#### 2.3.1 Hole Board

Mice are considered high novelty seekers (HNS) or low novelty seekers (LNS) according to whether their result is higher or lower than the mean of their group. The most usual measure is the frequency (number) of dips, although the latency to the first dip can also be used (*see* **Note 13**). Although the distribution is not completely bimodal (some mice had a similar number of head dips in LNS and HNS groups), they are clearly different with respect to the median score. To assess differences in the frequency of head-dipping between male adolescent HNS and LNS mice, a *t*-test is usually carried out. In all cases, this test reveals any significant differences between the two groups ($p < 0.01$).

In other studies, a mixed ANOVA is performed for each measure of the novelty-seeking test (number of dips and/or latency to the first dip). In function of the experimental design of the study, there is/are one, two, or three between-subject variables: *sex* (with two levels; males and females), *age* (with two levels; adolescents and young adults), or/and *pretreatment* (drug and saline). Sometimes, separate ANOVAs are performed for adolescent and adult animals with the variable "sex," with two levels (male and female), and for male and female animals with the variable "age," with two levels (adolescents and young adults).

Post hoc comparisons are performed with Bonferroni tests.

*2.3.2 Conditioned Place Preference*

For CPP, the time spent in the drug-paired compartment during pre- and postconditioning phases is analyzed by means of a mixed ANOVA with a *within variable—days*, with two levels (pre- and postconditioning test)—and one, two, three, or four between variables: *sex, age, pretreatment* (the same as explained in the previous section), and level of novelty-seeking (HNS and LNS).

Linear and logistic regression analyses may be employed to determine the association between the level of novelty-seeking and the development of CPP (using a CPP score, time spent in drug-paired compartment in Post-C minus time spent in the same compartment in Pre-C).

Differences between the time spent by mice in the drug- and saline-paired compartments in extinction and reinstatement tests after receiving priming doses are analyzed by means of Student's "*t*" tests. In some cases, the variable "days"—analyzed in the ANOVA described in the first paragraph of this section—has four levels: Pre-C, Post-C, extinction, and reinstatement. In addition, the time required for preference to be extinguished in each animal can be analyzed by means of the Kaplan–Meier test, with Breslow (generalized Wilcoxon) comparisons when appropriate.

In the ANOVAs, post hoc comparisons are performed with Bonferroni tests.

## 3    Notes

1. The age of animals plays an essential role in the effects observed in the hole board test. In rodents, adolescence covers the whole postnatal period, from weaning (21 postnatal days, PND 21) to adulthood (PND 60). Rodent adolescence has been classified in three periods: early adolescence (prepubescent or juvenile, from PND 21–34), middle adolescence (periadolescent, from PND 34–46), and late adolescence (young adult, from PND 46–59) [150]. Behavioral traits observed in adolescent humans [34] have also been observed in animal models. In comparison with their adult counterparts, adolescent rodents show increased exploration and risk-taking behavior [31, 36, 39, 165]. Juvenile rodents present greater hyperactivity and exploration in novel environments [36, 166], perform a higher number of explorations in the novel object recognition task and novel environment test of free-choice [66], and display higher locomotor activity in the initial minutes of exposure to an inescapable novel open field [63, 113]. Adolescent mice are also prone to present shorter latency to approach novelty and to explore a novel object/environment for longer periods than their young adult counterparts. In some studies, the hole board has been applied to adult mice (>PND 60), namely, to observe

changes in their behavior induced by environmental manipulations (e.g., exposure to stress).

2. It is important to consider the age of animals when selecting the dose to be used in the CPP paradigm. Generally, conditioned reward is enhanced in adolescents in comparison to adults [39]. For example, we have observed a clear influence of age on the rewarding effects of some drugs of abuse. Adolescent male mice conditioned with ethanol on PND 32–38 showed CPP with doses (1.25 and 2.5 g/kg) that were ineffective in young adult animals conditioned on PND 54–60 [167]. In contrast, we have seen that 1 mg/kg of cocaine is effective in inducing CPP in both adolescent (PND 29–32) and late adolescent (young adult) mice (PND 50–53), although the latter group showed reinstatement of the CPP induced by 25 mg/kg of cocaine with lower priming doses of this drug (12.5, 6.25, and 3.125 mg/kg) than early adolescent mice, in which reinstatement was achieved only with 12.5 mg/kg of MDMA [168]. In the case of MDMA, late adolescent mice are more sensitive to it's rewarding effects than early adolescent mice (PND 53–59 vs. PND 32–38, respectively, during conditioning). A low dose of MDMA (1.25 mg/kg) induces CPP in late adolescent mice only; furthermore, the CPP induced by 10 mg/kg of MDMA is reinstated in both age groups by 5 mg/kg of MDMA, but only late adolescent mice show further reinstatement with 2.5 mg/kg of MDMA [169].

   The influence of the novelty-seeking trait on drug-induced CPP also depends on the age of the animals. Previous reports by our group have shown that naïve HNS adolescent mice acquire CPP after conditioning with 1 mg/kg of cocaine, a dose that is ineffective in LNS animals [66, 114]. Conversely, when male adult rats are classified as HNS versus LNS, neither group develops CPP with 2.5 mg/kg of cocaine [107].

3. Sexual differences in the effects of drugs of abuse on the CPP paradigm have been reported; for instance, female rats develop CPP at lower doses than males [145]. Similarly, we have observed that ethanol induces CPP and reinstatement in young adult female mice at doses (2.5 g/kg) that are ineffective to induce these effects in male mice of the same age [167].

4. In CPP experiments we usually handle mice on each of the 3 days immediately prior to the preconditioning (Pre-C) phase in order to reduce their stress levels in response to experimental manipulations. Prevention of noise, a quiet environment throughout the laboratory and the habituation of animals to the room where the behavioral tests are performed is also critical if reliable results are to be achieved.

5. For CPP experiments, we employ a subthreshold dose of cocaine (or other drugs of abuse) in order to determine the sensitizing effect of adolescent drug treatments on the conditioned reinforcing effects of cocaine. If a dose that is ineffective in naïve mice can induce CPP in mice preexposed to a drug of abuse during adolescence, it can be assumed that the adolescent treatment has enhanced the vulnerability of mice to the rewarding effects of this drug. In other cases, to evaluate if adolescent drug exposure increases the probability of relapse to drug use after a period of abstinence, we administer a dose that induces CPP in naïve mice but does not produce reinstatement after drug priming (e.g., 6 mg/kg of cocaine) [66, 157].

6. Cannabinoid drugs are difficult to dissolve in physiological saline. It is recommendable to first mix the cannabinoid with a quantity of DSMO (for example, 3 mL) and a drop of Tween 80. Physiological saline is then added bit by bit. An electric shaker may be used to facilitate dissolution.

7. Intraperitoneal (i.p.) injection is an easy and effective way to administer drugs to mice (*see* Fig. 2). It is important to use a new needle for each animal (to reduce discomfort and risk of infection), although in the case of chronic treatments the syringe can be marked and maintained for later use in the same animal.

8. The novelty-seeking test must be performed at least 1 h after initiation of the dark phase of the cycle, since mice are nocturnal animals whose circadian phase of activity takes place at night. For the same reason, the experimental room should be only slightly illuminated.

9. It is recommended to place mice in the same corner of the cage at the beginning of the novelty-seeking test and it should be verified that a mouse of medium size cannot pass through the hole. Moreover, there are different frames with holes for rats or mice. This is especially important in the case of adolescent mice of some strains that are very small.

10. In a previous study carried out in our laboratory, we reported a higher predictive capacity of the hole board test for identifying "drug-vulnerable" individuals among adolescent mice of both sexes. HNS mice acquired CPP after conditioning with 1 mg/kg of cocaine, while LNS mice did not [114].

11. Our protocol of cocaine administration was adapted from that of Sillivan et al. [170] and has proved to induce long-lasting alterations in rodent behavior [170, 171]. It should be stressed that we use lower doses of cocaine to administered binges to mice that are subsequently to be conditioned with cocaine or MDMA, since we have observed that pretreatment with high

doses of cocaine decreases the subsequent rewarding effects of this drug [172].

12. Animals showing strong unconditioned aversion for any compartment (less than 33% of the session time) must be excluded from the rest of the protocol to ensure that the CPP procedure is unbiased in terms of initial spontaneous preference [157, 159]. Therefore, we determine which compartment will be paired with the drug (and which will be paired with the vehicle) for each mouse. We use a counterbalanced design by which half the animals in each group receive the treatment in one compartment (black) and the other half receive it in the other (white) compartment.

13. Latency data undergo logarithmic transformation before statistical analysis.

## 4 Advantages and Limitations of the Hole Board Test When Employed to Predict Vulnerability to Drug Addiction

A major limitation of studies in human beings is the impossibility of establishing causal relationships between events or facts. In the case of the novelty-seeking trait, we cannot determine whether a high novelty-seeking profile leads an individual to progress from recreational to compulsive drug consumption, or whether this behavioral profile is the consequence of drug abuse. Some studies have suggested that high novelty seekers have a differential sensitivity to drug reward at initiation of consumption that increases their vulnerability to develop addictive behavior [12, 173–176]; however, other studies challenge this putative relationship between sensation/novelty-seeking and addiction risk [18, 177–179]. In particular, it has been suggested that high novelty seekers are more prone to initiate and maintain a regular drug use, but that, without additional risk factors (such as a family history of addiction), they do not run an enhanced risk of developing drug addiction [179].

Animal models avoid the ethical problems of research with humans and have demonstrated themselves to be useful tools for increasing our understanding of the interacting factors that facilitate the development of drug addiction, including individual traits (such as a high novelty-seeking profile) and environmental events (e.g., early drug exposure or stressful experiences). In this sense, the hole board test allows us to assess the causality between a novelty-seeking profile and vulnerability to the development of addiction-like behavior in rodents. Another advantage of the hole board test is the validity of the model, since there is a close similarity between the behavioral patterns, physiological correlates, and psychobiological consequences observed in rodents and human beings with the novelty/sensation-seeking trait [23, 180].

The paradigm that we have described here allows researchers to study the interaction between two main factors that increase vulnerability to repeated drug consumption, that is, the behavioral predisposition of a high novelty-seeking profile and exposure to drugs of abuse during adolescence. The HNS phenotype may increase the probability that an adolescent will initiate drug experimentation and develop neuroadaptations that will lead later on to addiction in adulthood. Our research in mice has shown that the NS trait facilitates acquisition of the CPP induced by different drugs of abuse. Furthermore, we have observed that the HNS trait increases the effects of exposure to drugs of abuse during adolescence on the reinstatement and maintenance of cocaine CPP. Thus, adolescent exposure to psychostimulants, EtOH or cannabinoids induces more profound effects on the behavior of subjects with an HNS profile and their response to drugs of abuse. From a translational point of view, our paradigm models a subpopulation of adolescents which engages in drug use early in life and has a greater risk of developing abuse and addiction. Thus, there is no direct causal relationship between adolescent drug exposure and the later development of addiction; rather, our results suggest that there are subjects with particular behavioral traits, present prior to the onset of drug use, that are more vulnerable to addiction. For example, subjects with a higher propensity for sensation-seeking, in addition to a greater tendency to experiment with drugs, are likely to develop more neuroadaptations following drug exposure, thus resulting in the transition from voluntary to compulsive drug use.

Some of the limitations of our model are related to what the hole board and the CPP paradigms are really measuring. Besides novelty-seeking, the hole board test can be used to assess emotionality and anxiety in mice [79, 81]. In addition, the enhanced ability of HNS mice to develop CPP with low doses of psychostimulants or cannabinoids may be due to different factors; in comparison with LNS mice, those with an HNS profile may experience the drug as more rewarding, may enhance the attribution of incentive salience to the drug-associated cues and/or may acquire conditioned learning more efficiently.

The influence of the novelty-seeking trait on the rewarding effects of drugs of abuse needs to be studied in adolescent animals because novelty-seeking behavior and drug sensitivity are more pronounced in adolescent than in adult rodents. However, the influence of the novelty-seeking trait on drug consumption has generally been studied in adult rodents [12]. One limitation of the hole board is the fact that adolescent mice show a lower index of novelty-seeking (measured by the number of dips) than young adult mice (*see* Table 1); nevertheless, we have selected this model of novelty-seeking because it is the only one capable of predicting enhanced vulnerability to cocaine-induced CPP in HNS adolescent male mice. As of yet we have not evaluated sex differences in the way the novelty-seeking trait modulates the effects of adolescent drug exposure on the later development of addiction.

## 5 Future Directions and Conclusion

One priority of future research should be the identification of neurobiological substrates of the HNS profile and the associated vulnerability to drug addiction. Dopamine is the neurotransmitter most related with this behavioral trait. HNS animals present higher endogenous levels of dopamine, lower availability of D2/D3/D4 receptors and stronger responses to rewarding stimuli and reward-associated cues [181]. This characteristic striatal DA profile of HNS mice may contribute to their tendency to approach novel stimuli and to acquire drug-induced CPP with lower doses than LNS mice. Further knowledge of the role of other neurotransmitter systems and brain areas involved in novelty-seeking will no doubt contribute to the design of pharmacotherapies that reduce the risk of addiction in subjects with a more vulnerable behavioral profile.

A recent study in our laboratory has demonstrated that the novelty-seeking trait is a behavioral marker of vulnerability or resilience to the effects of stress on drug reward. In particular, young adult C57BL6 male mice suffering social stress (induced by exposure to repeated social defeat) are more sensitive to the effects of cocaine in the CPP paradigm during adulthood. However, we also observed that the long-term effects of social defeat were modulated by the behavioral response of mice to stress. Mice performing more dips in the hole board 24 h after the last defeat were later more vulnerable to the effects of stress on CPP; in fact, defeated HNS mice developed place conditioning with low doses of cocaine (which were ineffective in control and defeated LNS mice). The fact that defeated mice performing a lower number of dips in the hole board do not develop cocaine-induced CPP suggests they are resilient to the effects of stress. Future studies need to explore the interaction between the different variables that participate in vulnerability or resilience to stress.

Finally, as commented on before, sex differences in hole board behavior have been studied very little. Future studies should evaluate how novelty-seeking and other personality and environmental factors differentially modify the long-term effects of adolescent drug exposure in male and female subjects. The knowledge obtained in the following years will be of vital importance for drawing up the guidelines of specific preventive programs aimed at more vulnerable subjects.

In conclusion, our studies suggest that the high novelty-seeking endophenotype is a marker of susceptibility to the effects of environmental variables, such as adolescent drug exposure or stressful experiences, and that it increases the risk of drug addiction. Advances in knowledge of such endophenotypes will constitute the scientific basis for the development of new preventive strategies and effective individualized therapies aimed at individuals at risk of addiction that reduce drug consumption and mitigate this disorder.

## Acknowledgments

This work has been possible thanks to the grant PSI2017-83023 (Ministerio de Ciencia, Innovación y Universidades, Spain)

## References

1. Kuhn BN, Kalivas PW, Bobadilla AC (2020) Understanding addiction using animal models. Front Behav Neurosci 13:262. https://doi.org/10.3389/fnbeh.2019.00262

2. Bardo MT, Neisewander JL, Kelly TH (2013) Individual differences and social influences on the neurobehavioral pharmacology of abused drugs. Pharmacol Rev 65:255–290. https://doi.org/10.1124/pr.111.005124

3. Ersche KD, Turton AJ, Chamberlain SR, Müller U, Bullmore ET, Robbins TW (2012) Cognitive dysfunction and anxious-impulsive personality traits are endophenotypes for drug dependence. Am J Psychiatry 169:926–936. https://doi.org/10.1176/appi.ajp.2012.11091421

4. Gottesman II, Gould TD (2003) The endophenotype concept in psychiatry: etymology and strategic intentions. Am J Psychiatry 160:636–645. https://doi.org/10.1176/appi.ajp.160.4.636

5. Zukerman M (1979) Sensation seeking: beyond the optimal level of arousal. Erlbaum, Hillsdale, NJ

6. Zuckerman M (1994) Behavioral expressions and biosocial bases of sensation seeking. Cambridge University Press, Cambridge

7. Cloninger C (1987) The tridimensional personality questionnaire, version IV. Department of Psychiatry, Washington University School of Medicine, St. Louis, MO

8. Jupp B, Dalley JW (2014) Behavioral endophenotypes of drug addiction: etiological insights from neuroimaging studies. Neuropharmacology 76:487–497. https://doi.org/10.1016/j.neuropharm.2013.05.041

9. Kosten TA, Ball SA, Rounsaville BJ (1994) A sibling study of sensation seeking and opiate addiction. J Nerv Ment Dis 182:284–289. https://doi.org/10.1097/00005053-199405000-00006

10. Wills TA, Windle M, Cleary SD (1998) Temperament and novelty seeking in adolescent substance use: convergence of dimensions of temperament with constructs from Cloninger's theory. J Pers Soc Psychol 74:387. https://doi.org/10.1037/0022-3514.74.2.387

11. Ball S (2004) Personality traits, disorders, and substance abuse. In: Stelmack RM (ed) On the psychobiology of personality: essays in honor of Marvin Zuckerman. Elsevier Science, Amsterdam, pp 203–222. https://doi.org/10.1016/B978-008044209-9/50013-0

12. Nadal-Alemany R (2008) La búsqueda de sensaciones y su relación con la vulnerabilidad a la adicción y al estrés. Adicciones 20:59–72

13. Blanchard MM, Mendelsohn D, Stamp JA (2009) The HR/LR model: further evidence as an animal model of sensation seeking. Neurosci Biobehav Rev 33:1145–1154. https://doi.org/10.1016/j.neubiorev.2009.05.009

14. Piazza PV, Deroche V, Rougé-Pont F, Le Moal M (1998) Behavioral and biological factors associated with individual vulnerability to psychostimulant abuse. NIDA Res Monogr 169:105–133

15. Kreek MJ, Nielsen DA, Butelman ER, LaForge KS (2005) Genetic influences on impulsivity, risk taking, stress responsivity and vulnerability to drug abuse and addiction. Nat Neurosci 8:1450–1457. https://doi.org/10.1038/nn1583

16. Flagel SB, Akil H, Robinson TE (2009) Individual differences in the attribution of incentive salience to reward-related cues: implications for addiction.

Neuropharmacology 56:139–148. https://doi.org/10.1016/j.neuropharm.2008.06.027

17. Dalley JW, Everitt BJ, Robbins TW (2011) Impulsivity, compulsivity, and top-down cognitive control. Neuron 69:680–694. https://doi.org/10.1016/j.neuron.2011.01.020

18. Ersche KD, Turton AJ, Pradhan S, Bullmore ET, Robbins TW (2010) Drug addiction endophenotypes: impulsive versus sensation seeking personality traits. Biol Psychiatry 68:770–773. https://doi.org/10.1016/j.biopsych.2010.06.015

19. Silvia S, Martins SS, Storr CL, Alexandre PK, Chilcoat HD (2008) Adolescent ecstasy and other drug use in the National Survey of Parents and Youth: the role of sensation-seeking, parental monitoring and peer's drug use. Addict Behav 33(7):919–933. https://doi.org/10.1016/j.addbeh.2008.02.010

20. Stoops WW, Lile JA, Robbins CG, Martin CA, Rush CR, Kelly TH (2007) The reinforcing, subject-rated, performance, and cardiovascular effects of d-amphetamine: influence of sensation-seeking status. Addict Behav 32(6):1177–1188. https://doi.org/10.1016/j.addbeh.2006.08.006

21. Kelly TH, Robbins G, Martin CA, Fillmore MT, Lane SD, Harrington NG, Rush CR (2006) Individual differences in drug abuse vulnerability: d-amphetamine and sensation-seeking status. Psychopharmacology 189(1):17–25. https://doi.org/10.1007/s00213-006-0487-z

22. Stansfield KH, Kirstein CL (2007) Chronic cocaine or ethanol exposure during adolescence alters novelty-related behaviors in adulthood. Pharmacol Biochem Behav 86(4):637–642. https://doi.org/10.1016/j.pbb.2007.02.008

23. Bardo MT, Donohew RL, Harrington NG (1996) Psychobiology of novelty seeking and drug seeking behavior. Behav Brain Res 77:23–43

24. Jetha MK, Segalowitz SJ (2012) Adolescent brain development: implications for behavior. Academic Press, New York, NY

25. Calkins SD (2010) Psychobiological models of adolescent risk: implications for prevention and intervention. Dev Psychobiol 52:213–215. https://doi.org/10.1002/dev.20435

26. Laviola G, Pascucci T, Pieretti S (2001) d-Amphetamine-induced behavioural sensitization and striatal dopamine release in awake freely moving periadolescent rats. Pharmacol Biochem Behav 68:115–124. https://doi.org/10.1016/S0091-3057(00)00430-5

27. Walker QD, Kuhn CM (2008) Cocaine increases stimulated dopamine release more in periadolescent than adult rats. Neurotoxicol Teratol 30:412–418. https://doi.org/10.1016/j.ntt.2008.04.002

28. Badanich KA, Adler KJ, Kirstein CL (2006) Adolescents differ from adults in cocaine conditioned place preference and cocaine-induced dopamine in the nucleus accumbens septi. Eur J Pharmacol 550:95–106. https://doi.org/10.1016/j.ejphar.2006.08.034

29. Frantz KJ, O'Dell LE, Parsons LH (2007) Behavioral and neurochemical responses to cocaine in periadolescent and adult rats. Neuropsychopharmacology 32:625–637. https://doi.org/10.1038/sj.npp.1301130

30. Camarini R, Griffin WC, Yanke AB, dos Santos BR, Olive MF (2008) Effects of adolescent exposure to cocaine on locomotor activity and extracellular dopamine and glutamate levels in nucleus accumbens of DBA/2J mice. Brain Res 1193:34–42. https://doi.org/10.1016/j.brainres.2007.11.045

31. Spear LP (2011) Rewards, aversions and affect in adolescence: emerging convergences across laboratory animal and human data. Dev Cogn Neurosci 1(4):392–400

32. Spear LP (2013) Adolescent neurodevelopment. J Adolesc Health 52(2):S7–S13

33. Rodríguez-Arias M, Aguilar MA (2012) Polydrug use in adolescence. In: Belin D (ed) Addictions - from pathophysiology to treatment. InTech, Rijeka. https://doi.org/10.5772/47961

34. Steinberg L (2007) Risk taking in adolescence new perspectives from brain and behavioral science. Curr Dir Psychol 16:55–59. https://doi.org/10.1111/j.1467-8721.2007.00475.x

35. Kandel DB, Logan JA (1984) Patterns of drug use from adolescence to young adulthood: I. Periods of risk for initiation, continued use, and discontinuation. Am J Public Health 74:660–666. https://doi.org/10.2105/AJPH.74.7.660

36. Spear LP (2000) The adolescent brain and age-related behavioral manifestations. Neurosci Biobehav Rev 24:417–463. https://doi.org/10.1016/S0149-7634(00)00014-2

37. Crews FT, Boettiger CA (2009) Impulsivity, frontal lobes and risk for addiction. Pharmacol Biochem Behav 93:237–247. https://doi.org/10.1016/j.pbb.2009.04.018

38. Geier C, Luna B (2009) The maturation of incentive processing and cognitive control.

Pharmacol Biochem Behav 93:212–221. https://doi.org/10.1016/j.pbb.2009.01. 021

39. Schramm-Sapyta NL, Walker QD, Caster JM, Levin ED, Kuhn CM (2009) Are adolescents more vulnerable to drug addiction than adults? Evidence from animal models. Psychopharmacology 206:1–21. https://doi.org/10.1007/s00213-009-1585-5

40. Kuhn C, Johnson M, Thomae A, Luo B, Simon SA, Zhou G, Walker QD (2010) The emergence of gonadal hormone influences on dopaminergic function during puberty. Horm Behav 58:122–137. https://doi.org/10.1016/j.yhbeh.2009.10.015

41. Clark DB, Kirisci L, Tarter RE (1998) Adolescent versus adult onset and the development of substance use disorders in males. Drug Alcohol Depend 49:115–121. https://doi.org/10.1016/S0376-8716(97)00154-3

42. Kandel DB, Yamaguchi K, Chen K (1992) Stages of progression in drug involvement from adolescence to adulthood: further evidence for the gateway theory. J Stud Alcohol 53:447. https://doi.org/10.15288/jsa.1992.53.447

43. Chen C-Y, Storr CL, Anthony JC (2009) Early-onset drug use and risk for drug dependence problems. Addict Behav 34 (3):319–322. https://doi.org/10.1016/j.addbeh.2008.10.021

44. Horvath LS, Milich R, Lynam D, Leukefeld C, Clayton R (2004) Sensation seeking and substance use: a cross-lagged panel design. Individ Differ Res 2:175–183

45. Perkins KA, Lerman C, Coddington SB, Jetton C, Karelitz JL, Scott JA, Wilson AS (2008) Initial nicotine sensitivity in humans as a function of impulsivity. Psychopharmacology 200:529–544. https://doi.org/10.1007/s00213-008-1231-7

46. Fillmore MT, Ostling EW, Martin CA, Kelly TH (2009) Acute effects of alcohol on inhibitory control and information processing in high and low sensation-seekers. Drug Alcohol Depend 100:91–99. https://doi.org/10.1016/j.drugalcdep.2008.09.007

47. Kelly TH, Delzer TA, Martin CA, Harrington NG, Hays LR, Bardo MT (2009) Performance and subjective effects of diazepam and d-amphetamine in high and low sensation seekers. Behav Pharmacol 20:505–517. https://doi.org/10.1097/FBP.0b013e3283305e8d

48. Sax KW, Strakowski SM (1998) Enhanced behavioral response to repeated d-amphetamine and personality traits in humans. Biol Psychiatry 44:1192–1195. https://doi.org/10.1016/S0006-3223(98)00168-1

49. Hutchison KE, Wood MD, Swift R (1999) Personality factors moderate subjective and psychophysiological responses to damphetamine in humans. Exp Clin Psychopharmacol 7:493. https://doi.org/10.1037/1064-1297.7.4.493

50. Ballaz SJ (2009) Differential novelty detection in rats selectively bred for novelty-seeking behavior. Neurosci Lett 461(1):45–48. https://doi.org/10.1016/j.neulet.2009.05.066

51. Dere E, Huston JP, De Souza Silva MA (2007) The pharmacology, neuroanatomy and neurogenetics of one-trial object recognition in rodents. Neurosci Biobehav Rev 31:673–704. https://doi.org/10.1016/j.neubiorev.2007.01.005

52. Marquez C, Nadal R, Armario A (2005) Responsiveness of the hypothalamic-pituitary adrenal axis to different novel environments is a consistent individual trait in adult male outbred rats. Psychoneuroendocrinology 30:179–187. https://doi.org/10.1016/j.psyneuen.2004.05.012

53. Hefner K, Holmes A (2007) Ontogeny of fear-, anxiety- and depression-related behavior across adolescence in C57BL/6J mice. Behav Brain Res 176(2):210–215. https://doi.org/10.1016/j.bbr.2006.10.001

54. Piazza PV, Deminiere JM, Le Moal M, Simon H (1989) Factors that predict individual vulnerability to amphetamine self-administration. Science 245:1511–1513. https://doi.org/10.1126/science.2781295

55. Hooks MS, Jones GH, Smith AD, Neill DB, Justice JB Jr (1991) Response to novelty predicts the locomotor and nucleus accumbens dopamine response to cocaine. Synapse 9 (2):121–128. https://doi.org/10.1002/syn.890090206

56. Kabbaj M, Evans S, Watson SJ, Akil H (2004) The search for the neurobiological basis of vulnerability to drug abuse: using microarrays to investigate the role of stress and individual differences. Neuropharmacology 47(Suppl 1):111–122. https://doi.org/10.1016/j.neuropharm.2004.07.021

57. Dellu F, Piazza PV, Mayo W, Le Moal M, Simon H (1996) Novelty-seeking in rats--biobehavioral characteristics and possible relationship with the sensation-seeking trait in man. Neuropsychobiology 34(3):136–145. https://doi.org/10.1159/000119305

58. Davis BA, Clinton SM, Akil H, Becker JB (2008) The effects of novelty-seeking phenotypes and sex differences on acquisition of cocaine self-administration in selectively bred High-Responder and Low-Responder rats. Pharmacol Biochem Behav 90(3):331–338. https://doi.org/10.1016/j.pbb.2008.03.008

59. Larson EB, Carroll ME (2005) Wheel running as a predictor of cocaine self-administration and reinstatement in female rats. Pharmacol Biochem Behav 82:590–600. https://doi.org/10.1016/j.pbb.2005.10.015

60. Pawlak CR, Ho Y, Schwarting RK (2008) Animal models of human psychopathology based on individual differences in novelty-seeking and anxiety. Neurosci Biobehav Rev 32:1544–1568. https://doi.org/10.1016/j.neubiorev.2008.06.007

61. Harro J, Oreland L, Vasar E, Bradwejn J (1995) Impaired exploratory behaviour after DSP-4 treatment in rats: implications for the increased anxiety after noradrenergic denervation. Eur Neuropsychopharmacol 5:447–455. https://doi.org/10.1016/0924-977X(95)80003-K

62. Klebaur JE, Bevins RA, Segar TM, Bardo MT (2001) Individual differences in behavioral responses to novelty and amphetamine self-administration in male and female rats. Behav Pharmacol 12(4):267–275. https://doi.org/10.1097/00008877-200107000-00005

63. Philpot RM, Wecker L (2008) Dependence of adolescent novelty seeking behavior on response phenotype and effects of apparatus scaling. Behav Neurosci 122:861. https://doi.org/10.1037/0735-7044.122.4.861

64. Belin D, Berson N, Balado E, Piazza PV, Deroche-Gamonet V (2011) High-novelty-preference rats are predisposed to compulsive cocaine self-administration. Neuropsychopharmacology 36:569–579

65. Kliethermes CL, Crabbe JC (2006) Genetic independence of mouse measures of some aspects of novelty seeking. Proc Natl Acad Sci U S A 103:5018–5023. https://doi.org/10.1073/pnas.0509724103

66. Vidal-Infer A, Arenas MC, Daza-Losada M, Aguilar MA, Miñarro J, Rodríguez-Arias M (2012) High novelty-seeking predicts greater sensitivity to the conditioned rewarding effects of cocaine. Pharmacol Biochem Behav 102:124–132. https://doi.org/10.1016/j.pbb.2012.03.031

67. Cain ME, Saucier DA, Bardo MT (2005) Novelty seeking and drug use: contribution of an animal model. Exp Clin Psychopharmacol 13:367. https://doi.org/10.1037/1064-1297.13.4.367

68. Lalonde R (2002) The neurobiological basis of spontaneous alternation. Neurosci Biobehav Rev 26:91–104. https://doi.org/10.1016/S0149-7634(01)00041-0

69. Dember WN, Fowler H (1959) Spontaneous alternation after free and forced trials. Can J Psychol 13:151. https://doi.org/10.1037/h0083776

70. Douglas RJ, Isaacson RL (1965) Homogeneity of single trial response tendencies and spontaneous alternation in the T-maze. Psychol Rep 16:87–92. https://doi.org/10.2466/pr0.1965.16.1.87

71. Nicholls B, Springham A, Mellanby J (1992) The playground maze: a new method for measuring directed exploration in the rat. J Neurosci Methods 43:171–180. https://doi.org/10.1016/0165-0270(92)90026-A

72. Ennaceur A (2010) One-trial object recognition in rats and mice: methodological and theoretical issues. Behav Brain Res 215(2):244–254. https://doi.org/10.1016/j.bbr.2009.12.036

73. Klebaur J, Bardo M (1999) Individual differences in novelty seeking on the playground maze predict amphetamine conditioned place preference. Pharmacol Biochem Behav 63:131–136. https://doi.org/10.1016/S0091-3057(98)00258-5

74. Frick KM, Gresack JE (2003) Sex differences in the behavioral response to spatial and object novelty in adult C57BL/6 mice. Behav Neurosci 117:1283. https://doi.org/10.1037/0735-7044.117.6.1283

75. Ennaceur A, Aggleton JP (1997) The effects of neurotoxic lesions of the perirhinal cortex combined to fornix transection on object recognition memory in the rat. Behav Brain Res 88:181–193. https://doi.org/10.1016/S0166-4328(97)02297-3

76. Abreu-Villaça Y, Queiroz-Gomes FE, Dal Monte AP, Filgueiras CC, Manhães AC (2006) Individual differences in novelty-seeking behavior but not in anxiety response to a new environment can predict nicotine consumption in adolescent C57BL/6 mice. Behav Brain Res 167:175–182. https://doi.org/10.1016/j.bbr.2005.09.003

77. Mateos-García A, Roger-Sánchez C, Rodriguez-Arias M, Miñarro J, Aguilar M, Manzanedo C, Arenas M (2015) Higher sensitivity to the conditioned rewarding effects of cocaine and MDMA in High-Novelty-Seekers mice exposed to a cocaine binge during

adolescence. Psychopharmacology 232:101–113. https://doi.org/10.1007/s00213-014-3642-y

78. Boissier JR, Simon P (1962) The exploration reaction in the mouse. Preliminary note. Therapy 17:1225–1232

79. Takeda H, Tsuji M, Matsumiya T (1998) Changes in head-dipping behavior in the hole-board test reflect the anxiogenic and/or anxiolytic state in mice. Eur J Pharmacol 350:21–29. https://doi.org/10.1016/S0014-2999(98)00223-4

80. Saitoh A, Hirose N, Yamada M, Yamada M, Nozaki C, Oka T, Kamei J (2006) Changes in emotional behavior of mice in the hole board test after olfactory bulbectomy. J Pharmacol Sci 102:377–386. JST.JSTAGE/jphs/FP0060837 [pii]

81. Calabrese EJ (2008) An assessment of anxiolytic drug screening tests: hormetic dose responses predominate. CRC Crit Rev Toxicol 38:489–542. https://doi.org/10.1080/10408440802014238

82. Boissier JR, Simon P, Lwoff JM (1964) Use of a particular mouse reaction (hole board method) for the study of psychotropic drugs. Therapie 19:571–583

83. Kliethermes C, Kamens H, Crabbe J (2007) Drug reward and intake in lines of mice selectively bred for divergent exploration of a hole board apparatus. Genes Brain Behav 6:608–618. https://doi.org/10.1111/j.1601-183X.2006.00289.x

84. Bevins RA, Bardo MT (1999) Conditioned increase in place preference by access to novel objects: antagonism by MK-801. Behav Brain Res 99:53–60. https://doi.org/10.1016/S0166-4328(98)00069-2

85. Douglas LA, Varlinskaya EI, Spear LP (2003) Novel-object place conditioning in adolescent and adult male and female rats: effects of social isolation. Physiol Behav 80:317–325. https://doi.org/10.1016/j.physbeh.2003.08.003

86. Thomsen M, Caine SB (2007) Intravenous drug self-administration in mice: practical considerations. Behav Genet 37:101–118. https://doi.org/10.1007/s10519-006-9097-0

87. Arenas MC, Aguilar MA, Montagud-Romero S, Mateos-García A, Navarro-Francés CI, Miñarro J, Rodríguez-Arias M (2016) Influence of the novelty-seeking endophenotype on the rewarding effects of psychostimulant drugs in animal models. Curr Neuropharmacol 14(1):87–100. https://doi.org/10.2174/1570159x13666150921112841

88. Pierre PJ, Vezina P (1997) Predisposition to self-administer amphetamine: the contribution of response to novelty and prior exposure to the drug. Psychopharmacology 129 (3):277–284. https://doi.org/10.1007/s002130050191

89. Grimm JW, See RE (1997) Cocaine self-administration in ovariectomized rats is predicted by response to novelty, attenuated by 17-beta estradiol, and associated with abnormal vaginal cytology. Physiol Behav 61 (5):755–761. https://doi.org/10.1016/s0031-9384(96)00532-x

90. Piazza PV, Deroche-Gamonent V, Rouge-Pont F, Le Moal M (2000) Vertical shifts in self-administration dose-response functions predict a drug-vulnerable phenotype predisposed to addiction. J Neurosci 20 (11):4226–4232. https://doi.org/10.1523/JNEUROSCI.20-11-04226.2000

91. Kabbaj M, Norton CS, Kollack-Walker S, Watson SJ, Robinson TE, Akil H (2001) Social defeat alters the acquisition of cocaine self-administration in rats: role of individual differences in cocaine-taking behavior. Psychopharmacology 158(4):382–387. https://doi.org/10.1007/s002130100918

92. Mantsch JR, Ho A, Schlussman SD, Kreek MJ (2001) Predictable individual differences in the initiation of cocaine self-administration by rats under extended-access conditions are dose-dependent. Psychopharmacology 157 (1):31–39. https://doi.org/10.1007/s002130100744

93. Cummings JA, Gowl BA, Westenbroek C, Clinton SM, Akil H, Becker JB (2011) Effects of a selectively bred novelty-seeking phenotype on the motivation to take cocaine in male and female rats. Biol Sex Differ 2:3. https://doi.org/10.1186/2042-6410-2-3

94. Marinelli M, White FJ (2000) Enhanced vulnerability to cocaine self-administration is associated with elevated impulse activity of midbrain dopamine neurons. J Neurosci 20 (23):8876–8885. https://doi.org/10.1523/JNEUROSCI.20-23-08876.2000

95. Meyer AC, Rahman S, Charnigo RJ, Dwoskin LP, Crabbe JC, Bardo MT (2010) Genetics of novelty seeking, amphetamine self-administration and reinstatement using inbred rats. Genes Brain Behav 9 (7):790–798. https://doi.org/10.1111/j.1601-183X.2010.00616.x

96. Mitchell JM, Cunningham CL, Mark GP (2005) Locomotor activity predicts acquisition of self-administration behavior but not

cocaine intake. Behav Neurosci 119 (2):464–472. https://doi.org/10.1037/0735-7044.119.2.464

97. Beckmann JS, Marusich JA, Gipson CD, Bardo MT (2011) Novelty seeking, incentive salience and acquisition of cocaine self-administration in the rat. Behav Brain Res 216(1):159–165. https://doi.org/10.1016/j.bbr.2010.07.022

98. de la Peña I, Luck-Gonzales E, de la Peña JB, Kim BN, Hyun-Han D, Young-Shin C, Hoon-Cheong J (2015) Individual differences in novelty-seeking behavior in spontaneously hypertensive rats: enhanced sensitivity to the reinforcing effect of methylphenidate in the high novelty-preferring subpopulation. J Neurosci Methods 252:48–54. https://doi.org/10.1016/j.jneumeth.2014.08.019

99. Bird J, Schenk S (2013) Contribution of impulsivity and novelty-seeking to the acquisition and maintenance of MDMA self-administration. Addict Biol 18(4):654–664. https://doi.org/10.1111/j.1369-1600.2012.00477.x

100. Dickson PE, Ndukum J, Wilcox T, Clark J, Roy B, Zhang L, Li Y, Lin DT, Chesler EJ (2015) Association of novelty-related behaviors and intravenous cocaine self-administration in Diversity Outbred mice. Psychopharmacology 232(6):1011–1024. https://doi.org/10.1007/s00213-014-3737-5

101. Bienkowski P, Koros E, Kostowski W (2001) Novelty-seeking behaviour and operant oral ethanol self-administration in Wistar rats. Alcohol 36(6):525–528. https://doi.org/10.1093/alcalc/36.6.525

102. Nadal R, Armario A, Janak PH (2002) Positive relationship between activity in a novel environment and operant ethanol self-administration in rats. Psychopharmacology 162(3):333–338

103. Bardo MT, Bevins RA (2000) Conditioned place preference: what does it add to our preclinical understanding of drug reward? Psychopharmacology 153:31–43. http://www.ncbi.nlm.nih.gov/pubmed/11255927

104. Tzschentke TM (1998) Measuring reward with the conditioned place preference paradigm: a comprehensive review of drug effects, recent progress and new issues. Prog Neurobiol 56:613–672. https://doi.org/10.1016/S0301-0082(98)00060-4

105. Tzschentke TM (2007) Measuring reward with the conditioned place preference (CPP) paradigm: update of the last decade. Addict Biol 12:227–462. https://doi.org/10.1111/j.1369-1600.2007.00070.x

106. Aguilar MA, Rodríguez-Arias M, Miñarro J (2008) Neurobiological mechanisms of the reinstatement of drug-conditioned place preference. Brain Res Rev 59:253–277. https://doi.org/10.1016/j.brainresrev.2008.08.002

107. Gong W, Neill DB, Justice JB Jr (1996) Locomotor response to novelty does not predict cocaine place preference conditioning in rats. Pharmacol Biochem Behav 53(1):191–196

108. Kosten TA, Miserendino MJ (1998) Dissociation of novelty- and cocaine-conditioned locomotor activity from cocaine place conditioning. Pharmacol Biochem Behav 60 (4):785–791. https://doi.org/10.1016/s0091-3057(97)00388-2

109. Dietz D, Wang H, Kabbaj M (2007) Corticosterone fails to produce conditioned place preference or conditioned place aversion in rats. Behav Brain Res 181(2):287–291. https://doi.org/10.1016/j.bbr.2007.04.005

110. Capriles N, Watson S Jr, Akil H (2012) Individual differences in the improvement of cocaine-induced place preference response by the 5-HT2C receptor antagonist SB242084 in rats. Psychopharmacology 220 (4):731–740. https://doi.org/10.1007/s00213-011-2524-9

111. Robinet PM, Rowlett JK, Bardo MT (1998) Individual differences in novelty-induced activity and the rewarding effects of novelty and amphetamine in rats. Behav Process 44 (1):1–9. https://doi.org/10.1016/s0376-6357(98)00022-9

112. Erb SM, Parker LA (1994) Individual differences in novelty-induced activity do not predict strength of amphetamine-induced place conditioning. Pharmacol Biochem Behav 48 (3):581–586. https://doi.org/10.1016/0091-3057(94)90317-4

113. Mathews IZ, Morrissey MD, McCormick CM (2010) Individual differences in activity predict locomotor activity and conditioned place preference to amphetamine in both adolescent and adult rats. Pharmacol. Biochem Behav 95:63–71. https://doi.org/10.1016/j.pbb.2009.12.007

114. Arenas MC, Daza-Losada M, Vidal-Infer A, Aguilar MA, Miñarro J, Rodríguez-Arias M (2014) Capacity of novelty-induced locomotor activity and the hole-board test to predict sensitivity to the conditioned rewarding effects of cocaine. Physiol Behav 133:152–160. https://doi.org/10.1016/j.physbeh.2014.05.028

115. Brabant C, Quertemont E, Tirelli E (2005) Evidence that the relations between novelty-induced activity, locomotor stimulation and place preference induced by cocaine qualitatively depend upon the dose: a multiple regression analysis in inbred C57BL/6J mice. Behav Brain Res 158(2):201–210. https://doi.org/10.1016/j.bbr.2004.08.020

116. Shimosato K, Watanabe S (2003) Concurrent evaluation of locomotor response to novelty and propensity toward cocaine conditioned place preference in mice. J Neurosci Methods 128(1–2):103–110. https://doi.org/10.1016/s0165-0270(03)00153-5

117. Pelloux Y, Costentin J, Duterte-Boucher D (2004) Differential effects of novelty exposure on place preference conditioning to amphetamine and its oral consumption. Psychopharmacology 171(3):277–285. https://doi.org/10.1007/s00213-003-1584-x

118. Pelloux Y, Costentin J, Duterte-Boucher D (2006) Novelty preference predicts place preference conditioning to morphine and its oral consumption in rats. Pharmacol Biochem Behav 84(1):43–50. https://doi.org/10.1016/j.pbb.2006.04.004

119. Zheng X, Ke X, Tan B, Luo X, Xu W, Yang X, Sui N (2003) Susceptibility to morphine place conditioning: relationship with stress-induced locomotion and novelty-seeking behavior in juvenile and adult rats. Pharmacol Biochem Behav 75(4):929–935. https://doi.org/10.1016/s0091-3057(03)00172-2

120. Nadal R, Rotllant D, Márquez C, Armario A (2005) Perseverance of exploration in novel environments predicts morphine place conditioning in rats. Behav Brain Res 165(1):72–79. https://doi.org/10.1016/j.bbr.2005.06.039

121. Montagud-Romero S, Daza-Losada M, Vidal-Infer A, Maldonado C, Aguilar MA, Miñarro J, Rodríguez-Arias M (2014) The novelty-seeking phenotype modulates the long-lasting effects of intermittent ethanol administration during adolescence. PLoS One 9(3):e92576

122. Rodríguez-Arias M, Vaccaro S, Arenas MC, Aguilar MA, Miñarro J (2015) The novelty-seeking phenotype modulates the long-lasting effects of adolescent MDMA exposure. Physiol Behav 141:190–198

123. Rodríguez-Arias M, Roger-Sánchez C, Vilanova I, Revert N, Manzanedo C, Miñarro J, Aguilar MA (2016) Effects of cannabinoid exposure during adolescence on the conditioned rewarding effects of WIN 55212-2 and cocaine in mice: influence of the novelty-seeking trait. Neural Plast 2016:6481862. https://doi.org/10.1155/2016/6481862

124. Kravitz HM, Fawcett J, McGuire M, Kravitz GS, Whitney M (1999) Treatment attrition among alcohol-dependent men: is it related to novelty seeking personality traits? J Clin Psychopharmacol 19(1):51–56. https://doi.org/10.1097/00004714-199902000-00010

125. Evren C, Durkaya M, Evren B, Dalbudak E, Cetin R (2012) Relationship of relapse with impulsivity, novelty seeking and craving in male alcohol-dependent inpatients. Drug Alcohol Rev 31(1):81–90. https://doi.org/10.1111/j.1465-3362.2011.00303.x

126. Kankaanpää A, Meririnne E, Lillsunde P, Seppälä T (1998) The acute effects of amphetamine derivatives on extracellular serotonin and dopamine levels in rat nucleus accumbens. Pharmacol Biochem Behav 59:1003–1009

127. Ballaz SJ, Akil H, Watson SJ (2007) The 5-HT7 receptor: role in novel object discrimination and relation to novelty-seeking behavior. Neuroscience 149:192–202

128. Kerman IA, Clinton SM, Bedrosian TA, Abraham AD, Rosenthal DT, Akil H, Watson SJ (2011) High novelty-seeking predicts aggression and gene expression differences within defined serotonergic cell groups. Brain Res 1419:34–45

129. Kandel D (1975) Stages in adolescent involvement in drug use. Science 190(4217):912–914

130. Chadwick B, Miller ML, Hurd YL (2013) Cannabis use during adolescent development: susceptibility to psychiatric illness. Front Psychiatry 4:129

131. Newcomb MD, Bentler PM (1986) Cocaine use among adolescents: longitudinal associations with social context, psychopathology, and use of other substances. Addict Behav 11(3):263–273

132. DeSimone J (1998) Is marijuana a gateway drug? Eastern Econ J 24(2):149–164

133. O'Brien MS, Comment LA, Liang KY, Anthony JC (2012) Does cannabis onset trigger cocaine onset? A case-crossover approach. Int J Methods Psychiatr Res 21(1):66–75

134. Viola TW, Tractenberg SG, Wearick-Silva LE, de Oliveira Rosa CS, Pezzi JC, Grassi-Oliveira R (2014) Long term cannabis abuse and early-onset cannabis use increase the severity of cocaine withdrawal during detoxification and rehospitalization rates due to cocaine

dependence. Drug Alcohol Depend 144:153–159

135. Solinas M, Panlilio LV, Goldberg SR (2004) Exposure to Δ-9-tetrahydrocannabinol (THC) increases subsequent heroin taking but not heroin's reinforcing efficacy: a self-administration study in rats. Neuropsychopharmacology 29(7):1301–1311

136. Ellgren M, Spano SM, Hurd YL (2007) Adolescent cannabis exposure alters opiate intake and opioid limbic neuronal populations in adult rats. Neuropsychopharmacology 32 (3):607–615

137. Biscaia M, Fernandez B, Higuera-Matas A, Miguéns M, Viveros MA, García-Lecumberri C, Ambrosio E (2008) Sex-dependent effects of periadolescent exposure to the cannabinoid agonist CP-55,940 on morphine self-administration behaviour and the endogenous opioid system. Neuropharmacology 54(5):863–873

138. Higuera-Matas A, Soto-Montenegro ML, Del Olmo N, Miguéns M, Torres I, Vaquero JJ, Sánchez J, García-Lecumberri C, Desco M, Ambrosio E (2008) Augmented acquisition of cocaine self-administration and altered brain glucose metabolism in adult female but not male rats exposed to a cannabinoid agonist during adolescence. Neuropsychopharmacology 33(4):806–813

139. Panlilio LV, Zanettini C, Barnes C, Solinas M, Goldberg SR (2013) Prior exposure to THC increases the addictive effects of nicotine in rats. Neuropsychopharmacology 38 (7):1198–1208

140. Friedman AL, Jutkiewicz EM (2014) Effects of adolescent THC exposure on the behavioral effects of cocaine in adult Sprague-Dawley rats. Ph.D. thesis. http://deepblue. lib.umich.edu/bitstream/handle/2027.42/ 107721/amyfr.pdf?sequence=1& isAllowed=y

141. Manzanedo C, Aguilar MA, Rodríguez-Arias M, Navarro M, Minarro J (2004) Cannabinoid agonist-induced sensitisation to morphine place preference in mice. Neuroreport 15(8):1373–1377

142. Rodríguez-Arias M, Manzanedo C, Roger-Sanchez C, Do Couto BR, Aguilar MA, Minarro J (2010) Effect of adolescent exposure to WIN 55212-2 on the acquisition and reinstatement of MDMA-induced conditioned place preference. Prog Neuro-Psychopharmacol Biol Psychiatry 34 (1):166–171

143. Brenhouse HC, Andersen SL (2008) Delayed extinction and stronger reinstatement of cocaine conditioned place preference in adolescent rats, compared to adults. Behav Neurosci 122:460–465. https://doi.org/10. 1037/0735-7044.122.2.460

144. Brenhouse HC, Dumais K, Andersen SL (2010) Enhancing the salience of dullness: behavioral and pharmacological strategies to facilitate extinction of drug-cue associations in adolescent rats. Neuroscience 69:628–636. https://doi.org/10.1016/j. neuroscience.2010.05.063

145. Zakharova E, Leoni G, Kichko I, Izenwasser S (2009) Differential effects of methamphetamine and cocaine on conditioned place preference and locomotor activity in adult and adolescent male rats. Behav Brain Res 198:45–50. https://doi.org/10.1016/j.bbr. 2008.10.019

146. Aberg M, Wade D, Wall E, Izenwasser S (2007) Effect of MDMA (ecstasy) on activity and cocaine conditioned place preference in adult and adolescent rats. Neurotoxicol Teratol 29:37–46. https://doi.org/10.1016/j. ntt.2006.09.002

147. Balda MA, Anderson KL, Itzhak Y (2006) Adolescent and adult responsiveness to the incentive value of cocaine reward in mice: role of neuronal nitric oxide synthase (nNOS) gene. Neuropharmacology 51:341–349. https://doi.org/10.1016/j. neuropharm.2006.03.026

148. Campbell JO, Wood RD, Spear LP (2000) Cocaine and morphine-induced place conditioning in adolescent and adult rats. Physiol Behav 68:487–493. https://doi.org/10. 1016/S0031-9384(99)00225-5

149. Schramm-Sapyta NL, Pratt AR, Winder DG (2004) Effects of periadolescent versus adult cocaine exposure on cocaine conditioned place preference and motor sensitization in mice. Psychopharmacology 173:41–48. https://doi.org/10.1007/s00213-003-1696-3

150. Laviola G, Macrì S, Morley-Fletcher S, Adriani W (2003) Risk-taking behavior in adolescent mice: psychobiological determinants and early epigenetic influence. Neurosci Biobehav Rev 27(1–2):19–31. https://doi. org/10.1016/s0149-7634(03)00006-x

151. Sutcliffe JS, Marshall KM, Neill JC (2007) Influence of gender on working and spatial memory in the novel object recognition task in the rat. Behav Brain Res 177(1):117–125. https://doi.org/10.1016/j.bbr.2006.10. 029

152. Lynch WJ, Roth ME, Carroll ME (2002) Biological basis of sex differences in drug abuse: preclinical and clinical studies.

126    Claudia Calpe-López et al.

Psychopharmacology 164(2):121–137. https://doi.org/10.1007/s00213-002-1183-2

153. Quiñones-Jenab V (2006) Why are women from Venus and men from Mars when they abuse cocaine? Brain Res 1126(1):200–203. https://doi.org/10.1016/j.brainres.2006.08.109

154. Becker JB, Hu M (2008) Sex differences in drug abuse. Front Neuroendocrinol 29(1):36–47. https://doi.org/10.1016/j.yfrne.2007.07.003

155. Fattore L, Fadda P, Fratta W (2009) Sex differences in the self-administration of cannabinoids and other drugs of abuse. Psychoneuroendocrinology 34(Suppl 1):227–236. https://doi.org/10.1016/j.psyneuen.2009.08.008

156. Roth ME, Cosgrove KP, Carroll ME (2004) Sex differences in the vulnerability to drug abuse: a review of preclinical studies. Neurosci Biobehav Rev 28(6):533–546. https://doi.org/10.1016/j.neubiorev.2004.08.001

157. Maldonado C, Rodríguez-Arias M, Castillo A, Aguilar MA, Minarro J (2006) Gamma-hydroxybutyric acid affects the acquisition and reinstatement of cocaine-induced conditioned place preference in mice. Behav Pharmacol 17(2):119–131

158. Daza-Losada M, Rodríguez-Arias M, Aguilar MA, Miñarro J (2009) Acquisition and reinstatement of MDMA-induced conditioned place preference in mice pre-treated with MDMA or cocaine during adolescence. Addict Biol 14(4):447–456. https://doi.org/10.1111/j.1369-1600.2009.00173.x

159. Manzanedo C, Rodríguez-Arias M, Daza-Losada M, Maldonado C, Aguilar MA, Minarro J (2010) Effect of the CB1 cannabinoid agonist WIN 55212-2 on the acquisition and reinstatement of MDMA-induced conditioned place preference in mice. Behav Brain Funct 6:19

160. Navarro M, Carrera MRA, Fratta W, Valverde O, Cossu G, Fattore L, Chowen JA, Gómez R, del Arco I, Villanúa MA, Maldonado R, Koob GF, Rodríguez de Fonseca F (2001) Functional interaction between opioid and cannabinoid receptors in drug self-administration. J Neurosci 21(14):5344–5350

161. White AM, Kraus CL, Swartzwelder H (2006) Many college freshmen drink at levels far beyond the binge threshold. Alcohol Clin Exp Res 30:1006–1010

162. Moser P, Wolinsky T, Duxon M, Porsolt RD (2011) How good are current approaches to nonclinical evaluation of abuse and dependence? J Pharmacol Exp Ther 336:588–595. https://doi.org/10.1124/jpet.110.169979

163. Robbins TW (1978) The acquisition of responding with conditioned reinforcement: effects of pipradrol, methylphenidate, d-amphetamine, and nomifensine. Psychopharmacology 58:79–87. https://doi.org/10.1007/BF00426794

164. Manzanedo C, Aguilar MA, Rodríguez-Arias M, Miñarro J (2001) Effects of dopamine antagonists with different receptor blockade profiles on morphine-induced place preference in male mice. Behav Brain Res 121(1–2):189–197. https://doi.org/10.1016/s0166-4328(01)00164-4

165. Spear LP (2007) The developing brain and adolescent-typical behavior patterns: an evolutionary approach. In: Walker E, Romer D (eds) Adolescent psychopathology and the developing brain: integrating brain and prevention science. Oxford University Press, New York, NY. https://doi.org/10.1093/acprof:oso/9780195306255.003.0001

166. Doremus-Fitzwater TL, Varlinskaya EI, Spear LP (2009) Social and non-social anxiety in adolescent and adult rats after repeated restraint. Physiol Behav 97:484–494. https://doi.org/10.1016/j.physbeh.2009.03.025

167. Roger-Sánchez C, Aguilar MA, Rodríguez-Arias M, Aragon CM, Miñarro J (2012) Age-and sex-related differences in the acquisition and reinstatement of ethanol CPP in mice. Neurotoxicol Teratol 34(1):108–115. https://doi.org/10.1016/j.ntt.2011.07.011

168. Montagud-Romero S, Aguilar MA, Maldonado C, Manzanedo C, Miñarro J, Rodríguez-Arias M (2015) Acute social defeat stress increases the conditioned rewarding effects of cocaine in adult but not in adolescent mice. Pharmacol Biochem Behav 135:1–12. https://doi.org/10.1016/j.pbb.2015.05.008

169. García-Pardo MP, Rodríguez-Arias M, Maldonado C, Manzanedo C, Miñarro J, Aguilar MA (2014) Effects of acute social stress on the conditioned place preference induced by MDMA in adolescent and adult mice. Behav Pharmacol 25(5–6):532–546. https://doi.org/10.1097/FBP.0000000000000065

170. Sillivan SE, Black YD, Naydenov AV, Vassoler FR, Hanlin RP, Konradi C (2011) Binge cocaine administration in adolescent rats affects amygdalar gene expression patterns and alters anxiety-related behavior in adulthood. Biol Psychiatry 70(6):583–592.

https://doi.org/10.1016/j.biopsych.2011.03.035

171. Black YD, Maclaren FR, Naydenov AV, Carlezon WA Jr, Baxter MG, Konradi C (2006) Altered attention and prefrontal cortex gene expression in rats after binge-like exposure to cocaine during adolescence. J Neurosci 26 (38):9656–9665. https://doi.org/10.1523/JNEUROSCI.2391-06.2006

172. Manzanedo C, García-Pardo MP, Rodríguez-Arias M, Miñarro J, Aguilar MA (2012) Pre-treatment with high doses of cocaine decreases the reinforcing effects of cocaine in the conditioned place preference paradigm. Neurosci Lett 516(1):29–33. https://doi.org/10.1016/j.neulet.2012.03.044

173. Sargent JD, Tanski S, Stoolmiller M, Hanewinkel R (2010) Using sensation seeking to target adolescents for substance use interventions. Addiction 105:506–514. https://doi.org/10.1111/j.1360-0443.2009.02782.x

174. Nees F, Tzschoppe J, Patrick CJ, Vollstädt-Klein S, Steiner S, Poustka L, Banaschewski T, Barker GJ, Büchel C, Conrod PJ (2012) Determinants of early alcohol use in healthy adolescents: the differential contribution of neuroimaging and psychological factors. Neuropsychopharmacology 37:986–995. https://doi.org/10.1038/npp.2011.282

175. Spillane NS, Muller CJ, Noonan C, Goins RT, Mitchell CM, Manson S (2012) Sensation-seeking predicts initiation of daily smoking behavior among American Indian high school students. Addict Behav 37:1303–1306. https://doi.org/10.1016/j.addbeh.2012.06.021

176. Stephenson MT, Helme DW (2006) Authoritative parenting and sensation seeking as predictors of adolescent cigarette and marijuana use. J Drug Educ 36:247–270. https://doi.org/10.2190/Y223-2623-7716-2235

177. Corr PJ, Kumari V (2000) Individual differences in mood reactions to d-amphetamine: a test of three personality factors. J Psychopharmacol 14:371–377. https://doi.org/10.1177/026988110001400404

178. White TL, Lott DC, de Wit H (2006) Personality and the subjective effects of acute amphetamine in healthy volunteers. Neuropsychopharmacology 31:1064–1074. https://doi.org/10.1038/sj.npp.1300939

179. Ersche KD, Jones PS, Williams GB, Smith DG, Bullmore ET, Robbins TW (2013) Distinctive personality traits and neural correlates associated with stimulant drug use versus familial risk of stimulant dependence. Biol Psychiatry 74:137–144. https://doi.org/10.1016/j.biopsych.2012.11.016

180. Miczek KA, de Wit H (2008) Challenges for translational psychopharmacology research some basic principles. Psychopharmacology 199:291–301. https://doi.org/10.1007/s00213-008-1198-1194

181. Norbury A, Husain M (2015) Sensation-seeking: dopaminergic modulation and risk for psychopathology. Behav Brain Res 288:79–93. https://doi.org/10.1016/j.bbr.2015.04.015

# Chapter 5

## Cognitive Biases Associated with Vulnerability to the Development of Pathological Gambling

**Rafal Rygula, Justyna K. Hinchcliffe, and Karolina Noworyta**

### Abstract

Although gambling disorder is a serious social problem in modern societies, information about the cognitive traits that could determine vulnerability to this psychopathology is still scarce. In this chapter, we describe a behavioral protocol for preclinical experiments in rats that can be used for the assessment of the interaction between cognitive judgment bias measured as a stable and enduring behavioral trait and decision-making under risk.

For this, the rats are initially trained and screened in a series of ambiguous-cue interpretation tests for "pessimistic" and "optimistic" traits. Subsequently, the animals are retrained and retested in the rat slot machine task, which allows the assessment of decision-making under risk and the crucial feature of gambling-like behavior that has been investigated in rats and humans—the interpretation of "near-miss" outcomes as a positive (i.e., "clear win") situation.

Preclinical modeling of these cognitive distortions in animals can facilitate our understanding of their neurobiological bases and potentially stimulate novel treatment options.

**Key words** Rat, Animal model, Cognitive bias, Gambling, Ambiguous cue, Slot machine task, Optimism, Pessimism

## 1 Introduction

Pathological gambling is a complex and multifaceted disorder. Although various factors may lead to this disorder, one of the most profound appears to be the presence of specific cognitive biases and cognitive distortions. These biases refer to how the gambler thinks about randomness, chance, and skill [1–3], and foster an inappropriately high expectation of winning (optimism) during the game. Although a number of specific biases have been described in this context, arguably, the most classic cognitive distortions associated with pathological gambling include the "gambler's fallacy," which is a bias in the processing of random sequences; the "illusion of control," referring to the interpretation of skill involvement in situations that are governed by chance alone

María A. Aguilar (ed.), *Methods for Preclinical Research in Addiction*, Neuromethods, vol. 174,
https://doi.org/10.1007/978-1-0716-1748-9_5, © Springer Science+Business Media, LLC, part of Springer Nature 2022

[4, 5]; and hypersensitivity to the so-called near-miss outcomes, which are unsuccessful outcomes that are proximal to a major win [1].

The strong association of cognitive biases with the presentation of pathological gambling has led to the theory that distorted beliefs can play also a causative role in vulnerability to this disorder [6, 7]. Indeed, one of the cognitive biases that could critically determine vulnerability to gambling is overly optimistic judgment bias. According to Carver and Scheier [8], the generalized positive outcome expectancies of optimists result in persistence in attempting to accomplish goals in the face of adversity. Gambling is one of the domains in which persistence is unlikely to be consistently rewarded and in which optimism may be a liability [9]. In agreement with this assumption, recent research on optimism and gambling has found that optimists had more positive expectations for gambling than did pessimists and were less likely to reduce their betting after poor outcomes [9]. Despite these findings, the information concerning the interaction of optimism and gambling in both humans and rodents is still meagre. Although preclinical modeling of these cognitive distortions in animals can facilitate our understanding of their neurobiological bases and potentially stimulate novel treatment options, cognitive distortions have not been the subject of much preclinical behavioral research, probably due to difficulties in measurement.

The first successful attempts to model and measure cognitive judgment bias in animals occurred just over 15 years ago. In 2004, Harding and colleagues demonstrated the existence of judgment bias in rats using the ambiguous-cue interpretation (ACI) test [10]. In this behavioral paradigm, the rats are trained to discriminate between two stimuli: one stimulus that predicts a positive or favorable consequence (S+) and another stimulus that predicts a negative or less favorable consequence (S−). The discriminative stimuli acquire either a positive or negative valence, and the training continues until the animals demonstrate a stable, correct discrimination ratio. The ambiguous-cue test is composed of the presentation of two previously reinforced reference stimuli (S+ and S−) and additional ambiguous stimuli that are intermediate to S+ and S−. The response pattern to these additional intermediate stimuli (ambiguous cues) reflects the cognitive judgment bias in rats. The modified version of the ACI test [11–13] has been shown to be sensitive to various behavioral and pharmacological manipulations [14–17] and has been optimized to screen for the valence of cognitive judgment bias measured as a stable and enduring behavioral trait [18–21]. Cognitive judgment bias screening using multiple ACI tests [10, 11, 17] can be used to isolate groups of "optimistic" and "pessimistic" rats that consistently differ in their cognitive judgment bias over time [19, 20].

These "optimistic" and "pessimistic" traits can subsequently be correlated with the propensity to gamble, for example, in the rat slot machine task (rSMT) [22]. In this preclinical test, subjects respond to a series of three flashing lights, loosely analogous to the wheels of a slot machine, causing the lights to set to ON or OFF. A winning outcome is signaled if all three lights were illuminated. At the end of each trial, rats choose between responding on the "collect" lever, resulting in a reward on "win" trials but a time penalty on "loss" trials, or starting a new trial. Recent studies using this task revealed an interrelation between cognitive judgment bias and propensity to gamble in rats [12].

In this behavioral protocol, we describe an experiment that can be used for the assessment of the interaction between cognitive judgment bias measured as a stable and enduring behavioral trait in the ACI paradigm and sensitivity to "near-miss" outcomes in the rSMT.

# 2 Materials

## 2.1 Animals

In our laboratory, we routinely use male Sprague-Dawley rats (Charles River Laboratories, Sulzfeld, Germany) that weigh approximately 175–200 g at the beginning of experimentation (*see* **Note 1**). The animals are housed in groups of four in type IV Makrolon cages (59 × 38 × 20 cm, Ehret GMBH, Wandlitz, Germany), which have metal mesh lids and food hoppers and are lined with sawdust bedding (Abedot, Animalab, Poznan, Poland). The rats are housed in a temperature- (21 ± 1 °C) and humidity- (40–50%) controlled room under a 12/12-h light–dark cycle (lights on at 07:00 h). The behavioral training and testing are performed during the light phase of the light–dark cycle (between 07:00 h and 19:00 h).

## 2.2 Apparatus for Cognitive Judgment Bias Screening Using the ACI Paradigm

For the ACI training and testing in our laboratory, we use a set of standard operant conditioning chambers obtained from Med Associates St Albans, Vermont, USA. Each box is equipped with a light, a speaker, a liquid dispenser (set to deliver 0.01 cc of 20% sucrose solution), a grid floor through which scrambled electric shocks (0.5 mA) can be delivered, and two retractable levers. The levers are located at opposite sides of the feeder. All of the behavioral protocols are programmed in Med State notation code (Med Associates). The experimental procedures for the ACI test were originally described by Enkel and colleagues in 2010 [11]. For further equipment specification details, *see* Papciak and Rygula [23].

**2.3 Apparatus for Measuring the Propensity to Gamble Using the rSMT**

For the rSMT training and testing, we use another set of computer-controlled, operant conditioning boxes obtained from Coulbourn Instruments, Allentown, PA, USA. The boxes are enclosed within ventilated light- and sound-attenuating cabinets. Each box is fitted with a speaker, a house light, two retractable levers, and an array of three nose-poke apertures equally distributed on one of the walls. A stimulus light is displayed at the back of each hole, and to detect nose-poke responses, the infrared beams are set at the entry of each hole. A food magazine is located on the opposite wall with one retractable lever located on each side. The conditioning boxes are controlled by Graphic State 3.0 software (Coulbourn Instruments, Allentown, PA, USA).

# 3 Methods

**3.1 Rats**

**3.1.1 Handling and Daily Care**

The rats are habituated to the housing conditions for 2 weeks prior to any experimental work (*see* **Note 2**). They are tail marked with a unique number code to be easily identified and weighed weekly. The weights are monitored weekly against a standard animal growth curve; for example, for Sprague-Dawley rats, this growth curve is found in the "Charles River Laboratories Research Models and Services Catalogue." The animals are handled daily during the entire training and experimental period to reduce anxiety and exaggerated stress responses to human experimenters. During handling, rats are gently lifted by placing a hand around their shoulders, slowly stroked for a couple of minutes and released back into the home cage. The use of appropriate handling is crucial for all procedures and experimental manipulations to be carried out efficiently as well as for the reliability and reproducibility of the data obtained in experiments. During the entire experiment, the rats are mildly food restricted to approximately 90% of their free-feeding weights, which can be achieved by providing them with 15–20 g of food (standard laboratory chow, Labofeed, Kcynia, Poland) per rat per day. The maintenance of relevant percentage body weight must comply with the local and national guidelines for animal experimentation or the approved in vivo project license. The food restriction regime starts 1 week prior to ACI training. Water is freely available except during the test sessions (for more details, *see* Subheadings 3.3.1 and 3.4.1).

**3.2 Preparing the Operant Equipment and Daily Routines**

All behavioral training and testing are carried out in an experimental room, away from the animal holding room. All external olfactory factors (deodorant spray, perfume, etc.) and noise levels (mobile phones, radio, etc.) must be kept to a minimum. Before experimental work begins, the operant conditioning boxes are assembled respectively to the task requirement, and software controlling the chambers is programmed accordingly. To check the box's configuration and whether the output data are accurately recorded, it is advisable to run practice test trials for each operant conditioning box before commencing any training or testing sessions. Reward sucrose solution is freshly prepared prior to any behavioral sessions. Following experimentation, the operant conditioning boxes are cleaned daily using a wet cloth soaked in an odorless mild detergent–water solution.

### 3.3 Cognitive Judgment Bias Screening Using the Ambiguous-Cue Interpretation Paradigm

Before the training sessions, all animals are habituated to the experimental room for 30 min. In the meantime, the programs and functioning of the operant conditioning system are tested and liquid reward containers are filled with the previously prepared sucrose solution. Following habituation, each animal is placed inside a separate operant conditioning box, and the door of the sound-attenuating chamber is shut (*see* **Note 3**).

#### 3.3.1 Positive Tone Training

Positive tone training involves a lever training session where animals learn to press the left lever to obtain a sucrose solution in response to a tone (2 kHz or 9 kHz, 75 dB, 50 s, counterbalanced, *see* **Note 4**). Due to its association with a reward, this tone gains a positive valence and is defined as the "positive tone" (S+), and the associated lever is referred to as the "positive lever." To achieve a reliable performance criterion of a lever press for the reward (over 90% correct responses), it is advisable to use three training steps:

A. During the first step, presentations of the positive tone (each lasting 10 s) co-occur with the delivery of the sucrose solution. During the entire session, the house light is on, and both levers are retracted. The intertrial intervals (ITIs) last 10 s, and one training session lasts 30 min (equivalent to 90 reward deliveries), ending with the house light off. Following the training, the rats are returned to their home cages in the holding room, while the grid floors and pan trays are cleaned.

   There are no quantifiable criteria during this step. The training must be repeated until the rats learn the location of the reward and spend most of their time with their heads in the liquid dipper aperture. The animals can be observed through a spyhole installed in the sound-attenuating cabinet door.

B. During the second step, presentations of the positive tone (each lasting 50 s) co-occur with left lever extensions. Trials are separated by 10-s ITIs, while levers retract. Each lever press during the presentation of the tone is rewarded by the delivery of sucrose solution (liquid dipper arm rises for 10 s). During the entire session, the house light is on, except during the ITIs. The training session lasts for 30 min and ends with the house light off and lever retraction. Following the completion of training, the rats are returned to their home cages in the holding room. The grid floors and pan trays are cleaned. The number of lever presses is scored. The training continues until the rats reach the criterion of at least 200 responses per experimental session maintained over at least 3 training sessions.

C. **Step C** is similar to **step B**; however, the first lever press in a given trial initiates the reward delivery, retracts the lever, terminates the tone and initiates the ITI. The ITI lasts until the end of the 1-min trial. If the rat does not press the lever during the 50 s of tone presentation, the trial is scored as an "omission" and is followed by a 10-s ITI. During the entire session, the house light is on, except during the ITIs. The training session lasts for 30 min (or 30 trials) and ends with the house light being turned off and lever retraction. Following the training, the animals return to their home cages in the holding room. The numbers of lever presses and omissions are scored. The training continues until the animals attain the criterion of at least 90% correct responses per experimental session maintained over at least 3 training sessions.

*3.3.2 Negative Tone Training*

Negative tone training is composed of two steps, where rats are trained to press a lever that is located on the right side of the liquid dispenser to avoid an electric foot shock (0.5 mA, 10 s) when another tone (9 kHz or 2 kHz at 75 dB, 50 s, fully counterbalanced; *see* **Note 4**) predicts a punishment. This tone, as it is coupled with the punishment outcome, acquires a negative valence and is defined as the "negative tone," while the associated lever is referred to as the "negative lever." A stable avoidance performance is achieved in the following steps:

A. During the first step, presentations of the negative tone (each lasting 10 s) are accompanied by the occurrence of electric foot shock (0.2 mA) unless the rat presses the right lever "escape response," which terminates the electric shocks and tone. The ITI lasts 30 s. If an animal fails to press the lever during the tone, the shock terminates, the lever retracts and ITI begins. The trial is scored as an "escape omission." During the entire session, the house light is on, except during the ITIs. The training session lasts for 30 min and ends with the house light off and lever retraction. Following the completed session, the animals are returned to their home cages in the holding room. The number of lever presses of "escape responses" and "escape omissions" is scored. During training sessions, the shock intensity is gradually increased (0.1 mA increments) until the animals reach the criterion of at least 70% "escape responses" per experimental session maintained over at least 3 training sessions at a shock intensity of 0.5 mA.

B. During this step, the presentations of the negative tones predict the occurrence of the electric shocks (0.5 mA). The first tone is presented after 30 s and lasts for a maximum of 30 s. The electric foot shock begins 20 s after the tone onset and accompanies the tone for a maximum of 10 s. Pressing the negative lever before the shock onset terminates the tone, prevents the shock, retracts the lever and starts an ITI. This response is recorded as a "prevention response." Pressing the negative lever after the shock onset terminates the tone and shock, retracts the lever and begins an ITI. This response is recorded as an "escape response." If an animal fails to press the lever during the tone, the shock terminates, the lever retracts, the ITI starts and the trial is scored as an "escape omission." The delay between the tone onset and the electric shock delivery can be progressively increased from 1 s to 40 s, that is, 50 s of tone presentation that is accompanied for the last 10 s by an electric foot shock and followed by a 10 s ITI. During the entire session, the house light is on, except during the ITIs. The training sessions last for 30 min and end with the house light being turned off and lever retraction. Following completed sessions, the rats are returned to their home cages in the holding room. The total number of responses, including the number of "prevention responses," "escape responses" and "omissions," is scored. The animals are trained until they achieve the criterion of at least 70% "prevention responses" per experimental session maintained over at least 3 training sessions.

*3.3.3 Discrimination Training*

Discrimination training involves learning to discriminate between the positive and negative tones by correctly pressing the corresponding lever (as trained in the previous stages) to maximize the reward delivery and punishment avoidance. The tones, 20 positive and 20 negative, are presented pseudorandomly (preprogrammed sequence: 0, 1, 0, 1, 0, 1, 1, 0, 0, 1, where 0 = a positive and 1 = a negative tone) and separated by 10-s ITIs. A positive lever press in response to positive tone presentation results in reward delivery and begins the ITI. A negative lever press in response to negative tone presentation results in punishment avoidance and begins the ITI. An incorrect lever press, for example, response of pressing the left lever instead of the right lever during the negative tone presentation, as well as "escape responses" or "omissions," is considered failed trials. All rats must attain at least 70% correct responses for each lever to proceed to the ambiguous-cue interpretation test.

Stable discrimination levels can be achieved in two training steps:

A. In this step, the tones, 20 positive and 20 negative, are presented alternately and separated by 10 s ITIs. Each trial consists of only one lever extension with the associated tone presentation (for up to 50 s). Pressing the positive lever during the positive tone presentation results in an instant reward delivery and begins the ITI. This result is scored as a "reward." If the rat fails to press the positive lever following the positive tone presentation, the lever retracts and a 10 s ITI starts. This result is scored as a "reward omission." Pressing the negative lever during the negative tone presentation and before the shock onset (during the first 40 s of the negative tone presentation) results in negative tone termination and begins the ITI. This result is scored as a "prevention response." If the rat fails to press the negative lever during the negative tone presentation but presses it after the shock onset (during the last 10 s of the tone presentation), this terminates the shock and begins the ITI. This result is scored as an "escape response." If the rat completely fails to press the negative lever during the negative tone presentation and fails to press it after the shock onset, this result is scored as an "escape omission." The session continues for 40 min, which is equal to 40 completed trials (the presentation of 20 positive and 20 negative tones). During the entire session, the house light is on, except during the ITIs. The training session ends with the house light being turned off and lever retraction. Following the discrimination training, the rats are returned to their home cages in the holding room. The following behavioral parameters are recorded and scored: the total number of responses, number of "rewards," number of "reward omissions," number of "prevention responses," number of "escape responses" and number of "escape omissions." The training continues until the rats reach the criterion of at least 70% correct responses for each lever per experimental session maintained over at least 3 training sessions before proceeding to **step B** of the discrimination training.

B. This step is similar to **step A**, but the presentation of each tone co-occurs with the extension of both levers. The tones (20 positive and 20 negative) are presented in pseudorandomized order and separated by 10-s ITIs. The positive lever press during the positive tone presentation (50 s) results in a reward delivery and begins the ITI and is scored as a "reward." If the rat fails to press the positive lever during the positive tone presentation, the lever retracts, the ITI starts and the trial is scored as a "reward omission." During positive tone presentation, if the rat presses the incorrect lever, that is, the negative lever, both levers retract, the ITI begins and the trial is scored as a "reward mistake." The negative lever press during the negative tone presentation and before the shock onset (within the first 40 s) terminates the tone, starts the ITI and is scored as a "prevention response." If the rat fails to press the negative lever during the negative tone presentation but does it after the shock onset (within the last 10 s of the tone presentation), the shock is terminated, the ITI starts and the trial is scored as an "escape response." If the rat completely fails to press the negative lever during the negative tone presentation and fails to press it after the shock onset, this result is scored as an "escape omission." If in the time of the negative tone presentation the rat presses the positive lever, it results in the retraction of this lever, but the other lever remains extended and this is scored as a "punishment mistake." The animal may still press the second, correct negative lever to terminate the tone and to avoid the shock later, but this response is not scored. During the entire session, the house light is on, except during the ITIs. The session continues for 40 min, which equals 40 completed trials (presentation of 20 positive and 20 negative tones), and ends with the house light off and lever retraction. Following the training, the rats are returned to their home cages in the holding room. The following behavioral parameters are recorded and scored: the total number of responses, number of "rewards," number of "reward omissions," number of "prevention responses," number of "escape responses," number of "escape omissions," number of "reward mistakes" and number of "punishment mistakes." Rats must achieve at least 70% correct responses for each lever maintained over 3 consecutive training sessions to proceed to the ambiguous-cue testing.

The most common problem during the training of rats in the ACI paradigm is achieving the stable criterion in the discrimination training. Usually, the animals learn faster to obtain a reward in response to the positive tone. They learn slower to associate the negative tone and level to avoid the punishment, or they fail to respond (*see* **Note 5**).

*3.3.4 Ambiguous-Cue Interpretation Testing*

The ambiguous-cue testing session consists of 20 positive, 20 negative, and 10 intermediate (ambiguous) tone presentations. The frequency of the intermediate tones is set to 5 kHz at 75 dB. The ambiguous tone frequency was chosen based on the protocol previously described by Enkel and colleagues [11], and several experiments from our laboratory confirmed this tone to act as intermediate in terms of the response pattern [13, 15, 16, 20, 23]. The tones are presented in a pseudorandomized order (0, 1, 0, 2, 1, 0, 0, 1, 2, 1, where 0 = a positive tone, 1 = a negative tone and 2 = an ambiguous tone) and are separated by 10-s ITIs. Any lever press during the ambiguous tone stops the tone presentation but has no further consequences. During the ACI testing session, similar to stage B of the discrimination training, the response of pressing the positive lever during the positive tone presentation results in an instant reward delivery, begins the ITI and is scored

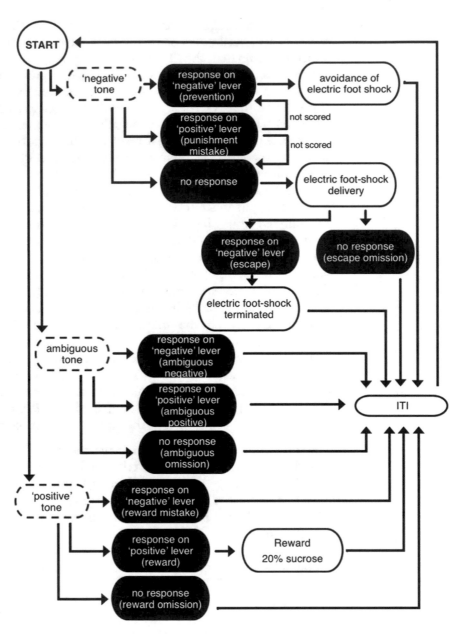

**Fig. 1** Schematic diagram of events during an ACI test

as a "reward." If the rat fails to press the positive lever during the positive tone presentation, the lever retracts, the ITI starts and the trial is scored as a "reward omission." If the rat presses the incorrect lever, that is, the negative lever, during the positive tone presentation, both levers are retracted, the ITI is initiated, and the result is scored as a "reward mistake." A negative lever press during the negative tone presentation and before the shock onset (during the first 40 s) terminates the tone, begins the ITI and is scored as a "prevention response." If the rat fails to press the negative lever during the negative tone presentation but responds after the shock onset, it terminates the shock and starts the ITI. This trial is scored as an "escape response." If the rat completely fails to press the negative lever during the negative tone presentation, before or after the shock onset, this trial is scored as an "escape omission." If the rat presses the incorrect lever, that is, the positive lever during the negative tone presentation, it results in the retraction of the positive lever, but the negative lever remains available. This trial is scored as a "punishment mistake." The animal may still press the other, correct negative lever to terminate the tone and to avoid the shock later, but this response is not scored.

All rats' responses to the ambiguous tones are scored separately. If during the presentation of the ambiguous tone the rat presses the positive lever, it is scored as an "ambiguous positive response." If during the presentation of the ambiguous tone the rat presses the negative lever, it is scored as an "ambiguous negative response." If the rat fails to press any lever, this is scored as an "ambiguous omission." Any lever press during the ambiguous tone presentation terminates the tone, retracts both levers and begins the ITI, and it is not further reinforced (trial has no outcome). For a schematic diagram of the events during an ACI test, please *see* Fig. 1.

During the entire session, the house light is on, except during the ITIs. The session lasts 50 min (20 positive, 20 negative, and 10 ambiguous tone trials) and ends with the house light off and both levers retracted.

Following the testing session, the rats are returned to their home cages in the holding room. The grid floors and operant boxes are cleaned. The following behavioral parameters are recorded and scored: the total number of responses, number of "rewards," number of "reward omissions," number of "prevention responses," number of "escape responses," number of "escape omissions," number of "reward mistakes," number of "punishment mistakes," number of "ambiguous positive responses," number of "ambiguous negative responses," and number of "ambiguous omissions."

*3.3.5 Data Analysis*

The lever responses to each tone (positive tones: numbers of "rewards," "reward mistakes" and "reward omissions"; ambiguous tones: numbers of "ambiguous positive responses," "ambiguous negative responses" and "ambiguous omissions"; and negative tones: numbers of "prevention responses," "punishment mistakes," "escape responses" and "escape omissions") are analyzed as the proportion of the overall number of possible responses to a given tone. The cognitive judgment bias index can be calculated when the proportion of negative responses to the ambiguous cues is subtracted from the proportion of positive responses, which results in values ranging between −1 and 1. Values above 0 indicate an overall positive judgment and an "optimistic" interpretation of the ambiguous cue, whereas values below 0 indicate an overall negative judgment and a "pessimistic" interpretation of the ambiguous cue. The omissions are analyzed separately. When analyzing the omissions of the responses to the negative tone, the proportion of the "escape omissions" is added to the proportion of the "escape responses," as the latter are, in fact, also omissions. Normal, healthy animals can show a high degree of interindividual variability; hence, their scores of cognitive bias index can vary between −1 and 1.

*3.3.6 Assessment of Cognitive Judgment Bias as a Stable and Enduring Behavioral Trait*

To assess the cognitive judgment bias as an enduring behavioral trait, the animals undergo a series of ten consecutive ACI tests that are carried out with 1-week intervals (*see* **Note 6**). Based upon the average cognitive bias index that is acquired from these ten ACI tests, the rats can be divided into two subgroups: those displaying trait "optimism" and those displaying trait "pessimism" (*see* Fig. 2). The animals are classified as "optimistic" if the average cognitive bias index from the ten screening tests equals or is above 0. The animals with an average cognitive bias index from the ten screening tests below 0 are classified as "pessimistic."

**3.4 Assessment of Gambling-like Behaviors Using the Rat Slot Machine Task**

Because during the ACI training and testing the levers acquired positive and negative valence (considering the association with a reward or punishment outcome), the rSMT is conducted in another type of operant conditioning boxes (Coulbourn Instruments, Allentown, PA, USA) to avoid any potential confounding factors. Additionally, the sweet sucrose solution serving as a reward in the ACI paradigm is replaced with reward pellets (45 mg, Bio-Serv, New Jersey, USA). All training and testing stages described in this protocol are adapted from Winstanley et al. [22] and have been described elsewhere [12].

*3.4.1 Preparations Before the rSMT Training Sessions*

Before the training sessions, similar to the ACI paradigm, all animals are habituated to the experimental room for 30 min. The functioning of the operant conditioning system (i.e., software, levers, and pellet dispenser) should be tested prior to any training and testing session (*see* **Note 7**).

*3.4.2 Behavioral Training for the rSMT*

Training sessions last 30 min, and animals are tested once a day. This is accomplished in several steps.

A. The first step involves habituation to the testing chambers over two sessions, during which the rats are placed in the operant chamber with a house light turned on and with reward pellets (45 mg, Bio-Serv, New Jersey, USA) located in the food magazine and nose-poke apertures.

B. In the second step, the rats are retrained to press the levers to earn the food reward in a fixed ratio 1 schedule of

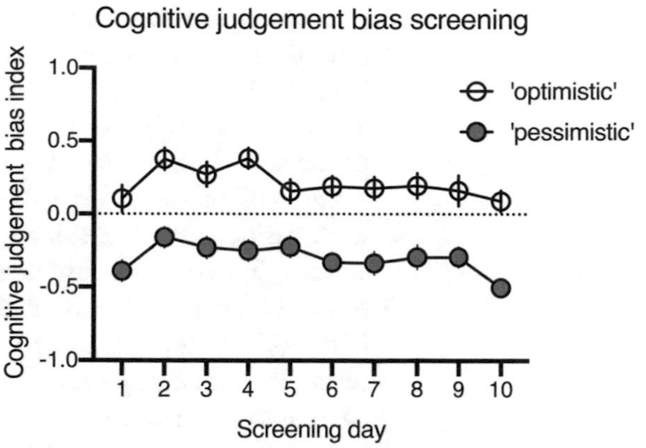

**Fig. 2** Example of cognitive judgement bias screening results. The mean ± SEM cognitive bias index of the animals classified as 'optimistic' (open circles, $N = 18$) and of those classified as 'pessimistic' (filled circles, $N = 20$) across ten ACI tests

reinforcement. Only one lever is presented during the session. When the animal has made more than 50 lever presses in a session, training is repeated on the other lever. The order of lever presentation is counterbalanced between the animals. Training to press both levers is important for further training as one of the levers serves as the "collect" lever. By pressing it, the rat indicates a choice that suggests that the animal assumed the trial was the "win" trial. The other retractable lever is a "roll" lever. By pressing it, the rat indicates a decision not to collect the reward but rather to initiate or continue the game. Once the animals have learnt to press the levers, the proper training for the slot machine task (**steps C, D, and E**) begins.

C. In the third step, rats learn to press the right lever in order to initiate every trial and to nose poke when a flashing light is displayed in the array to obtain a reward. At the beginning of the task, the right "roll" lever is extended and each response on this lever activates the light inside hole 2 to flash at a frequency of 2 Hz. A single nose-poke response in this aperture changes the frequency of the light inside to continuous display, and to enhance the status of the aperture, a 15 kHz tone sounds. Following a nose poke, the magazine light illuminates and one pellet is dispensed into the magazine. The collection of the food pellet terminates the light and extends the "roll" lever into the chamber to enable the rat to initiate the next trial. Animals can proceed to the next step after achieving the criterion of more than 50 completed trials within the 30-min session.

D. In the fourth training step, rats are required to learn that the status of hole 2 (ON or OFF) determines the output (rewarded or punished) of pressing the collect lever. After trial initiation, the animals are again required to nose poke into hole 2. In one-half of the trials, the nose poke into hole 2 causes the light to set to ON and a 15 kHz tone to be presented. These trials are assigned as "win" trials. On the other half of the trials following a nose-poke response, the light inside hole 2 is set to OFF and a 4 kHz tone is set to sound. These trials are assigned as "loss" trials. After a response in hole 2, either the right "roll" lever and/or the left "collect" lever are presented. The lever press triggers the retraction of both levers. Pressing the "roll" lever cancels the current trial, and a new trial starts with the light flashing in hole 2 (irrespective of the trial type: "win" or "loss"). The outcome of a collect lever press depends on the trial type. On "win" trials, responding on the "collect" lever results in the arrival of two food pellets followed by the "roll" lever extension, enabling the animal to initiate a new trial. Response on the "collect" lever during "loss" trials is punished by a 10 s time out period, when no reward can be obtained and no new trials can be initiated. Directly after the time out, the "roll" lever is extended again and the rat can initiate a new trial. Training sessions carry on until rats press on the "collect" lever in ≥80% of "win" trials and ≤20% of "loss" trials.

E. In the last training step, a response in an additional hole (hole 3) is required before the levers are presented. The trial setup is identical to **step D**, except that after the tone has sounded following a response in hole 2, hole 3 begins to flash. The rats are required to nose poke into this hole, after which the light is set to ON or OFF, the corresponding tone is played and both levers are presented. When both lights are set to ON, a "win" is signaled and a reward (five sugar pellets) is delivered. All other trial types (OFF–ON; ON–OFF; OFF–OFF) are considered "loss" trials, where responding on the "collect" lever results in a 10 s time out. Training continues for five sessions, after which the animals are moved on to the full task (*see* Subheading 3.4.3).

*3.4.3 Testing Rats in the rSMT*

Following the completion of successive versions of the slot machine program, which gradually increase in complexity, the proper rSMT testing commences.

During the test session, the rat has to press the "roll" lever, which prompts its retraction and initiates the trial by flashing the hole 1 light at a frequency of 2 Hz (Fig. 3). When the rat nose-pokes into the hole 1, depending on the schedule, the light inside the hole is either turned ON or turned OFF for the rest of the trial. The light ON or OFF status of the hole is accompanied by either a

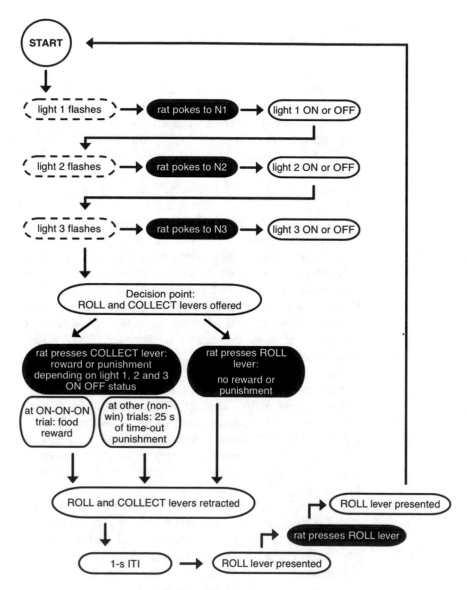

**Fig. 3** Schematic diagram of events during an rSMT

high (15 kHz at 68 dB) or a low (4 kHz at 68 dB) acoustic tone, which acts as an additional cue. The sounds last for 1 s.

Following the hole 1 step, the light in hole 2 begins to flash. When the rat makes a nose-poke response into hole 2, as in the previous step, the light inside is either turned ON or turned OFF, depending on the schedule, and a high or low corresponding tone is presented for 1 s. Then, the light inside the hole 3 begins to flash, and the sequence of events (depending on the schedule, light ON or OFF accompanied with a corresponding acoustic signal) is repeated, similar to the procedure for holes 1 and 2. At this moment, the animal is exposed to and, by nose-poking, has responded to the series of three stimuli. For a schematic diagram of the events during an rSMT, *see* Fig. 3.

To increase the number of meaningful trials, the original procedure that was described by Winstanley and colleagues [22] is modified by reducing the number of possible 1- and 2-light pattern outcomes to (ON–OFF–OFF and OFF–OFF–ON) and (ON–ON–OFF and OFF–ON–ON), respectively. In this way each trial can have only 6 out of 8 possible patterns: ON–ON–ON, ON–ON–OFF, OFF–ON–ON, ON–OFF–OFF, OFF–OFF–ON, and OFF–OFF–OFF. This gives one state with 3 lights ON (ON–ON–ON, a rewarded, "clear win"), two states with two lights ON (ON–ON–OFF or OFF–ON–ON, a "near miss"), two states with 1 light ON (ON–OFF–OFF, OFF–OFF–ON, a "near loss") and 1 state with all lights OFF (OFF–OFF–OFF, a "clear loss").

The frequency of all six light patterns is predefined in a pseudorandom order and each pattern is presented once per block of six trials so that the animals can obtain a food reward only once in each block. The pseudorandom order of trials in each block is different so that animals cannot learn the sequence. Each rat is required to interpret the given pattern as a "win" or "loss." This is accomplished at the "decision point" by extending both the "roll" and the "collect" levers. On the clear "win" trials, during which all three lights are set to ON, pressing the "collect" lever results in a delivery of five sucrose pellets. On any other pattern, with at least one light being turned OFF, responding on the "collect" lever results in a penalty of 25 s of time out, during which the animal cannot quickly start a new trial. However, pressing the "roll" lever on any trial cancels the potential reward or the time out so that the rat can immediately initiate a new trial.

The rat's optimal strategy is to respond with the "collect" lever only on "win" trials to obtain the food reward and to respond with a "roll" lever on every "loss" trial to avoid the time out and initiate a new trial. There is no time requirement to respond on the lever or the nose poke. Rats are trained once a day, 5 days per week, until they establish a stable trial performance ($\geq$25 trials per session and $\leq$ 50% collect responses on clear loss outcomes (OFF–OFF–OFF) during five consecutive sessions.

*3.4.4 Data Analysis*

The main analyzed parameters include the proportion of pressing the "collect" lever in response to "clear win," "near miss," "near loss," and "clear loss" outcomes. Other parameters include the latency to press the "collect" lever following "clear win," "near miss," "near loss," and "clear loss" outcomes and the average numbers of trials finished per session. The results are analyzed as an average of the five rSMT tests. For analysis, the data from 2-light trials (ON–ON–OFF and OFF–ON–ON) and 1-light trials (ON–OFF–OFF and OFF–OFF–ON) can be pooled (*see* **Note 8**).

**3.5  Experimental Design**

The experimental design is schematically presented in Fig. 4. After attaining stable discrimination performance in the ACI paradigm (more than 70% correct responses to each tone over 3 consecutive days), each rat undergoes the cognitive judgment bias screening procedure as previously described (*see* Subheading 3.3.6). After the "optimistic" or "pessimistic" traits of the individual animals have been established, the rats are retrained for the rSMT in another set of experimental boxes and subsequently tested for 5 consecutive days on the rSMT (*see* Subheading 3.4.3). The mean results from these five tests are compared between the groups of animals that were previously classified as "optimistic" and "pessimistic."

# 4  Notes

1. Prior to conducting experiments with laboratory animals, experimenters need to undertake the relevant training and obtain personal license for animal experimentation (check national policies). All experimental procedures must be first approved and conducted in accordance with the national and institutional ethics committees/review groups and must follow the officially approved procedures and guidelines for the care, husbandry, and use of experimental animals.

2. A sufficient amount of time should be allowed for proper habituation and nonaversive handling of the rats by the experimenter. Properly handled rats are more relaxed and easily accept seek human contact, which directly translates into their state of wellbeing. Handling stress can be an undesirable source of variation in experiments that deal with affective and cognitive processes.

3. Each rat should always be trained and tested in the same operant box, for example, rat # 1 in box number 1 and rat # 7 in box number 7, and the animals should be tested in the same order (e.g., test rat # 6 after testing rat # 5 and never the other way around).

4. The cue-outcome combinations between the experimental animals should be fully counterbalanced prior to the first training session so that for half of the rats, the 9 kHz tone predicts punishment and the 2 kHz tone predicts reward and vice versa for the other half.

5. To increase the number of "prevention responses," reduce the value of the reward sucrose solution (depending on the rat's individual performance, decrease the percentage of the sucrose solution used as a reward, e.g., from 20% to 10% or 5%). For a short time, one can also alter the ratio of reward- to punishment-associated trials in the discrimination session, for

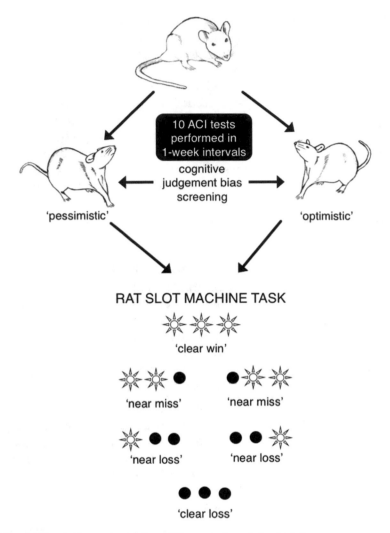

**Fig. 4** Schematic representation of the experimental schedule

example, from the initial 0.5:0.5 (20 rewarded and 20 punished) to 0.4:0.6 (16 rewarded and 24 punished) or 0.3:0.7 (12 rewarded and 28 punished). Typically, approximately 90% of trained animals accomplish the criterion of 70% correct responses to both reference cues and can be further tested in the ACI paradigm.

6. The rats should not be tested on consecutive days as this may result in a loss of ambiguity for the intermediate tone because the animals will learn to associate it with a lack of outcome, which could influence their subsequent choices and lead to false conclusions. Repeated testing can be accomplished by separating two ACI test sessions with a discrimination session (e.g., an ACI test on day 1, a discrimination test on day 2, and an ACI test on day 3).

The rats should not be trained or tested more than once on the same day due to large consumption of the sucrose solution and potential confounding with their satiety levels, leading to a reduced performance and false-negative results. Satiety level may confound the outcomes of the ACI test; hence, the rats should be fed and tested at the same times each day (e.g., a 17:00 h feeding and a 09:00 h testing). If there are different cohorts of rats in the same holding room, all of the animals should be fed at the same time, if possible, to avoid frustration in the hungry animals.

7. As animals consume a large number of reward pellets, frequent cleaning of the pellet dispenser, tubing, and the magazine is advised. No broken or halved pellets should be used. Pellet dust and debris can easily build up, leading to blockage and unreliable pellet delivery.

8. As responding to different light patterns in 1- and 2-light trials usually reveals no significant differences, the data within the 2-light trials (ON–ON–OFF and OFF–ON–ON) and within the 1-light trials (ON–OFF–OFF and OFF–OFF–ON) can be pooled to simplify the further analyses.

## Acknowledgments

This work was supported by the statutory funds of the Maj Institute of Pharmacology Polish Academy of Sciences.

## References

1. Clark L (2010) Decision-making during gambling: an integration of cognitive and psychobiological approaches. Philos Trans R Soc Lond B Biol Sci 365:319–330

2. Ladouceur R, Dube D, Giroux I et al (1995) Cognitive biases in gambling—American roulette and 6/49-lottery. J Soc Behav Pers 10:473–479

3. Ladouceur R, Sylvain C, Boutin C et al (2001) Cognitive treatment of pathological gambling. J Nerv Ment Dis 189:774–780

4. Langer EJ (1975) Illusion of control. J Pers Soc Psychol 32:311–328

5. Thompson SC, Armstrong W, Thomas C (1998) Illusions of control, underestimations, and accuracy: a control heuristic explanation. Psychol Bull 123:143–161

6. Cocker PJ, Winstanley CA (2015) Irrational beliefs, biases and gambling: exploring the role of animal models in elucidating vulnerabilities for the development of pathological gambling. Behav Brain Res 279:259–273

7. Jacobsen LH, Knudsen AK, Krogh E et al (2007) An overview of cognitive mechanisms in pathological gambling. Nord Psychol 59:347–361

8. Carver CS, Scheier MF (2014) Dispositional optimism. Trends Cogn Sci 18:293–299

9. Gibson B, Sanbonmatsu DM (2004) Optimism, pessimism, and gambling: the downside of optimism. Personal Soc Psychol Bull 30:149–160

10. Harding EJ, Paul ES, Mendl M (2004) Animal behaviour: cognitive bias and affective state. Nature 427:312

11. Enkel T, Gholizadeh D, Von Bohlen O, Halbach O et al (2010) Ambiguous-cue interpretation is biased under stress- and depression-like states in rats. Neuropsychopharmacology 35:1008–1015

12. Rafa D, Kregiel J, Popik P et al (2016) Effects of optimism on gambling in the rat slot machine task. Behav Brain Res 300:97–105

13. Rygula R, Szczech E, Papciak J et al (2014) The effects of cocaine and mazindol on the cognitive judgement bias of rats in the ambiguous-cue interpretation paradigm. Behav Brain Res 270:206–212

14. Organtzidis H, Hales C, Hinchcliffe J et al (2019) Different methods for assessing cognitive affective biases in rats using the judgement bias task. Pharmacol Rep 71:1312–1313

15. Papciak J, Popik P, Fuchs E et al (2013) Chronic psychosocial stress makes rats more 'pessimistic' in the ambiguous-cue interpretation paradigm. Behav Brain Res 256:305–310

16. Rygula R, Papciak J, Popik P (2014) The effects of acute pharmacological stimulation of the 5-HT, NA and DA systems on the cognitive judgement bias of rats in the ambiguous-cue interpretation paradigm. Eur Neuropsychopharmacol 24:1103–1111

17. Rygula R, Pluta H, Popik P (2012) Laughing rats are optimistic. PLoS One 7:e51959

18. Rafa D, Kregiel J, Popik P et al (2015) Effects of optimism on gambling in the rat slot machine task. Behav Brain Res 300:97–105

19. Rygula R, Golebiowska J, Kregiel J et al (2015) Effects of optimism on motivation in rats. Front Behav Neurosci 9:32

20. Rygula R, Papciak J, Popik P (2013) Trait pessimism predicts vulnerability to stress-induced anhedonia in rats. Neuropsychopharmacology 38:2188–2196

21. Rygula R, Popik P (2016) Trait "pessimism" is associated with increased sensitivity to negative feedback in rats. Cogn Affect Behav Neurosci 16:516–526

22. Winstanley CA, Cocker PJ, Rogers RD (2011) Dopamine modulates reward expectancy during performance of a slot machine task in rats: evidence for a 'near-miss' effect. Neuropsychopharmacology 36:913–925

23. Papciak J, Rygula R (2017) Measuring cognitive judgement bias in rats using the ambiguous-cue interpretation test. Curr Protoc Neurosci 78:9.57.1–9.57.22

# Part II

## Environmental Variables of Vulnerability to Addiction

# Chapter 6

# Extended Drug Access and Escalation of Drug Self-Administration

**Florence Allain, Ndeye Aissatou Ndiaye, and Anne-Noël Samaha**

## Abstract

In this chapter we first describe the intravenous drug self-administration technique in rats, with a focus on cocaine. Where relevant, we also describe how self-administration procedures can be adapted for use in both female and male rats. We then discuss some of the pharmacokinetic variables that can influence the development of behavioral features of cocaine addiction. These variables include the speed of intravenous drug delivery, the amount and temporal pattern (intermittency) of intake. In this context, we present and compare different self-administration procedures that have been used to model DSM-like features relevant to addiction in rats. These procedures include Short-Access, Long-Access, and Intermittent-Access cocaine self-administration, and variations therein. We highlight that some procedures (i.e., Long-Access) are best suited to study changes in cocaine intake over time. Others (i.e., Intermittent-Access) are especially effective to study increases in incentive motivation for cocaine over time. Work comparing these procedures supports two important conclusions. First, excessive/escalating cocaine intake is not a necessary prerequisite to produce the increased incentive motivation that defines the addicted state. Second, Intermittent-Access cocaine self-administration might not only better model human patterns of cocaine intake, but might also be uniquely suited to study the cocaine-induced changes in neurobiology, psychology, and behavior involved in the addiction process.

**Key words** Intravenous drug self-administration, Rat, Male, Female, Cocaine, Long-access, Intermittent-access, Addiction

## 1 Introduction

### 1.1 Self-Administration: An Operant Conditioning Procedure

Self-administration of a reward involves voluntarily performing a behavioral response to obtain this reward. The nature of the behavioral response (or operant response) can vary across experiments and species (e.g., pecking for a bird, nose-poking, pressing a lever or pulling a chain for a rodent). However, across experiments and species, the concept remains the same; a subject is placed in a specialized cage/apparatus and can earn rewards when the predetermined operant response is emitted.

Learning the association between a behavioral response and an outcome is termed operant or instrumental conditioning. The

María A. Aguilar (ed.), *Methods for Preclinical Research in Addiction*, Neuromethods, vol. 174,
https://doi.org/10.1007/978-1-0716-1748-9_6, © Springer Science+Business Media, LLC, part of Springer Nature 2022

outcome can either have positive reinforcing/rewarding properties (i.e., pleasant/wanted effects) or negative reinforcing properties (i.e., removal of something unpleasant, which is a pleasant outcome). Both positive and negative reinforcers *increase* the probability of the preceding response. This is in contrast to when an operant response produces a punishing outcome. This is unpleasant and so punishment *decreases* the likelihood that the animal will emit the preceding behavioral response in the future.

In the context of this chapter, we focus on drug self-administration procedures, which exploit the positive reinforcing properties of drugs to study voluntary drug intake and the transition to addiction. The concept of operant conditioning was developed and studied extensively by Burrhus Frederic Skinner, who taught pigeons to peck at a predetermined target in order to get food [1, 2]. The delivery of food is contingent upon the operant response, and the presentation of food reinforces that response. This is the basic concept that is exploited in drug self-administration studies. Drugs, just like food, water, safety, and sex, have rewarding and reinforcing properties, such that they will support the learning and performance of operant behaviors. In reference to Skinner's work on this type of learning, the specialized cages used for operant conditioning tasks are still widely known as *Skinner boxes*.

In this chapter, we will describe the type of operant conditioning where drugs (specifically cocaine) positively reinforce a behavior (lever pressing in rats).

## 1.2 Discovering that Nonhuman Laboratory Animals Voluntarily Self-Administer Drugs

Humans self-administer drugs across the globe, and they have done so across history. Evidence that animals other than humans voluntarily self-administer drugs in a laboratory context comes from a study conducted by S. D. S. Spragg in 1940. In this study, Spragg chronically treated chimpanzees with morphine. After developing signs of abstinence from morphine, the chimpanzees learned and performed the behaviors that produced the drug. For instance, they voluntarily selected a drug-associated box containing a syringe of morphine, they selected the right stick to open this box, and they would also pull Spragg into the test room associated with morphine administration [3]. In parallel, when chimpanzees were food-deprived instead of morphine-deprived, they would select the food-associated box containing a banana. The conclusion was that nonhuman animals, at least chimpanzees, would perform behavioral responses to obtain a drug of abuse. Importantly, it was also concluded that nonhuman animals would do this only if they developed physical dependence to the drug and were then kept from it for a time. As we will see next, work in rats refuted this conclusion.

In 1962 James R. Weeks published an article entitled *Experimental Morphine Addiction: Method for automatic intravenous injections in unrestrained rats* [4]. The study used a Skinner box where freely moving female rats could press a lever to get an intravenous injection of morphine. This was done by implanting a catheter into the jugular vein of the rats [5, 6]. Weeks showed that physically morphine-dependent rats would learn to press the lever to self-administer morphine. This was a major breakthrough in the field, because it confirmed that species other than primates would voluntarily self-administer a drug of abuse in the laboratory. The finding was replicated 2 years later in morphine-dependent monkeys [7]. Next, it was shown that even nondependent animals would self-administer morphine [8, 9], such that "regarding the 10 µg/kg dose, none of the monkeys took enough morphine to become physically dependent" [9]. The authors concluded that "morphine can act as a reinforcer, independent of its property to produce physical dependence" [9]. The technique has also been extended to other substances with addictive potential including psychostimulant drugs [8, 10–15].

Most drugs self-administered by humans are also self-administered by rats and monkeys [16]. In this context, we point interested readers to [17] for an informative discussion of how appetitive versus consummatory drug-directed behaviors can also be studied and interpreted in laboratory animals. While the nature of the operant behaviors leading to drug differs from humans (e.g., exploiting the many ways to gather money, finding a point of purchase) to rats (e.g., pressing a lever, nose-poking), voluntary drug self-administration is seen across species. This provides the basis for powerful animal models to study the biological, psychological and behavioral effects of voluntary drug intake.

### 1.3 From Drug Self-Administration to the Study of Drug Addiction

While humans and laboratory animals voluntarily take drugs, this is not tantamount to drug addiction [18]. Around 10–15% of individuals that consume drugs will develop a substance use disorder [19]. This indicates that addiction is not an inevitable consequence of drug use, even when drug use is chronic. Therefore, a key question in addiction research is why some individuals can keep control over their drug consumption, while others make the transition to addiction. Understanding the neurobiological, psychological and behavioral changes involved in this transition is fundamental to predicting and preventing addiction. Thus, it is of great importance to have preclinical models of drug addiction that can be translated to the human condition [20]. As in humans, just because a laboratory rat reliably self-administers a drug does not make this rat useful to study addiction. There exist no reliable biomarkers of drug addiction. Instead, diagnosis is based on the presence/absence of defined psychological and behavioral features, as outlined in the Diagnostic and Statistical Manual of Mental Disorders [21]. The number of features also determines the severity of the disorder, from mild to moderate to severe [21]. A similar approach, based on the number of addiction-relevant behavioral features, can be used to identify rats with an addiction-like phenotype [21–24].

Many factors modulate the development of addiction-like features and the transition to addiction. These include individual and environmental variables (e.g., genetics, age, biological sex, stress exposure, and others [25–29]). In addition, *how* a drug is consumed can also be decisive. Research shows that the *speed of drug delivery*, the *amount*, and the *temporal pattern* of drug intake (i.e., intermittency) each determine the transition to addiction. Below, in describing the intravenous drug self-administration technique, we highlight findings from this literature, showing that such pharmacokinetic factors determine the response to drugs of abuse. This discussion will focus particularly on cocaine, as the effects of speed, amount, and pattern of intake have been studied most extensively with this drug. Where relevant, we also highlight differences in outcome between females and males.

## 2    Protocol

### 2.1    Materials and Methods for Food Self-Administration in Rats

#### 2.1.1    The Skinner Box

Figure 1 represents an operant conditioning chamber also called a *Skinner box*. Rats are generally housed in a room distinct from the testing room where the operant chambers are located. However, some laboratories have housed rats in the operant chambers, where they also have free access to food and water [30, 31].

Operant conditioning chambers are generally placed inside sound-attenuating cubicles with a ventilating fan that also serves to attenuate external noise. Each chamber is equipped with a metal grid floor. A pan is placed below this grid floor. The pan can be kept empty, or filled with bedding or cardboard to facilitate cleaning between test sessions. The chambers contain a houselight (left wall in Fig. 1) and two 4-cm wide levers. A discrete cue light is also located above each lever. In between the two levers, rats can have access to a food/liquid receptacle where food/liquid rewards can be delivered. Note that the location of each of the above elements in an operant chamber (food receptacle, levers, houselight, cue lights, etc.) can differ across laboratories.

#### 2.1.2    Lever Presses Reinforced by Food

- Food self-administration can be used to hasten later acquisition of drug self-administration behavior. Training laboratory rats to press a test lever to self-administer food is generally quick and easy.

- As a sample procedure, rats can learn to press a lever (the active lever) to obtain a food pellet in the receptacle. Sessions can last 1 h/day, under FR1 (Fixed Ratio 1, one lever press delivers one food pellet). The ratio requirement can then be increased if

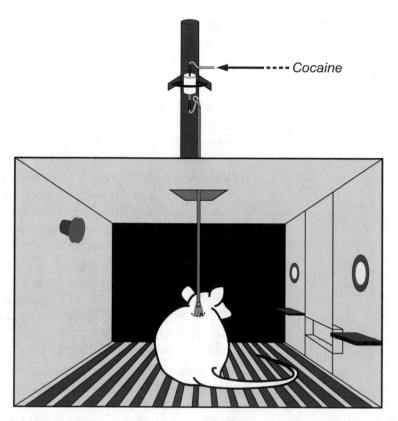

**Fig. 1** Schematic representation of a Skinner box adapted for cocaine self-administration in rats. Dimensions can vary, but a typical chamber has the following dimensions; 31.8 cm × 25.4 cm × 26.7 cm. On the right wall: An active and an inactive lever. Pressing the active lever produces an intravenous cocaine injection through tubing in the tether system. Pressing the inactive lever has no programmed consequences. On the left wall: a houselight which is turned on to signal the beginning of a self-administration session and then turned off at the end

needs be. At all times, pressing the inactive lever has no programmed consequence.

- Rats are food restricted during this step (e.g., 15 g/day in adult male rats and 13 g/day in adult female rats [32]) to promote food seeking during the task. Thus, acquisition of reliable food self-administration behavior is generally quick for the majority of rats. Acquisition criteria can include taking >10 pellets/session, on two successive sessions. In our laboratory, we generally allow a maximum of 100 pellets/session, and most rats generally self-administer all 100 (*see* **Note 1**).

## 2.2 Materials and Methods for Intravenous Drug Self-Administration in Rats

### 2.2.1 Catheters

- Catheters are easy to make in-house. The following materials are required; cannulae (C313G-5UP; stainless-steel; 5-mm upward projection; 11-mm length; 22 Gauge; Plastics1; Fig. 2), 100–200 μl pipette tips (Fig. 2), silastic tubing (Inner $\emptyset$ = 0.51 mm; Outer $\emptyset$ = 0.94 mm; Dow Corning, CAT. NO. 508-002; Fig. 2), polyolefin heat-shrink tubing (Inner $\emptyset$ = 1.17 mm; Recovered $\emptyset$ = 0.58 mm; Alpha Wire, FIT 221 3/64; Fig. 2) and Nylon mesh (Bard® Mesh, Monofilament polypropylene mesh, REF. NO. 0112660; Fig. 2).

- To build a catheter, the canula should be bent at a ~135° angle. Bending the cannula at this angle is especially important. First, this angle ensures that the catheter is placed in the right position under the skin during the surgery. Second, this angle will minimize rat discomfort.

- The canula is fixed to a piece of silastic tubing (~15 cm for a catheter to be used in a male rat and ~13.5 cm for a female) after immersion of the silastic tubing in Chloroform for distension. The resulting canula/silastic tubing complex is then reinforced with a piece of heat-shrink tubing that will cover the canula after being heated with a soldering iron.

- The canula with tubing is then inserted into a cut pipette tip. The pipette tip is then filled with dental cement to consolidate the catheter, a piece of nylon mesh is placed on top of the dental cement, and a drop of dental cement is added to secure this piece of mesh.

- Two silicone bubbles are placed at the extremity of the silastic tubing (Fig. 2). For female rats: one bubble is placed at ~2.1–2.4 cm from the beveled extremity of the tubing (~3 cm for male rats) and the second bubble at ~1.7 cm above the first bubble (same in male rats). Silicone bubbles will serve to secure the catheter within the vein during surgery. Finally, the extremity of the silastic tubing is cut at a beveled angle to facilitate its insertion into the vein.

### 2.2.2 Implanting a Catheter into the Jugular Vein of a Rat

- The experimenter can choose to implant into the left or the right jugular vein.

- Rats are first anesthetized under isoflurane (5% for induction and 2–3% for maintenance) and injected with analgesic and antibiotic. Lubricating ointment is also applied to both eyes before the start of the surgery.

- Skin in the back and above the jugular vein is shaved and disinfected (EtOH 70% and Chlorhexidine). An incision is made in the skin on the rat's back in between the scapulae. Another incision is made in the skin above the jugular vein beats. A cleaned catheter (first immersed in Cidex OPA and then rinsed

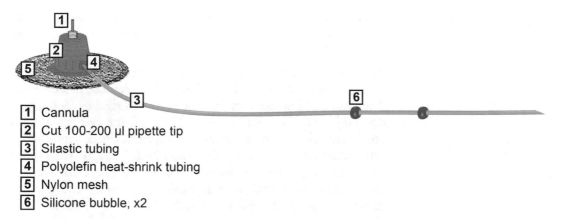

1 Cannula
2 Cut 100-200 µl pipette tip
3 Silastic tubing
4 Polyolefin heat-shrink tubing
5 Nylon mesh
6 Silicone bubble, x2

**Fig. 2** Schematic representation of an intravenous catheter. Materials needed to build a catheter are numbered and defined

with tap water) is subcutaneously transferred from the rat's back to the jugular vein area under the front paw ipsilateral to the vein.

- The beveled extremity of the catheter is inserted into the jugular vein up until the first silicone bubble on the catheter.

- With suture thread, the first silicone bubble is tied to the vein and the second to the chest muscle. The catheter is linked to a syringe of saline during the procedure and blood/saline is pulled/pushed to check good position of the catheter into the vein. Good position in the vein is confirmed when one can draw blood from the catheter during surgery.

- On the rat's back, the nylon mesh is flattened subcutaneously to avoid potential discomfort (*see* **Note 2**).

*2.2.3 Intravenous Cocaine Self-Administration*

- Materials needed for cocaine self-administration are the same as those needed for food self-administration, except that rats are now catheterized, operant responding is reinforced with cocaine, and each operant chamber is equipped with a syringe pump and a tether apparatus. The tether is a Tygon® (Polymers) tubing (Inner $\emptyset = 0.51$ mm; Outer $\emptyset = 1.52$ mm; Cole-Parmer® REF. NO. 06419-01) which is protected by a metal spring linking the rat's catheter to the syringe filled with drug (Fig. 1). The tether system has to be adjusted such that it does not pull on the rat's catheter. Females can also show more cocaine-induced psychomotor activity than males do. As such, it is important to frequently monitor female rats when they are in the operant cages, to avoid breaks to the tether apparatus. In both sexes, tether apparatus should be verified visually after each cocaine self-administration session, to confirm that it is intact before starting the following session.

- After jugular catheterization, rats are given approximately 6–8 days of recovery in males and 10–14 days in females where they are monitored every day (*see* **Note 3**).

- Rats can easily learn to self-administer cocaine during 1-h sessions (1 session/day) under an FR1 schedule of reinforcement. The ratio requirement can then be increased during training, as needed for experimental purposes [32–34]. Increasing the ratio requirement often promotes greater discrimination between the active versus inactive levers. To avoid potential overdose, a time-out period (e.g., 20 s) can follow each cocaine injection. The cue light above the active lever is turned on during the cocaine injection. Inactive lever presses have no programmed consequences.

- Care should be taken in interpreting data from the first cocaine self-administration session after food training, because rats are still adapting to food being substituted by cocaine. For reliable acquisition of cocaine self-administration at a dose of 0.25 mg/kg/injection, acquisition criteria can include taking at least six injections per 1-h session at regular intervals (injections should be spaced out throughout the session) and pressing twice more the active versus the inactive lever. Acquisition criteria depend on the drug studied and the dose used. For example, cocaine self-administration under fixed ratio follows an inverted U-shaped dose–response curve [35]. This suggests that the injection criterion should be adjusted according to dose.

- The data are mixed, but some studies suggest that the rate of acquisition of intravenous cocaine self-administration can differ between male and female rats. The dose of cocaine used and prior operant experience can explain such differences between studies [36]. Some studies report that female rats can acquire cocaine self-administration more readily than male rats can [37, 38]. Still, other studies report that the sexes might not differ in acquisition of cocaine self-administration behavior when test conditions promote rapid acquisition [e.g., higher cocaine doses, food restriction, operant pretraining; [36] and references therein]. We also point the reader to recent studies showing no sex differences in the rate of acquisition of intravenous cocaine self-administration [32, 39].

## 3   Behavioral Features Relevant to Cocaine Addiction in Rats and Representative Results

Taking a drug, even regularly, is not necessarily reflective of drug addiction. For this reason, preclinical intravenous cocaine self-administration procedures have been developed [22–24] so that they promote the development of DSM-like criteria observed in humans [21].

These DSM-based criteria include the following.

- *The stimulant is often taken in larger amounts or over a longer period than was intended.*

- *A need for markedly increased amounts of the stimulant to achieve intoxication or desired effect.*

These signs of cocaine addiction in humans can be modeled in the laboratory rat using self-administration procedures that promote excessive and escalating cocaine intake over time. Below we will describe some of the variables that can influence how much cocaine a rat will self-administer over time. These variables include (but are not exclusive to) the following.

### 3.1 Dose and Schedule of Cocaine Reinforcement

- The dose of cocaine self-administered per infusion is determined by both the concentration of the cocaine solution made available during each self-administration session and by the length of each self-administered infusion. The experimenter determines both variables.

- The rate of cocaine self-administration can vary depending on the dose of cocaine available and the schedule of reinforcement [14]. Under a fixed ratio schedule of drug reinforcement (which is the most widely used in self-administration studies [40]), the rate of cocaine self-administration follows an inverted U-shaped dose-response curve [35, 41, 42]. Thus, the number of cocaine infusions taken will increase at lower cocaine doses and decrease at higher cocaine doses. This can make data interpretation difficult. Rats could be taking more as dose is increased because there is increased reinforcing efficacy, but as dose is increased further, rats might now take *less* because of greater satiety at higher doses [43].

- Under a progressive ratio (PR) of drug reinforcement, the ratio required to obtain the next injection increases exponentially [43–45]. While under FR, rate of cocaine intake follows an inverted U-shaped dose–response curve [35], one can observe a linear dose-response curve under PR [46]. Thus, PR can have some advantages over FR when analyzing the reinforcing properties of cocaine across a range of doses [43]. Under a PR schedule, cocaine becomes exponentially harder to obtain during the test session. During a test session of fixed duration, there comes a point where rats no longer earn infusions because reaching the next ratio is too demanding. The last ratio reached is called the "breakpoint" and it is an index of how motivated the rat is to get cocaine. This ratio offers an opportunity to compare drugs and compare experimental treatments. However, under a progressive ratio schedule, self-administration behavior can be different with opiates versus psychostimulant drugs [43]. This has led to the conclusion that "[...] no single schedule can capture what appear to be fundamental differences between distinct classes of drugs. Rather, several schedules of reinforcement may be required to characterize the multidimensional properties of drugs." [43].

- Under both FR and PR, cocaine intake can either escalate over time [47–50] or remain stable, especially at lower doses [32].

**3.2 Self-Administration Session Length**

- After reliable acquisition of cocaine self-administration during 1-h daily sessions, session length can be increased to 6 h/day. Compared to Short Access session (ShA, 1–2 h/day), Long Access sessions (LgA, 6+ h/day) promote escalation of cocaine intake in rats [49, 50]. While ShA-rats can keep stable levels of cocaine intake for more than 3 months [50], LgA-rats can escalate their intake, and do so within 2–7 LgA sessions ([49–58]; For representative results *see* Fig. 3a, black *versus* white curves; Adapted from [53]).

- LgA cocaine self-administration procedures do not always produce escalation of intake [32, 53, 59, 60, 61], and this depends upon the cocaine dose used, the speed of intravenous cocaine delivery and biological sex [32, 53, 60, 62].

- Female rats can self-administer much more cocaine than male rats over LgA sessions and are also more likely to escalate their drug use ([32, 63]; For representative results *see* Fig. 3b; Adapted from [32]; *see* **Note 4**).

**3.3 Speed of Intravenous Cocaine Delivery**

- The speed of cocaine delivery through the tether system (Fig. 1) and into the intravenous vein can be manipulated by using different syringe-pump motors during the self-administration session. Different syringe-pump motor models allow for a different number of rotations per minute (RPM). This in turn determines the amount of fluid that is delivered in a fixed time period (e.g., during a 5-s infusion). Accordingly, concentration of the cocaine solution has to be adjusted as a function of the target infusion volume, and the duration of each infusion. As an example, a 3.33 RPM motor (fast) can be used to deliver ~150 μl of drug over 5 s but a 0.1 RPM motor (slower) can be used to deliver ~80 μl of drug over 90 s [33]. Both speeds deliver the same amount of cocaine. Thus, for rats weighing 400 g, at 0.25 mg/kg/injection of cocaine, 0.1 mg of drug will be transported through the tether system either over 5 s or over 90 s. It is of particular importance when studying the speed of cocaine delivery to remove the dead volume through the tether system before starting each self-administration session. A dead volume

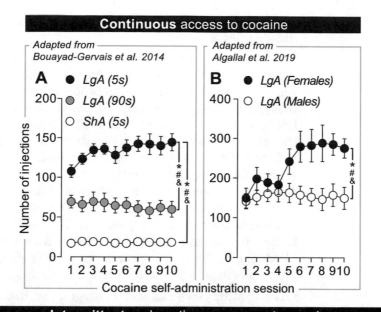

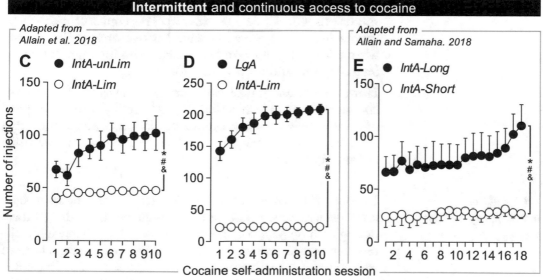

**Fig. 3** Comparison of different cocaine self-administration procedures in rats and their effects on cocaine intake over time. (**a**) Intake data from male rats given Long-Access (LgA) versus Short-Access (ShA) self-administration sessions (Black versus white curve, respectively) and rats given access to rapid cocaine infusions (LgA-5s) versus more sustained cocaine infusions during each Long-Access session (LgA-90s; Black versus gray curve, respectively). (**b**) Intake data from female and male rats given Long-Access cocaine self-administration sessions. (**c**) Intake data from male rats given Intermittent-Access to an unlimited number of cocaine Infusions (IntA-unLim) versus male rats given Intermittent-Access to a limited number of cocaine infusions (IntA-Lim). (**d**) Intake data from male rats given Long-Access cocaine self-administration sessions versus male rats given Intermittent-Access to a limited number of cocaine infusions. (**e**) Intake data from male rats given 6-h (IntA-Long) or 2-h (IntA-Short) Intermittent-Access self-administration sessions. *$P$s < 0.05, main effect of Group. #$P$s < 0.05, main effect of Session. &$P$s ≤ 0.05, Session × Group interaction

in the system will have a higher impact for cocaine injections delivered more slowly.

- For a drug like cocaine, comparing effects of fast versus slower cocaine infusions is of interest clinically, because humans consume cocaine via different routes of administration (slow routes; orally, intranasally, versus faster routes; inhalation or intravenous injection), and because the route of cocaine administration is thought to influence the susceptibility to cocaine addiction [64, 65]. Thus, varying the speed of cocaine delivery from ~5 to ~90 s in self-administration studies is relevant notably because this range of speeds causes a right-ward shift in cocaine Tmax in the brain with slower injections without causing a significant downward shift in Cmax [33, 66, 67]. As such, using this range, one can study the effects of differences in the rate of rise of cocaine in the brain, without the confounding influence of differences in achieved concentration in the brain.

- Of note, many preclinical studies suggest that laboratory animals will reliably self-administer cocaine over a range of injection speeds [12, 48, 53, 61, 62, 68–73]. This concords with observations in humans, where cocaine has reinforcing efficacy even when taken using slow routes of administration (e.g., intranasal intake vs. inhalation or i.v. injection; [64]).

- However, cocaine intake will increase as speed of intravenous cocaine delivery increases [53, 61, 62] and increasing the speed of cocaine delivery also promotes escalation of intake over time ([53, 62]; For representative results *see* Fig. 3a, black *versus* gray curves; Adapted [53]).

### 3.4 Pattern/Intermittency of Drug Access

- In contrast to Long-Access cocaine self-administration procedures which achieve continuously high levels of drug intake during each session, cocaine intake in humans is likely intermittent during each bout of intoxication [74, 75]. During a bout of consumption, experienced drug users consume cocaine intermittently, in a pattern that would generate spikes and troughs in brain drug concentrations [23, 41, 74]. In contrast, brain concentrations of cocaine remain steady during a LgA session [75]. To model the intermittency of cocaine intake in human users, and to produce intermittent spikes and troughs in brain cocaine concentrations, a new Intermittent Access (IntA) procedure has been developed, whereby rats are given access to cocaine for short 5–6 min bouts intercalated with longer 25–26 min bouts where cocaine is not available [32, 33, 75–78].

- IntA rats self-administer less cocaine than LgA rats do [32, 51, 75, 76]. However, both LgA and IntA rats can escalate their cocaine intake over sessions [34, 39, 51, 75, 77, 79].

- In our laboratory, to study the relationship between escalation of cocaine intake and other behavioral features relevant to addiction in rats, we have used IntA cocaine self-administration procedures where rats do not escalate their intake [33, 34, 51]. In one IntA group, we limited the number of cocaine injections (2–4 injections) per 5–6 min drug period: this group is called IntA-Lim. In a second IntA group, we manipulated the session length which is known to be important for the expression of escalation when the schedule is continuous [49]. This new IntA group was named IntA-Short [34] and rats had access to cocaine intermittently for 2 h/day (Note that IntA-unLim and IntA-Long refer to the same 6-h IntA group where cocaine is freely available during the 5–6 min drug periods).

- IntA-Lim rats show stable cocaine intake over sessions while IntA-unLim rats escalate their drug use over time (For representative results *see* Fig. 3c; Adapted from [51]). IntA-Lim rats can also be compared with the more traditional LgA rats where, as expected, LgA rats escalate their cocaine use over time while IntA-Lim rats maintain low and stable levels cocaine intake over sessions (For representative results *see* Fig. 3d; Adapted from [51]). Comparisons between these procedures is helpful to study the effects of different amounts and different temporal patterns of cocaine self-administration. IntA-short rats do not escalate their cocaine intake over time while IntA-Long rats do (For representative results *see* Fig. 3e; Adapted from [34]).

- As seen using LgA procedures, where female rats can self-administer more cocaine than male rats do [32, 63], IntA also promotes sex differences in the behavioral response to cocaine [32, 39, 80].

## 4    Different Cocaine Self-Administration Procedures and Effects on Cocaine Use in Rats

We have described different variables that can influence the development of both excessive and escalating cocaine self-administration in rats using different self-administration procedures. In the next section, we highlight the effects these procedures have on other, DSM-based behavioral features of cocaine addiction [21].

These features include the following.

- *A persistent desire or unsuccessful efforts to cut down or control stimulant use.*

- *A great deal of time is spent in activities necessary to obtain the stimulant, use the stimulant, or recover from its effects.*

- *Craving, or a strong desire or urge to use the stimulant.*

All three features can be interpreted as reflecting an excessive motivation to procure and consume cocaine. In rats, the motivation to self-administer cocaine can be measured under a progressive ratio schedule of drug reinforcement [44, 45, 81] or using a within-session threshold procedure which measures demand for cocaine [42, 82–85].

Table 1 presents a summary of how the different procedures we have described above (ShA, LgA, IntA and its variations) influence the amount of cocaine taken, escalation of intake over time, and incentive motivation to take the drug. The table also provides a summary of how these different procedures compare with respect to the amount of cocaine intake and the degree of incentive motivation for cocaine that they produce, respectively.

**Table 1**

**Relationships between the amount of cocaine intake, escalation of intake and incentive motivation to take cocaine across different self-administration procedures**

| Comparisons between procedures | Amount of cocaine intake | Escalation of cocaine intake | Motivation for cocaine |
|---|---|---|---|
| LgA versus ShA | LgA > ShA | LgA only | LgA > ShA |
| LgA (5 s) versus LgA (90 s) | LgA (5 s) > LgA (90 s) | LgA only | LgA (5 s) > LgA (90 s) |
| LgA versus IntA unLim | LgA > IntA-unLim | Both | IntA-unLim > LgA |
| LgA versus IntA Lim | LgA > IntA-Lim | LgA only | IntA-Lim > LgA |
| LgA (90 s) versus IntA Lim | LgA (90 s) > IntA-Lim | Neither | IntA-Lim > LgA (90 s) |
| IntA unLim versus IntA Lim | IntA unLim > IntA-Lim | IntA unLim only | IntA unLim = IntA-Lim |
| IntA unLim versus ShA | IntA unLim = ShA | _____ | IntA unLim > ShA |
| IntA Long versus IntA Short | IntA Long > IntA Short | IntA Long only | IntA Long = IntA Short |

This table illustrates that large amounts of cocaine intake and/or escalation of intake can be poor predictors of the incentive motivation to take cocaine. *See* text for details. *ShA* short access, where self-administration sessions last 1–2 h; *LgA* long access, where self-administration sessions last 6+ h; *5 s* where each intravenous cocaine infusion is delivered over 5 seconds; *90s* where each intravenous cocaine infusion is delivered over 90 seconds; *IntA-Unlim* where an unlimited number of infusions is available intermittently during sessions lasting 4–6 h; *IntA-Lim* where a limited number of infusions is available intermittently during sessions lasting 4–6 h. ">" and "=" indicate the effect a procedure produces on the variable of interest relative to the other procedure. This table includes all relevant comparisons that have been published in the literature to date. The shorter horizontal line indicates that the two procedures have not been compared in published work to date

**4.1  Short-Access Versus Long-Access**

- *Amount of cocaine intake/escalation of cocaine self-administration.* LgA rats self-administer more cocaine than ShA rats do, and LgA rats also escalate their drug use over sessions while ShA rats often maintain stable intake over sessions [49, 50, 52–58].

- *Motivation to self-administer cocaine.* LgA rats respond more for cocaine under a PR schedule than ShA rats do [56–58, 86, 87].

**4.2  Long-Access 5 s Versus Long-Access 90 s**

- *Amount of cocaine intake/escalation of cocaine self-administration.* During LgA sessions, rapid intravenous cocaine injections promote escalation of cocaine intake [53, 62].

- *Motivation to self-administer cocaine.* During LgA sessions, rapid intravenous cocaine injections also promote higher responding for the drug under a PR schedule [53, 61].

*In the following, we compare different procedures, where all used rapid intravenous cocaine infusions (1–5 s)—except for Sect. 4.5.*

**4.3  Intermittent-Access Versus Long-Access**

- *Amount of cocaine intake/escalation of cocaine self-administration.* LgA rats self-administer more cocaine than IntA do [32, 51, 75, 76]. Both IntA rats and LgA rats can escalate their drug use over sessions [34, 39, 51, 75, 77, 79].

- *Motivation to self-administer cocaine.* In spite of self-administering less cocaine than LgA rats do, IntA rats show more motivation to take cocaine, as measured using behavioral economics procedures [75].

**4.4  Intermittent-Access Limited Versus Long-Access**

- *Amount of cocaine intake/escalation of cocaine self-administration.* IntA-Lim rats have access to a limited number of cocaine infusions/session. LgA rats can self-administer eightfold more cocaine than IntA-Lim rats can [51]. LgA rats also escalate their drug use over sessions while IntA-Lim maintain low and stable levels of drug intake over time [51].

- *Motivation to self-administer cocaine.* IntA-Lim rats respond more for cocaine under a PR schedule [51].

**4.5  Intermittent-Access Limited Versus Long-Access (90 s)**

- *Amount of cocaine intake/escalation of cocaine self-administration.* In a recent study, we compared two self-administration procedures that combined variation in two pharmacokinetics variables: the speed of cocaine delivery and the pattern of drug intake. We compared LgA rats self-administering cocaine infusions delivered over 90 s to IntA-Lim rats self-administering cocaine infusions delivered over 5 s. Cocaine intake is highest in LgA-90s rats, though they do not escalate their intake over time [88].

- *Motivation to self-administer cocaine.* The IntA-Lim rats respond more for the drug under PR than LgA-90s rats do [88].

### 4.6 Intermittent-Access Limited Versus Intermittent-Access Unlimited

- *Amount of cocaine intake/escalation of cocaine self-administration.* IntA-Lim rats have intermittent access to a limited number of cocaine infusions per session. These rats generally self-administer all allotted infusions, but they cannot escalate their intake over time [51]. In contrast, IntA-unLim rats have intermittent access to an unlimited number of cocaine infusions, and these rats escalate their intake over time [51].

- *Motivation to self-administer cocaine.* In spite of having taken significantly less cocaine in the past, IntA-Lim rats show a similar level of incentive motivation for cocaine as IntA-unLim rats do, as measured by responding under PR [51].

### 4.7 Intermittent-Access Versus Short-Access

- *Amount of cocaine intake/escalation of cocaine self-administration.* IntA rats (with no limit on the number of infusions they can take) self-administer a similar amount of cocaine than ShA rats do [75, 76].

- *Motivation to self-administer cocaine.* IntA show more motivation to take cocaine, as measured using behavioral economics procedures [75].

### 4.8 Intermittent-Access Long Versus Intermittent-Access Short

- *Amount of cocaine intake/escalation of cocaine self-administration.* IntA-Long rats (who have 6-h sessions) self-administer more cocaine than IntA-Short rats do (2-h sessions). IntA-Long rats also escalate their cocaine use over time while IntA-Short rats do not [34].

- *Motivation to self-administer cocaine.* IntA-Long rats and IntA-short rats respond similarly for cocaine under a PR schedule [34]. This suggests that under IntA, session length is not a critical factor to promote the development of an excessive motivation for cocaine.

### 4.9 Role of Excessive/Escalating Cocaine Self-Administration in the Development of Other Behavioral Features of Cocaine Addiction

Based largely on data from laboratory rats, it is widely believed that excessive and escalating drug intake is uniquely sufficient to promote a drug-addiction phenotype [22, 89, 90]. This idea is based in great part on studies comparing ShA and LgA procedures, and showing that LgA promotes both excessive/escalating cocaine self-administration and increased incentive motivation to take the drug. Indeed, when cocaine access is continuous within sessions, giving rats longer self-administration sessions (LgA vs. ShA) allows them to take more drug and can also produce escalation, and this excessive/escalating intake is thought to be both necessary and uniquely effective in producing other behavioral features of addiction. As we have reviewed above, during LgA sessions, increasing the speed of i.v. cocaine delivery (from delivering cocaine over 90 s to 5 s) also promotes both escalation and increased motivation to obtain cocaine (ShA versus LgA, and LgA-5s versus LgA-90s, Table 1).

However, when rats are given intermittent, rather than continuous cocaine access during each self-administration session, excessive/escalating cocaine self-administration and incentive motivation for the drug are dissociable. Specifically, increased incentive motivation for cocaine develops even in the absence of excessive/escalating cocaine intake (IntA versus LgA, IntA-Lim versus LgA, IntA-Short versus IntA-Long, Table 1). Moreover, limiting cocaine intake during IntA self-administration does not prevent IntA experience from producing enhanced incentive motivation for the drug (IntA-Lim, IntA-Short, Table 1).

Altogether, these studies highlight the importance of assessing more than one behavioral feature of addiction in preclinical models. Excessive and escalating cocaine intake can sometimes predict the development of other addiction-relevant signs, but there is no reliable relationship.

## 5 Notes

1. In some cases, a few rats do not reliably acquire food self-administration task. This is a rare occurrence. If this happens, rats can be put back into the operant chambers for an overnight food self-administration session.

2. Female rats have thinner skin than male rats do. The skin around the catheter in females can take more time to heal. Because of this, it is especially important to make external sutures around the catheter in females to completely close the incision around the catheter port. Also, we have observed that females tend to scratch themselves more intensely and frequently than males do. As such, we recommend cutting female rats' claws every 2 weeks during experimentation to prevent damages to the catheter.

3. Females need more recovery, because their skin is thinner and requires more time to heal. For studies comparing males and females directly, all rats can be given 10–14 days of recovery.

4. During LgA sessions, while rats of both sexes should be monitored, it is especially important to frequently monitor female rats to assess (1) the need to replace cocaine-containing syringes during the 6-h session and (2) the potential damages to the tether system that can be caused by hyperlocomotion induced by excessive/escalating cocaine self-administration in females.

## 6 Conclusions

In this chapter, we briefly outlined the discovery of voluntary drug self-administration in laboratory animals, both primates and rodents. We then described the surgical and behavioral training and testing techniques needed to implement intravenous drug self-administration studies in the rat. The techniques are fit for use in both female and male rats, and we have noted procedural differences according to biological sex of the subjects where appropriate. Investigating the effects of voluntary cocaine self-administration in animals of both sexes is critical, because the response to drugs can be different in females versus males. For example, women report taking more cocaine [91], and they can be more vulnerable to relapse after abstinence [92, 93]. Women also progress faster from initial cocaine use to entering treatment [91, 94], and they enter treatment at a younger age [95]. We have also described two DSM-based features of drug addiction that can be modeled in the rat; excessive/escalating cocaine intake and motivation for the drug. We have discussed how these two features can be dissociable, with some cocaine self-administration procedures enhancing intake, and others being particularly effective in enhancing incentive motivation for cocaine. Because cocaine use in humans is likely intermittent, IntA cocaine self-administration experience in rats offers a unique opportunity to study the neurobiological, psychological, and behavioral processes involved in addiction.

## References

1. Catania A, Laties V (1999) Pavlov and Skinner: two lives in science (an introduction to B. F. Skinner's "Some responses to the stimulus 'Pavlov'"). J Exp Anal Behav 72:455–461

2. Skinner BF (1938) The behavior of organisms. Appleton-Century-Crofts, New York

3. Spragg SDS (1940) Morphine addiction in chimpanzees, Comparative psychology monographs, vol 15. Johns Hopkins Press, Baltimore, pp 1–132

4. Weeks JR (1962) Experimental morphine addiction: method for automatic intravenous injections in unrestrained rats. Science 138:143–144

5. Davis JD (1966) A method for chronic intravenous infusion in freely moving rats. J Exp Anal Behav 9:385–387

6. Weeks JR, Davis JD (1964) Chronic intravenous cannulas for rats. J Appl Physiol 19:540–541

7. Thompson T, Schuster CR (1964) Morphine self-administration, food-reinforced, and avoidance behaviors in rhesus monkeys. Psychopharmacologia 5:87–94

8. Deneau G, Yanagita T, Seevers MH (1969) Self-administration of psychoactive substances by the monkey. Psychopharmacologia 16:30–48

9. Woods JH, Schuster CR (1968) Reinforcement properties of morphine, cocaine, and SPA as a function of unit dose. Int J Addict 3:231–237

10. Deneau G, Yanagita T, Seevers MH (1964) Psychogenic dependence to a variety of drugs in the monkey. Pharmacologia 6:182–182

11. Pickens R (1968) Self-administration of stimulants by rats. Int J Addict 3:215–221

12. Pickens R, Dougherty J, Thompson T (1969) Effects of volume and duration of infusion on cocaine reinforcement with concurrent activity recording. In: NAS-NRC (ed) Minutes of the

meeting of the committee on problems of drug dependence. NAS-NRC, Washington, DC, pp 5805–5811

13. Pickens R, Harris WC (1968) Self-administration of d-amphetamine by rats. Psychopharmacologia 12:158–163

14. Pickens R, Thompson T (1968) Cocaine-reinforced behavior in rats: effects of reinforcement magnitude and fixed-ratio size. J Pharmacol Exp Ther 161:122–129

15. Deneau GA, Inoki R (1967) Nicotine self-administration in monkeys. Ann N Y Acad Sci 142:277–279

16. Schuster CR, Thompson T (1969) Self administration of and behavioral dependence on drugs. Annu Rev Pharmacol 9:483–502

17. Roberts DC, Gabriele A, Zimmer BA (2013) Conflation of cocaine seeking and cocaine taking responses in IV self-administration experiments in rats: methodological and interpretational considerations. Neurosci Biobehav Rev 37:2026–2036

18. UNODC. United Nations Office on Drugs and Crime (2017) World drug report 2017

19. Anthony JC, Warner LA, Kessler RC (1994) Comparative epidemiology of dependence on tobacco, alcohol, controlled substances, and inhalants: basic findings from the national comorbidity survey. Exp Clin Psychopharmacol 2:244–268

20. Spanagel R (2017) Animal models of addiction. Dialogues Clin Neurosci 19:247–258

21. APA. DSM V (2013) Diagnostic and statistical manual of mental disorders. American Psychiatric Association

22. Ahmed SH (2012) The science of making drug-addicted animals. Neuroscience 211:107–125

23. Kawa AB, Allain F, Robinson TE, Samaha AN (2019) The transition to cocaine addiction: the importance of pharmacokinetics for preclinical models. Psychopharmacology 236:1145–1157

24. Roberts DC, Morgan D, Liu Y (2007) How to make a rat addicted to cocaine. Prog Neuro-Psychopharmacol Biol Psychiatry 31:1614–1624

25. Duaux E, Krebs MO, Loo H, Poirier MF (2000) Genetic vulnerability to drug abuse. Eur Psychiatry 15:109–114

26. Goldman D, Oroszi G, Ducci F (2005) The genetics of addictions: uncovering the genes. Nat Rev Genet 6:521–532

27. Kreek MJ, Nielsen DA, Butelman ER, LaForge KS (2005) Genetic influences on impulsivity, risk taking, stress responsivity and vulnerability to drug abuse and addiction. Nat Neurosci 8:1450–1457

28. Sinha R (2008) Chronic stress, drug use, and vulnerability to addiction. Ann N Y Acad Sci 1141:105–130

29. Brady KT, Randall CL (1999) Gender differences in substance use disorders. Psychiatr Clin North Am 22:241–252

30. Roberts DC, Brebner K, Vincler M, Lynch WJ (2002) Patterns of cocaine self-administration in rats produced by various access conditions under a discrete trials procedure. Drug Alcohol Depend 67:291–299

31. Caprioli D, Celentano M, Paolone G et al (2008) Opposite environmental regulation of heroin and amphetamine self-administration in the rat. Psychopharmacology 198:395–404

32. Algallal H, Allain F, Ndiaye NA, Samaha AN (2020) Sex differences in cocaine self-administration behavior under long access versus intermittent access conditions. Addict Biol 25:e12809

33. Allain F, Roberts DC, Levesque D, Samaha AN (2017) Intermittent intake of rapid cocaine injections promotes robust psychomotor sensitization, increased incentive motivation for the drug and mGlu2/3 receptor dysregulation. Neuropharmacology 117:227–237

34. Allain F, Samaha AN (2018) Revisiting long-access versus short-access cocaine self-administration in rats: intermittent intake promotes addiction symptoms independent of session length. Addict Biol 24:641–651

35. Sizemore GM, Gaspard TM, Kim SA et al (1997) Dose-effect functions for cocaine self-administration: effects of schedule and dosing procedure. Pharmacol Biochem Behav 57:523–531

36. Lynch WJ, Taylor JR (2004) Sex differences in the behavioral effects of 24-h/day access to cocaine under a discrete trial procedure. Neuropsychopharmacology 29:943–951

37. Carroll ME, Morgan AD, Lynch WJ et al (2002) Intravenous cocaine and heroin self-administration in rats selectively bred for differential saccharin intake: phenotype and sex differences. Psychopharmacology 161:304–313

38. Hu M, Crombag HS, Robinson TE, Becker JB (2004) Biological basis of sex differences in the propensity to self-administer cocaine. Neuropsychopharmacology 29:81–85

39. Kawa AB, Robinson TE (2019) Sex differences in incentive-sensitization produced by intermittent access cocaine self-administration. Psychopharmacology 236:625–639

40. Spealman RD, Goldberg SR (1978) Drug self-administration by laboratory animals: control

by schedules of reinforcement. Annu Rev Pharmacol Toxicol 18:313–339

41. Allain F, Minogianis EA, Roberts DC, Samaha AN (2015) How fast and how often: the pharmacokinetics of drug use are decisive in addiction. Neurosci Biobehav Rev 56:166–179

42. Oleson EB, Roberts DC (2009) Behavioral economic assessment of price and cocaine consumption following self-administration histories that produce escalation of either final ratios or intake. Neuropsychopharmacology 34:796–804

43. Arnold JM, Roberts DC (1997) A critique of fixed and progressive ratio schedules used to examine the neural substrates of drug reinforcement. Pharmacol Biochem Behav 57:441–447

44. Hodos W (1961) Progressive ratio as a measure of reward strength. Science 134:943–944

45. Richardson NR, Roberts DC (1996) Progressive ratio schedules in drug self-administration studies in rats: a method to evaluate reinforcing efficacy. J Neurosci Methods 66:1–11

46. French ED, Lopez M, Peper S et al (1995) A comparison of the reinforcing efficacy of PCP, the PCP derivatives TCP and BTCP, and cocaine using a progressive ratio schedule in the rat. Behav Pharmacol 6:223–228

47. Morgan D, Liu Y, Roberts DC (2006) Rapid and persistent sensitization to the reinforcing effects of cocaine. Neuropsychopharmacology 31:121–128

48. Liu Y, Roberts DC, Morgan D (2005) Sensitization of the reinforcing effects of self-administered cocaine in rats: effects of dose and intravenous injection speed. Eur J Neurosci 22:195–200

49. Ahmed SH, Koob GF (1998) Transition from moderate to excessive drug intake: change in hedonic set point. Science 282:298–300

50. Ahmed SH, Koob GF (1999) Long-lasting increase in the set point for cocaine self-administration after escalation in rats. Psychopharmacology 146:303–312

51. Allain F, Bouayad-Gervais K, Samaha AN (2018) High and escalating levels of cocaine intake are dissociable from subsequent incentive motivation for the drug in rats. Psychopharmacology 235:317–328

52. Ferrario CR, Gorny G, Crombag HS et al (2005) Neural and behavioral plasticity associated with the transition from controlled to escalated cocaine use. Biol Psychiatry 58:751–759

53. Bouayad-Gervais K, Minogianis EA, Levesque D, Samaha AN (2014) The self-administration of rapidly delivered cocaine promotes increased motivation to take the drug: contributions of prior levels of operant responding and cocaine intake. Psychopharmacology 231:4241–4252

54. Mandt BH, Copenhagen LI, Zahniser NR, Allen RM (2015) Escalation of cocaine consumption in short and long access self-administration procedures. Drug Alcohol Depend 149:166–172

55. Aujla H, Martin-Fardon R, Weiss F (2008) Rats with extended access to cocaine exhibit increased stress reactivity and sensitivity to the anxiolytic-like effects of the mGluR 2/3 agonist LY379268 during abstinence. Neuropsychopharmacology 33:1818–1826

56. Wee S, Mandyam CD, Lekic DM, Koob GF (2008) Alpha 1-noradrenergic system role in increased motivation for cocaine intake in rats with prolonged access. Eur Neuropsychopharmacol 18:303–311

57. Orio L, Edwards S, George O et al (2009) A role for the endocannabinoid system in the increased motivation for cocaine in extended-access conditions. J Neurosci 29:4846–4857

58. Hao Y, Martin-Fardon R, Weiss F (2010) Behavioral and functional evidence of metabotropic glutamate receptor 2/3 and metabotropic glutamate receptor 5 dysregulation in cocaine-escalated rats: factor in the transition to dependence. Biol Psychiatry 68:240–248

59. Kippin TE, Fuchs RA, See RE (2006) Contributions of prolonged contingent and noncontingent cocaine exposure to enhanced reinstatement of cocaine seeking in rats. Psychopharmacology 187:60–67

60. Mantsch JR, Yuferov V, Mathieu-Kia AM et al (2004) Effects of extended access to high versus low cocaine doses on self-administration, cocaine-induced reinstatement and brain mRNA levels in rats. Psychopharmacology 175:26–36

61. Minogianis EA, Levesque D, Samaha AN (2013) The speed of cocaine delivery determines the subsequent motivation to self-administer the drug. Neuropsychopharmacology 38:2644–2656

62. Wakabayashi KT, Weiss MJ, Pickup KN, Robinson TE (2010) Rats markedly escalate their intake and show a persistent susceptibility to reinstatement only when cocaine is injected rapidly. J Neurosci 30:11346–11355

63. Roth ME, Carroll ME (2004) Sex differences in the escalation of intravenous cocaine intake following long- or short-access to cocaine self-administration. Pharmacol Biochem Behav 78:199–207

64. Hatsukami DK, Fischman MW (1996) Crack cocaine and cocaine hydrochloride. Are the differences myth or reality? JAMA 276:1580–1588

65. Chen CY, Anthony JC (2004) Epidemiological estimates of risk in the process of becoming dependent upon cocaine: cocaine hydrochloride powder versus crack cocaine. Psychopharmacology 172:78–86

66. Minogianis EA, Shams WM, Mabrouk OS et al (2019) Varying the rate of intravenous cocaine infusion influences the temporal dynamics of both drug and dopamine concentrations in the striatum. Eur J Neurosci 50:2054–2064

67. Samaha AN, Li Y, Robinson TE (2002) The rate of intravenous cocaine administration determines susceptibility to sensitization. J Neurosci 22:3244–3250

68. Balster RL, Schuster CR (1973) Fixed-interval schedule of cocaine reinforcement: effect of dose and infusion duration. J Exp Anal Behav 20:119–129

69. Crombag HS, Ferrario CR, Robinson TE (2008) The rate of intravenous cocaine or amphetamine delivery does not influence drug-taking and drug-seeking behavior in rats. Pharmacol Biochem Behav 90:797–804

70. Kato S, Wakasa Y, Yanagita T (1987) Relationship between minimum reinforcing doses and injection speed in cocaine and pentobarbital self-administration in crab-eating monkeys. Pharmacol Biochem Behav 28:407–410

71. Panlilio LV, Goldberg SR, Gilman JP et al (1998) Effects of delivery rate and non-contingent infusion of cocaine on cocaine self-administration in rhesus monkeys. Psychopharmacology 137:253–258

72. Schindler CW, Cogan ES, Thorndike EB, Panlilio LV (2011) Rapid delivery of cocaine facilitates acquisition of self-administration in rats: an effect masked by paired stimuli. Pharmacol Biochem Behav 99:301–306

73. Woolverton WL, Wang Z (2004) Relationship between injection duration, transporter occupancy and reinforcing strength of cocaine. Eur J Pharmacol 486:251–257

74. Beveridge TJR, Wray P, Brewer A et al (2012) Analyzing human cocaine use patterns to inform animal addiction model development. Published abstract for the College on problems of drug dependence annual meeting, Palm Springs, CA

75. Zimmer BA, Oleson EB, Roberts DC (2012) The motivation to self-administer is increased after a history of spiking brain levels of cocaine. Neuropsychopharmacology 37:1901–1910

76. Calipari ES, Ferris MJ, Zimmer BA et al (2013) Temporal pattern of cocaine intake determines tolerance vs sensitization of cocaine effects at the dopamine transporter. Neuropsychopharmacology 38:2385–2392

77. Kawa AB, Bentzley BS, Robinson TE (2016) Less is more: prolonged intermittent access cocaine self-administration produces incentive-sensitization and addiction-like behavior. Psychopharmacology 233:3587–3602

78. Zimmer BA, Dobrin CV, Roberts DC (2011) Brain-cocaine concentrations determine the dose self-administered by rats on a novel behaviorally dependent dosing schedule. Neuropsychopharmacology 36:2741–2749

79. Kawa AB, Valenta AC, Kennedy RT, Robinson TE (2019) Incentive and dopamine sensitization produced by intermittent but not long access cocaine self-administration. Eur J Neurosci 50:2663–2682

80. Nicolas C, Russell TI, Pierce AF et al (2019) Incubation of cocaine craving after intermittent-access self-administration: sex differences and estrous cycle. Biol Psychiatry 85:915–924

81. Depoortere RY, Li DH, Lane JD, Emmett-Oglesby MW (1993) Parameters of self-administration of cocaine in rats under a progressive-ratio schedule. Pharmacol Biochem Behav 45:539–548

82. Bentzley BS, Fender KM, Aston-Jones G (2013) The behavioral economics of drug self-administration: a review and new analytical approach for within-session procedures. Psychopharmacology 226:113–125

83. Bentzley BS, Jhou TC, Aston-Jones G (2014) Economic demand predicts addiction-like behavior and therapeutic efficacy of oxytocin in the rat. Proc Natl Acad Sci U S A 111:11822–11827

84. Hursh SR (1991) Behavioral economics of drug self-administration and drug abuse policy. J Exp Anal Behav 56:377–393

85. Oleson EB, Richardson JM, Roberts DC (2011) A novel IV cocaine self-administration procedure in rats: differential effects of dopamine, serotonin, and GABA drug pre-treatments on cocaine consumption and maximal price paid. Psychopharmacology 214:567–577

86. Paterson NE, Markou A (2003) Increased motivation for self-administered cocaine after escalated cocaine intake. Neuroreport 14:2229–2232

87. Wee S, Orio L, Ghirmai S et al (2009) Inhibition of kappa opioid receptors attenuated

increased cocaine intake in rats with extended access to cocaine. Psychopharmacology 205:565–575

88. Minogianis EA, Samaha AN (2020) Taking rapid and intermittent cocaine infusions enhances both incentive motivation for the drug and cocaine-induced gene regulation in corticostriatal                    regions. bioRxiv:2020.04.22.055715

89. Benowitz NL, Henningfield JE (1994) Establishing a nicotine threshold for addiction. The implications for tobacco regulation. N Engl J Med 331:123–125

90. Jonkman S, Pelloux Y, Everitt BJ (2012) Drug intake is sufficient, but conditioning is not necessary for the emergence of compulsive cocaine seeking after extended self-administration. Neuropsychopharmacology 37:1612–1619

91. Griffin ML, Weiss RD, Mirin SM, Lange U (1989) A comparison of male and female cocaine abusers. Arch Gen Psychiatry 46:122–126

92. Elman I, Karlsgodt KH, Gastfriend DR (2001) Gender differences in cocaine craving among non-treatment-seeking individuals with cocaine dependence. Am J Drug Alcohol Abuse 27:193–202

93. McKay JR, Rutherford MJ, Cacciola JS et al (1996) Gender differences in the relapse experiences of cocaine patients. J Nerv Ment Dis 184:616–622

94. McCance-Katz EF, Carroll KM, Rounsaville BJ (1999) Gender differences in treatment-seeking cocaine abusers—implications for treatment and prognosis. Am J Addict 8:300–311

95. Kosten TA, Gawin FH, Kosten TR, Rounsaville BJ (1993) Gender differences in cocaine use and treatment response. J Subst Abus Treat 10:63–66

# Chapter 7

# Preclinical Models of Relapse to Psychostimulants Induced by Environmental Stimuli

## Anna Maria Borruto, Ana Domi, Laura Soverchia, Esi Domi, Hongwu Li, and Nazzareno Cannella

## Abstract

A major aim of addiction research is the understanding of the pathophysiological profile of relapse risk and the development of treatments for relapse prevention. Exposure to drug-paired environmental stimuli elicits craving and increases the likelihood to relapse. Therefore, scholars in the addiction field have developed several preclinical models of cued relapse in order to study the biological and pharmacological background of this phenomenon. Here we provide an overview of the nowadays available models of cued relapse to psychostimulant seeking. We begin describing the models of relapse induced by drug-contingent and discriminative stimuli, and then we give an overview of the models of context-induced relapse. Finally, we illustrate the models of incubation of cue-induced psychostimulant craving. For each relapse model we provide technical details, a step by step protocol, and troubleshooting tips. The researcher interested in studying the contribution of environmental stimuli to relapse will find here the tools to choose the optimal method to answer their question, and technical details necessary to the methodological implementation of their research.

**Keywords** Addiction, Psychostimulant, Cue relapse, Environmental stimuli, Conditioned stimuli, Discriminative stimuli, Drug context, Incubation of craving, Reinstatement model, Drug seeking

## 1 Introduction

Psychostimulants use disorder, like other addictions, is associated with emotional disfunction and susceptibility to relapse [1–3]. During consumption, the incentive and emotional values of drugs are transferred to drug-related stimuli like paraphernalia and environmental contexts that by this association become able to evoke conditioned psychophysiological effects, triggering craving for the drug, and hence increasing the risk to relapse [4–7]. Thus, understanding the contribution of drug-associated cues to relapse at neurobiological and pharmacological level is a key step toward the

Anna Maria Borruto and Ana Domi contributed equally.

María A. Aguilar (ed.), *Methods for Preclinical Research in Addiction*, Neuromethods, vol. 174,
https://doi.org/10.1007/978-1-0716-1748-9_7, © Springer Science+Business Media, LLC, part of Springer Nature 2022

development of treatment aimed at relapse prevention. Therefore, several preclinical protocols of cued reinstatement of drug-seeking have been developed to help studying the neurobiology and pharmacology of cued relapse. In this chapter, we present an overview on the technical implementation of the nowadays available cued-relapse models of psychostimulant seeking in rodents. Although here we focus and give specific instructions on the implementation of cued relapse models of psychostimulant seeking, the principle and main guidelines described are applicable to each drug of abuse.

Cue-induced reinstatement of psychostimulant seeking is a phenomenon studied by using different models: the conditioned stimulus-induced (CS) reinstatement, discriminative stimulus (DS)-induced reinstatement, contextual cues induced reinstatement (ABA model), the ABA-renewal model, and the incubation of drug craving model [4, 8]. All these reinstatement protocols for psychostimulant drugs involve a surgical phase (in which rodents are implanted with an indwelling venous catheters) and three main experimental phases: (1) self-administration training, (2) extinction (not in the incubation model), and (3) the reinstatement test [9].

## 2    Animals and Material

### 2.1    Animals

Cue induced reinstatement can be studied in rat as well as in mice [10–13]. Both species are valuable means to pursue research goals. Mice are more suitable to develop genetically engineered models [14], while rats are more suitable to model complex behavioral phenomena, but recent advances are improving genome editing technologies in the rat as well [15]. Mice and rat models of cue induced reinstatement of psychostimulant seeking are methodologically similar, and the same principles and similar techniques are applied to both species. Here we will focus mostly, but not exclusively, on rat protocols. Those readers specifically interested in mice protocol can refer to previously published work [16–18].

Strain is one of the most influential factors in cue-induced reinstatement models. The strains of rats mostly used are the Sprague-Dawley, Wistar, and Long-Evans [19–21]. Sprague-Dawley and Wistar rats are both outbred albino strains while Long-Evans is an outbred hooded rat strain. It is well known that strain can interact with two other parameters, such as housing (single-housed vs. group-housed) and the use of food pretraining before starting psychostimulant self-administration [19, 22]. According to a meta-analysis [19], housing may interact with strain to affect reinstatement levels. Single housing enhances reinstatement levels in Wistar and Long-Evans rats, whereas it decreases reinstatement in Sprague–Dawley rats. Moreover, reinstatement levels are increased in Wistar rats exposed to a food training, whereas in Long-Evans and Sprague–Dawley rats reinstatement is negatively correlated with food training. According to this meta-analysis, "Wistar and Long-Evans rats may be generally more sensitive to environmental influences, while Sprague–Dawley rats appear to be more stable" [19].

Incubation of cue-induced cocaine craving is one of the most robust compared to other drugs of abuse [4, 23]. Age, housing conditions, and gender can influence the expression of cue-induced incubation of cocaine craving. For example, a study conducted by Li et al., showed that time-dependent increase in cue-induced reinstatement (incubation) was less pronounced in rats that took cocaine as adolescents compared with adults [24]. The effect of environmental enrichment instead is still unclear, as contrasting results have been reported [25, 26]. A limited number of studies have investigated sex differences in incubation of cocaine craving although a greater magnitude of incubation has been recorded in female Sprague-Dawley rats. Working with behavior in female rats, the hormonal cycle is an important variable to be paid attention for. Indeed, estrous females exhibit an enhanced association between cocaine effects and environmental cues [27] and they show exacerbated incubation of cocaine craving [28].

## 2.2 Equipment

To study cue-induced reinstatement of psychostimulant seeking, animals must be trained to intra-venous drug self-administration, and during self-administration training environmental stimuli must be presented in contingency with drug delivery and availability. The following tools are the minimum requirement:

1. Operant chambers enclosed in sound and light attenuating cubicles equipped with ventilation systems for air renewal. Operant chambers should include at least one operandum, *but the availability of a second operandum is strongly recommended* (*see* below about inactive responding). Rat response at the operandum is required to deliver the drug bolus. In classically designed apparatuses, there are at least two operanda. One operandum is defined as "active" and it is associated with drug delivery. The other is defined as inactive and response at the inactive operandum has no scheduled consequences. The inactive operandum is used to control for nonspecific behavior produced by the animal. An operandum can be a lever, a hole for nose poking or, more rarely in rodents, a touchscreen. The operant chamber itself represent a contextual cue and it is associated with drug availability. However, for cue-induced reinstatement studies, these chambers are equipped with cue-lights, house-lights, and tone generators to produce discrete and contextual cues associated with drug delivery and availability respectively (*see* below).

   Pictures representing standard self-administration chamber equipment are presented in Fig. 1.

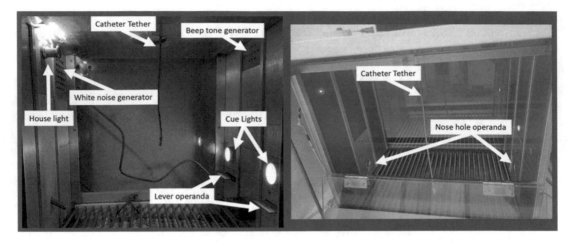

**Fig. 1** Representation of a standard self-administration chamber equipped with levers (right panel) and nose hole (left panel) operanda

2. A drug delivery system composed by an infusion pump, swivel, counterbalanced arm assembly and tether. A syringe containing the drug-solution is mounted on the infusion pump. The syringe is connected to the rat's catheter back-mount by a tube passing through the swivel and running inside the coil tether. The swivel allows the rats to move around freely while connected to the delivery system. The tether is necessary to protect the delivery tube from rat nibbling and gnawing, and to screw the delivery system to the catheter, avoiding disconnection. The size of the syringe is usually 5–20 ml, and it is chosen based on the intended bolus volume, pump speed, and infusion length. As an example, in our laboratory we have pumps running at 3.33 rpm speed, activating the pump for 5 s and using a 10 ml syringe we obtain an infusion volume of 0.1 ml.

**2.3 Intra-venous Catheter Implant and Maintenance**

An indwelling catheter is implanted in the animal jugular vein, usually the right one. The distal end of the catheter is inserted inside the jugular vein (generally 10 mm in mice and 40 mm in rats) and it usually reaches the right atrium. The tube exiting the vein runs subcutaneously over the shoulder and the intra-scapular region. The proximal end is connected to a bent metallic cannula embedded into a back-mount made of dental cement or plastic. Part of the cannula protrudes from the animal's back, exposing the catheter entrance to be connected to the drug-delivery system. This part of the cannula is usually surrounded by a threaded plastic ferrule that allows screwing the tether to the back-mount for a more stable and secure connection. Often, researchers isolate rodents to avoid gnawing on each other's plastic ferrule; however, to avoid social isolation, a metallic sleeve can be used to protect the plastic ferrules.

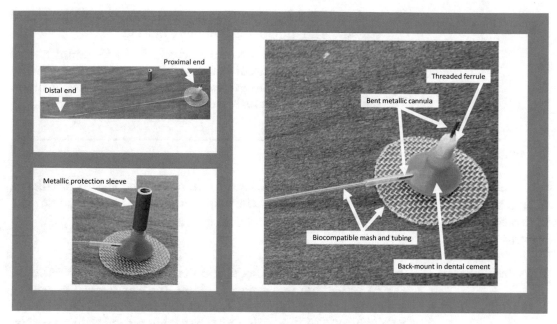

**Fig. 2** Representation of a custom-made intra-venous catheter

A custom-made intravenous catheter is represented in Fig. 2.

A heparinized solution is used to flush the catheter before and after drug self-administration training sessions, though some operator does it only once, either before or after the session. Catheters are flushed by 0.1 ml of solution. The heparin concentration varies according to the heparin formula used. For instance, sodium heparin is commonly used at a concentration ranging between 20 and 100 International Units (UI)/ml, whereas recently in our lab we switched to nadroparin at 5 UI/ml and found no difference in catheter patency maintenance.

**2.4**
**Troubleshooting Tips**

1. The catheter obviously represents an open way for germs to the circulation system and animals are therefore exposed to septicemia. To limit this risk, beside standard hygienic precautions, self-administration boxes should be cleaned daily and drug-delivery lines cleaned and filled with germ stabilizer when not in use (ask your veterinary for the best solution available in your facility).

2. Some researcher adds an antibiotic at low concentration to the heparin solution [29, 30] to help controlling the bacterial load. In addition, in our experience this helped to prolong the catheter patency, especially in mice [31, 32].

3. At least 3 days of recovery are generally left after surgery, however, whenever possible we suggest waiting at least 1 week. Beside the obviously better quality of recovery, we

noticed that animals with a longer recovery generally acquire self-administration better and more quickly.

4. There is no need to flush the catheters when self-administration sessions are not run, but when flushing pay attention to avoid solution spilling from the catheter before capping it, if blood remains inside the tube it will coagulate and clog the catheter.

# 3    Methods

## 3.1  Cue-Induced Reinstatement of Psychostimulant Seeking

The first step is to train the rats to self-administer the drug. This can be done with or without the help of a food pretraining phase.

### 3.1.1  Food Pretraining

In order to facilitate the acquisition of active operandum responding to self-administer a reward, some student begin by training rodents to respond for a food reinforcer (often a 45 mg pellet of palatable food) during a single 16-h overnight food training session [33–35] or during several daily sessions until a stable active responding is achieved [36–40]. When adoption of a food pretraining phase is desired or necessary, beside the devices described above, the operant chambers must be equipped with a food pellet dispenser, usually located between the two operanda. Food-training sessions are normally conducted under fixed ratio-1(FR1) contingency; meaning that each response at the active operandum results in the delivery of one pellet of food. Similarly, to experimental sessions described later, response at the inactive operandum should have no programmed consequences. It is important that the conditioned stimuli (CS) that will be associated with the psychostimulant are never presented during the food training sessions in order to avoid ambiguous cueing that could affect the interpretability of the reinstatement tests.

Food pretraining is necessarily associated with fasting. Animals are normally maintained at 90% of their free-feeding body-weight [41] during food self-administration training because this increases their exploratory behavior, and therefore the chance to respond to the operandum, thus facilitating the association between the responding at the operandum and pellet delivery. This clearly speeds up the acquisition of self-administration, but it is also perceived by many as a criticism to the use of food pretraining because of the obvious implication of the stress and motivation systems. It is important to clarify that food-pretraining is not a mandatory step. Indeed, often rats are trained directly to psychostimulant self-administration and they are perfectly able to acquire operandum responding whether lever presses or nose pokes are used as operant behavior [29, 42–48]. However, also those students who prefer not to use food-pretraining can find it useful when training is started in aged animals or when they are working with strains or in conditions in which the animal expresses little exploratory behavior or little capacity to make associations. Indeed, exploratory behavior and events association are key elements to guarantee a probabilistic responding at an operandum and a prompt association with the reward delivery respectively. Without these two elements no animal can learn how to self-administer any reinforcer.

*3.1.2 Protocol of CS-Induced Reinstatement*

CS-induced reinstatement protocols are intended to model the contribution of environmental stimuli explicitly paired with drug consumption to relapse in order to study the neurobiology of such association. Animals are trained to an instrumental response for drug reinforcement which is explicitly paired with the presentation of a discrete stimulus (i.e., light, tone, or a combination of tone and light). Association of multiple cues (i.e., light plus tone) with the drug reinforcement was reported to be more effective than a singular cue (light or tone alone) in inducing the reinstatement of drug seeking [49]. These CS repeatedly associated to drug reinforcement acquire an incentive value and can induce reinstatement of drug seeking [50–55]. After self-administration training, during which drug reinforcements are paired with CS, animals are subjected to extinction training, in which active operandum responding no longer results in drug delivery and CS presentations, causing a drop in seeking. Reinstatement test begins the day after the last extinction session, under conditions identical to self-administration, except that the drug is not delivered. Thus, responses at the active operandum is reinstated by the presentation of drug-paired CS in absence of the drug itself. The availability of an inactive operandum plays a crucial role in the interpretation of the results, as this operandum functions as negative control. Indeed, reinstatement test should not increase responses at the inactive operandum, otherwise the increase observed at the active operandum cannot be considered a reinstatement of drug-seeking but rather a generalized increase in activity that is reflected in unspecific increase in operandum responding [56]. Consistently, treatment effect on the inactive operandum could indicate an unspecific effect. This being said, also when there is a reinstatement effect on the inactive operandum, this is usually negligible compared to the active operandum responses that include both goal-directed and generalized activity. Therefore, in some case a change score analysis has been adopted by subtracting the number of inactive responses to the number of active responses to obtain goal-directed responses [57, 58].

The evaluation of the inactive operandum responses is a crucial point on data interpretation that applies to all procedures described in this chapter.

**Schematic of a general  cue-induced reinstatement protocol**

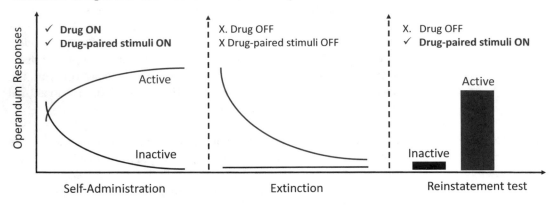

**Fig. 3** Didactic illustration of a general cue-induced reinstatement protocol. During self-administration training response at the drug and cue reinforced active operandum (red line) increases over time while responses at the non-reinforced inactive operandum decreases (black line). During extinction, operanda responses are reinforced by neither the drug nor the drug-paired stimuli and therefore responding at the active operandum approaches inactive response level, i.e. it extinguishes. On reinstatement test, re-exposure to drug-paired stimuli, in the absence of the drug, reinstates drug-seeking, i.e. it increases responding at the active operandum (red bar) while inactive operandum responses remain at extinction level (black bar)

A schematic of a general cue-induced reinstatement protocol is represented in Fig. 3.

**Self-Administration Training Protocol**

Self-administration training sessions should be run on a daily base, until a stable baseline and clear active/inactive operandum response discrimination is achieved. A week-end break in training is a common practice, but it can be avoided if necessary. Rats tend to have a high response to psychostimulants, and therefore a time-out period of 20–40 s is usually scheduled after each infusion to limit overdose risk. During timeout, responses at either operanda have no programmed consequences. For the same reasons, also a maximum number of infusions allowed is sometime used. This maximum number of infusions allowed varies between papers and it commonly ranges between 10 and 35. Training usually begin in FR1 contingency, but with psychostimulants reinforcement schedule is usually increased to FR3 or FR5, as these drugs induce a high level of responding.

A standard self-administration training session lasts for 2 h, but 1, 3, or 6 h sessions are also commonly used.

There are several cueing schedules that are currently used in association with drug delivery and a detailed description of each single cueing schedule would hardly be comprehensive. However, general guidelines can be drawn [59–64]. Classically, drug delivery is associated with a discrete cue that is normally composed by a cue-light, often associated with a beep-tone, activated contingently with drug-infusion. Beside the discrete cues, discriminative/

contextual cues are sometime used and activated throughout the session length to signal drug availability. These can be a house-light enlightened throughout the session or an intermittent beep-tone or white noise.

**Self-Administration Protocol Steps**

1. Load drug syringes on the infusion pumps.

2. Check the proper functioning of the infusion pumps, operanda, stimulus lights (house and cue lights) and tones.

3. Select the desired self-administration program.

4. Move animals from the vivarium to the self-administration room and start connecting the animals to the infusion system. Before connecting the rats, it is important to activate the infusion pumps manually and let solution dropping until all air bubbles are removed from the infusion tubes. To connect the animal, take each animal individually and remove the cap from the catheter pedestal. Flush the catheter with 0.1 ml of heparinized solution to check patency. Connect the catheter pedestal to the catheter connection/infusion tether. It is important to make sure that the connection is tight, otherwise, rodents may disconnect and data cannot be used for analyses.

5. Start the self-administration training session.

6. At the end of the self-administration training session, take out the rodents from the operant chambers. Flush the catheter with 0.1 ml of heparinized solution to flush the drug solution from the catheter dead volume, and protect the catheter pedestal with the cap.

7. Return the animals to the vivarium.

8. Record data and analyze training progression on a daily base. Once training criteria are met, rodents can enter the extinction phase. Commonly, self-administration training is completed when a plateau is reached and there is $\pm 10\%$ variability in number of reinforcement earned for at least three consecutive self-administration sessions [39, 59, 60, 65–68] or, depending on the session length and compound dose, it is continued until each animal obtain the self-administration criterion of $N$ sessions (often 10) with at least $X$ infusions (often 10) per session with 75% discrimination between the active and inactive operanda maintained for more than 3 consecutive days [61, 69–73].

*Extinction Training*

Extinction training begin after a stable acquisition of self-administration is acquired. During extinction, operand responses are no longer reinforced neither with the drug nor with CS. This yield to a drop in responding that will be the necessary condition to observe reinstatement of drug seeking.

Extinction sessions usually have the same duration of self-administration sessions or they are shortened to 1 h. A widely used extinction criterion is the mean number of active operandum responses remaining below 15–20% of baseline self-administration level for at least 3 consecutive days [39, 69, 72, 74, 75]. Another commonly used criterion is no more than 10–15 responses per hour for three consecutive sessions [60, 61].

*Extinction protocol steps*

1. Repeat **steps 2–4** of self-administration protocol described above, with the difference that the program for the extinction phase differs from that used in the self-administration training as there is no drug delivery and cue presentation.

2. Rats can be tethered but there is no need to connect the infusion tube as no solution is infused.

3. Start the extinction session.

4. At the end of the extinction session, return the animal to the vivarium.

5. Record data and analyze them on a daily base to monitor the achievement of extinction criteria.

**Reinstatement Test**

Once extinction criteria are met, the reinstatement test can be run. On reinstatement test animals are exposed to the same condition of self-administration sessions, except that the drug is not delivered. Reexposure to CS induces a statistically significant increase in active operandum responding, which is taken as a measure of drug seeking induced by CS.

As discussed above, it is important to observe the responses at the inactive operandum, that should remain at extinction level if the effect of CS are specific in inducing drug-seeking.

*Reinstatement protocol steps*

1. Repeat **steps 2–4** of self-administration protocol described above, with the difference that no syringe is loaded on the infusion pump.

2. Load the appropriate operant conditioning software program (the same used for self-administration.

3. Rats can be tethered but there is no need to connect the infusion tube as no solution is infused.

4. Start the reinstatement session.

5. At the end of the session, return the animal to the vivarium.

6. Record data and analyze them.

**CS-Induced Reinstatement Troubleshooting Tips**

1. In healthy animals, bodyweight tend to increase at young age and plateau or slightly increase in adult. Monitor bodyweight

on a weekly base as a proxy of health status. Ask your facility veterinary for immediate advice if a drop is observed.

2. During self-administration it is important to monitor the trend of infusion and operandum responding at individual level. A sudden increase in inactive operandum responding or decrease in infusion might indicate that the catheter is no longer patent. To verify this, several steps can be taken: first, check that the catheter is not clogged; second, look if there is a leakage of solution from the back-mount area (in this case sometime the tether can be wet after the session); third, try to withdraw blood from the catheter (keep in mind that while being able to withdraw blood is a sign of catheter patency, not being able to do it does not mean the catheter is not patent); fourth, if all these tests are negative and doubt on catheter patency still persists, test it by the injection of a fast acting barbiturate [45, 76], immediate loss of reflexes is a sign of catheter patency.

3. Psychostimulants are characterized by a stable inter-individual within-session pattern of intake [77–79]. This variable is a helpful proxy of self-administration stability and a loss of stability could indicate possible loss of catheter patency or anticipate health problems.

4. It is advisable to run extinction sessions continuously, that is, with no week-end breaks. This because a break of even 1 day could induce an increase in responding when the animal is returned to the operant apparatus, the so-called Monday effect, delaying the achievement of extinction criteria.

*3.1.3  Protocol of Drug Discrimination and Reinstatement of Seeking Induced by Discriminative Stimuli*

The CS-induced reinstatement of drug-seeking model can be enriched by the implementation of a discriminative phase. In this phase rats are trained to discriminate between CS predictive of the drug of abuse (S$^+$), and CS predictive of a nonreinforcing solution (S$^-$). The experimental procedure consists of three different phases: discrimination training, extinction and reinstatement. The discrimination phase is similar to the self-administration training described above but it includes two different kind sessions: (1) in the reinforced session, rodents are trained to self-administer the psychostimulant drug in the presence of a discriminative stimulus predictive of drug availability (S$^+$) (e.g., a specific odor, tone or cue light); (2) in the nonreinforced sessions, rats are subjected to saline self-administration, and saline infusions are associated with a different discriminative stimulus (S$^-$, drug nonavailability) [8, 37, 80, 81]. After the discrimination phase, animals undergo extinction sessions in the absence of discriminative stimuli. On the reinstatement test phase, drug seeking is assessed in the presence of the S$^+$ and the S$^-$ using a repeated testing design. A specific reinstatement of active operand responding induced by S$^+$ but not S$^-$ indicates that stimuli specifically associated with the drug of abuse, and not

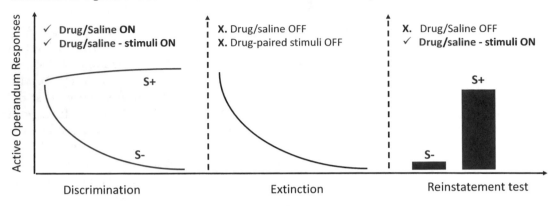

**Fig. 4** Didactic illustration of a general discriminative cue-induced reinstatement protocol. Subjects are initially trained to cocaine self-administration in the presence of environmental stimuli (S+) predictive of drug availability and associated with drug delivery (not represented in the picture). Then, during discrimination phase drug self-administration sessions in S+ environmental conditions (red line) are alternated with self-administration of a non-reinforcing solution (blue line) in a different environmental setup (S−). This causes a decrease in active operandum responding in S− sessions, i.e. a discrimination between drug-reinforced and non-reinforced sessions. During extinction, operanda responses are reinforced by neither the solutions delivery nor S+/S− stimuli and therefore responding at the active operandum extinguishes (black line). On reinstatement test, re-exposure to S− condition should not increase active operandum responding compared to extinction (blue bar) whereas exposure to S+ stimuli reinstates drug-seeking (red bar). Inactive operandum pattern of responding is not represented, but it follows the same trend described in Fig. 3

others, reinstates drug seeking [37, 38, 81]. In comparison with the more classical CS protocol described above, the added value of the discriminative stimuli protocol is the addition of nonreinforcer stimulus control that further strengths the interpretation of cue-induced reinstatement of drug seeking. This allow the experimenter to disentangle the biological adaptation consequential to drug-related stimuli from those associated with experimental handling.

A schematic of a general discriminative cue-induced reinstatement protocol is represented in Fig. 4.

### Training Animals to Self-Administer and Discriminate Drugs over Saline

The self-administration training can follow different protocols however, there is always an initial training phase where rats are subjected only to drug self-administration. The training phase is maintained until rats acquire a stable self-administration. The training phase is followed by a discrimination phase, where drug and saline self-administration sessions are alternated in pseudorandom order. This phase is maintained until rats discriminate between drug and saline sessions, that is, saline responding significantly drops while drug responding is maintained stably high. The key point is that drug and saline infusions are predicted by different discriminative stimuli and associated with different discrete CS. For instance, in the seminal works published by Weiss laboratory [37, 38, 81] they used a white noise active during the drug session as a discriminative stimulus for cocaine availability and the infusion-contingent illumination of the cue light above the active operand as discrete stimulus. In saline sessions they used the illumination of the house-light during the session as a discriminative stimulus for nonreinforcing solution availability and the presentation of a tone cue (7 kHz, 70 dB) as discrete stimulus associated to saline. During the discrimination phase, they run three 1-h sessions a day, 40–60 min apart from each other, in which only one solution and associated cues are available in pseudorandom order.

Self-administration is maintained until rats stabilize responding for the drug of abuse ($\pm$10% for at least 3 consecutive days) and responding for saline is stably low ($\leq$5 responses for at least three successive sessions) [36–38, 81].

*Discrimination protocol steps:*

1. The methods to run self-administration are identical to those described in Subheading 3.1.2 except that the solutions self-administered are alternatively the psychostimulant and the saline solution. Saline and psychostimulant solutions are associated with clearly different environmental stimuli.

Extinction.

As described in Subheading 3.1.2.

Reinstatement Test

The reinstatement test is similar to the self-administration training and to as described in Subheading 3.1.2, except that the first day after the last extinction session, rats are exposed to $S^-$ condition, that is, to the same setup of saline self-administration of the discrimination phase. The day after, rats are exposed to $S^+$ condition, that is, to the same setup of drug self-administration of the discrimination phase. If rats discriminated well between saline and drug associated cues, $S^+$ but not $S^-$ should reinstate active operandum responding which indicates that cues have been discriminated and reinstatement is given exclusively by drug-paired stimuli [37, 38, 63, 64, 81]. A further advantage of this protocol is that experimental challenges can be tested also against $S^-$ induced responding giving a measure of the specificity of treatment effects [82].

The same troubleshooting tips described above apply here.

3.1.4 Protocol of Drug-Context Induced Reinstatement

In the classical ABA model (also known as context-induced reinstatement), rodents are trained to self-administer drugs in one context (Context A), in which explicit CS presentations are not programmed to occur but it is the context itself in which the animal experience the drug that functions as drug-paired cue. After training, animals are subjected to extinction in a different context (Context B). On reinstatement test day, the subjects are reexposed to the previously drug-paired context (Context A), which results in rein-statement of responding [34, 83, 84].

Technically speaking, self-administration training, extinction and reinstatement session are not different from what described in Subheading 3.1.2. Rodents are trained to self-administer cocaine in operant conditioning chambers (Context A) containing a specific set of visual, auditory, olfactory, and tactile contextual stimuli, but no discrete cues are contingently paired with drug delivery at any point during testing [85–89]. For extinction training, the chamber (Context B) contains a different set of visual, auditory, olfactory, and tactile contextual stimuli compared to the Context A [85–89]. Reinstatement test is made by reexposing the animal to context A in the absence of the drug. Normally, in this protocol a group of animals is maintained in context B on reinstatement day to control that the reinstatement observed in context A is not due to factors independent of conditioning [85–89].

The same troubleshooting tips described above apply here.

*3.1.5 ABA-Renewal Protocol*

The ABA-renewal model works with the same characteristics of context-induced reinstatement model, except that during self-administration training in the Context A, drug delivery is contingently paired with a discrete CS [90]. After self-administration training, conditioned response is extinguished in a distinctly different context (Context B) but in the presence of the previously drug-paired CS. On reinstatement test day, the subjects are reexposed to the previously drug-paired context (Context A), which results in reinstatement of responding under the CS presentation; this model is called ABA renewal [91, 92]. Beside the ABA-renewal model, two models of renewal have been established on the basis of the context in which extinction and reinstatement phases take place: ABC and AAB renewal [92–96]. In the ABC renewal model acquisition, extinction, and testing take place in three different contexts. In the AAB renewal model, acquisition and extinction occur in the same context, but testing occurs in a different one [97, 98]. The protocol for renewal models (ABA, ABC, AAB) follow the same steps described for contextual-induced reinstatement described above [83, 84, 99–105].

The same troubleshooting tips described above apply here.

*3.2 Incubation of Psychostimulant Craving*

Human addicts exposed to drug-associated cues experience drug craving [106, 107]. This craving becomes sensitized during the first few weeks of abstinence [108]. This sensitization of craving during abstinence has been named "incubation" [52]. Cue-induced reinstatement models have been used also to study incubation of craving during abstinence. Incubation of drug seeking in response to cocaine-paired cues in the rat was first described by Grimm and colleagues as a behavioral phenomenon in which rats experience a time-dependent increase in cue-elicited responding at the active operand as they progress into abstinence [52].

Classical procedures of cue-induced incubation of craving involves three phases: training, forced abstinence (withdrawal) and incubation test (cue-induced drug seeking). Following the initial training where rodents learn to self-administer cocaine, animals enter the abstinence phase in which they are left undisturbed in their home cages for a defined period. After the abstinence phase, rodents enter the test phase in which they are subjected to cue-reinforced responding. Reinstatement test can be performed either as cue-extinction test or as within session extinction/cue-reinstatement test (*see* below for detail) [46, 109–111].

### 3.2.1 Training Animals to Drug Self-Administration.

Training is technically identical to the CS-induced reinstatement protocols in Subheading 3.1.2. However, while in the protocols described above rats are usually trained in 1–2 h long sessions, for incubation protocol 6-h sessions are commonly, though not always, used [23, 52, 112–115]. Evidences indicate that the phenomenon of incubation of drug craving is more robust following an extended-access self-administration procedure (6 h) than after a limited period of time (1–2 h) [23]. Nevertheless, incubation of cocaine craving is reported to occur also after short access training [59, 113, 116]. Most of the studies adopted a training phase of 10 daily 6-h sessions but in some work rats were trained for at least 2 weeks [25, 52, 117–119]. However, Halbout and colleagues demonstrated that incubation could be observed also after a single self-administration session [10].

### 3.2.2 Abstinence Phase

After training rats are subjected to an abstinence phase, during which they remain in their home cage and are not exposed to self-administration apparatuses. The environmental conditions in which the abstinence phase is held influence incubation of craving. For instance, it has been demonstrated that social housing and environmental enrichment attenuates incubation of craving. However, if the enriched environment is discontinued the attenuation of craving disappears [25, 26]. Another manipulation affecting incubation of craving during the withdrawal period is motor training. Recent studies have demonstrated that aerobic exercise such as wheel training can attenuate incubation of cocaine-craving [117, 120–123].

### 3.2.3 Cue-Induced Incubation of Drug-Craving

After a period of withdrawal, incubation of craving induced by drug-paired cues is evaluated in animals by either cue-extinction or cue-induced reinstatement tests [46, 52, 111]. In order to give a measure of the increasing craving, tests are usually repeated 24 h after the last self-administration session, and at least a second time into withdrawal. The increase over time of cue-induced responding

**Schematic of a general incubation of cue-induced craving protocol**

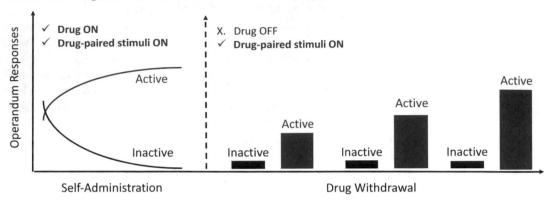

**Fig. 5** Didactic illustration of a general incubation of cue-induced craving protocol. During self-administration training response at the drug and cue reinforced active operandum (red line) increases over time while responses at the non-reinforced inactive operandum decreases (black line). During abstinence subjects are repeatedly tested for cue-induced drug-seeking. The increase of active (red bar) but not inactive (black bar) responding over time is used as a measure of incubation of drug craving

at the active operandum is considered a measure of increasing craving [23, 46, 52, 111].

- *Cue-extinction procedure*

  In cue-extinction procedure, rats are exposed for 1-h to the same conditions of self-administration except that the drug is not delivered, the ratio of seeking is given by the number of active operandum responses [52, 111].

- *Cue-induced reinstatement procedure*

  This test is run adopting a within-session extinction/reinstatement procedure [67]. On test day animals are subjected to several hours (6 to 8) of extinction in the absence of the CS previously associated with the drug until extinction criterion is met (less than 10–20 responses in the last hour, depending on the reference paper). Following extinction, the animal is subjected to cue-induced reinstatement for 1 h where response to the active operandum is reinforced by CS as in self-administration sessions except that the drug is not delivered. Reinstatement is measured as the difference in active operandum responses between the last hour of extinction and the cue-reinstatement part of the session [46, 52, 124].

  A schematic of a general incubation of cue-induced craving protocol is represented in Fig. 5.

*3.2.4 Incubation of Craving Troubleshooting Tips*

1. In the extinction/reinstatement test we suggest starting the reinstatement as soon as the extinction criterion is met and not waiting the 6 h to avoid over-extinction that could cause the rat not responding during reinstatement.

2. In the extinction/reinstatement test not all the animals necessarily extinguish their lever pressing in only 6 h. In this case allow these animals up to 2 additional hours of extinction. If they still do not meet the criteria exclude these animals.

# 4 Conclusions

Here we have revised the practical implementation of several protocols to study cue induced reinstatement of psychostimulant seeking. The choice of the method to use depends on the specific question the experimenter wants to answer. CS-induced reinstatement allows to study the contribution of discrete drug-contingent cues on relapse. The use of discriminative cues gives the possibility to disentangle the experience of the cue itself from the incentive value that the cue acquires following the association with drug reinforcement. The contextual reinstatement models allow to study the effect of drug-context and, in the "renewal" version, to disentangle the interaction between contextual and contingent cueing. Finally cue-induced reinstatement can be used in the incubation model to study increase of drug craving during abstinence.

# Acknowledgments

This work was supported by ERAB EA1840 to N.C.

# References

1. Koob GF, Volkow ND (2010) Neurocircuitry of addiction. Neuropsychopharmacology 35 (1):217–238. https://doi.org/10.1038/npp.2009.110

2. Koob GF, Volkow ND (2016) Neurobiology of addiction: a neurocircuitry analysis. Lancet Psychiatry 3(8):760–773. https://doi.org/10.1016/S2215-0366(16)00104-8

3. Everitt BJ, Robbins TW (2016) Drug addiction: updating actions to habits to compulsions ten years on. Annu Rev Psychol 67:23–50. https://doi.org/10.1146/annurev-psych-122414-033457

4. Pickens CL, Airavaara M, Theberge F et al (2011) Neurobiology of the incubation of drug craving. Trends Neurosci 34 (8):411–420. https://doi.org/10.1016/j.tins.2011.06.001

5. Courtney KE, Schacht JP, Hutchison K et al (2016) Neural substrates of cue reactivity: association with treatment outcomes and relapse. Addict Biol 21(1):3–22. https://doi.org/10.1111/adb.12314

6. Moeller SJ, Beebe-Wang N, Woicik PA et al (2013) Choice to view cocaine images predicts concurrent and prospective drug use in cocaine addiction. Drug Alcohol Depend 130 (1–3):178–185. https://doi.org/10.1016/j.drugalcdep.2012.11.001

7. Prisciandaro JJ, Myrick H, Henderson S et al (2013) Prospective associations between brain activation to cocaine and no-go cues and cocaine relapse. Drug Alcohol Depend 131(1–2):44–49. https://doi.org/10.1016/j.drugalcdep.2013.04.008

8. Fuchs RA, Lasseter HC, Ramirez DR et al (2008) Relapse to drug seeking following prolonged abstinence: the role of environmental stimuli. Drug Discov Today Dis Models 5(4):251–258. https://doi.org/10.1016/j.ddmod.2009.03.001

9. Beardsley PM, Shelton KL (2012) Prime-, stress-, and cue-induced reinstatement of

extinguished drug-reinforced responding in rats: cocaine as the prototypical drug of abuse. Curr Protoc Neurosci Chapter 9:Unit 9.39. https://doi.org/10.1002/0471142301.ns0939s61

10. Halbout B, Bernardi RE, Hansson AC et al (2014) Incubation of cocaine seeking following brief cocaine experience in mice is enhanced by mGluR1 blockade. J Neurosci 34(5):1781–1790. https://doi.org/10.1523/JNEUROSCI.1076-13.2014

11. Mead AN, Zamanillo D, Becker N et al (2007) AMPA-receptor GluR1 subunits are involved in the control over behavior by cocaine-paired cues. Neuropsychopharmacology 32(2):343–353. https://doi.org/10.1038/sj.npp.1301045

12. Orejarena MJ, Lanfumey L, Maldonado R et al (2011) Involvement of 5-HT2A receptors in MDMA reinforcement and cue-induced reinstatement of MDMA-seeking behaviour. Int J Neuropsychopharmacol 14(7):927–940. https://doi.org/10.1017/S1461145710001215

13. Georgiou P, Zanos P, Ehteramyan M et al (2015) Differential regulation of mGlu5 R and MuOPr by priming- and cue-induced reinstatement of cocaine-seeking behaviour in mice. Addict Biol 20(5):902–912. https://doi.org/10.1111/adb.12208

14. Fowler CD, Kenny PJ (2012) Utility of genetically modified mice for understanding the neurobiology of substance use disorders. Hum Genet 131(6):941–957. https://doi.org/10.1007/s00439-011-1129-z

15. Yoshimi K, Mashimo T (2018) Application of genome editing technologies in rats for human disease models. J Hum Genet 63 (2):115–123. https://doi.org/10.1038/s10038-017-0346-2

16. Yan Y, Nabeshima T (2009) Mouse model of relapse to the abuse of drugs: procedural considerations and characterizations. Behav Brain Res 196(1):1–10. https://doi.org/10.1016/j.bbr.2008.08.017

17. Caine SB, Negus SS, Mello NK (1999) Method for training operant responding and evaluating cocaine self-administration behavior in mutant mice. Psychopharmacology 147 (1):22–24. https://doi.org/10.1007/s002130051134

18. Rocha BA (1999) Methodology for analyzing the parallel between cocaine psychomotor stimulant and reinforcing effects in mice. Psychopharmacology 147(1):27–29. https://doi.org/10.1007/s002130051136

19. Oberhofer J, Noori HR (2019) Quantitative evaluation of cue-induced reinstatement model for evidence-based experimental optimization. Addict Biol 24(2):218–227. https://doi.org/10.1111/adb.12588

20. Kuhn BN, Kalivas PW, Bobadilla AC (2019) Understanding addiction using animal models. Front Behav Neurosci 13:262. https://doi.org/10.3389/fnbeh.2019.00262

21. Deiana S, Fattore L, Spano MS et al (2007) Strain and schedule-dependent differences in the acquisition, maintenance and extinction of intravenous cannabinoid self-administration in rats. Neuropharmacology 52(2):646–654. https://doi.org/10.1016/j.neuropharm.2006.09.007

22. Clemens KJ, Caille S, Cador M (2010) The effects of response operand and prior food training on intravenous nicotine self-administration in rats. Psychopharmacology 211(1):43–54. https://doi.org/10.1007/s00213-010-1866-z

23. Lu L, Grimm JW, Hope BT et al (2004) Incubation of cocaine craving after withdrawal: a review of preclinical data. Neuropharmacology 47(Suppl 1):214–226. https://doi.org/10.1016/j.neuropharm.2004.06.027

24. Li C, Frantz KJ (2009) Attenuated incubation of cocaine seeking in male rats trained to self-administer cocaine during periadolescence. Psychopharmacology 204(4):725–733. https://doi.org/10.1007/s00213-009-1502-y

25. Thiel KJ, Painter MR, Pentkowski NS et al (2012) Environmental enrichment counters cocaine abstinence-induced stress and brain reactivity to cocaine cues but fails to prevent the incubation effect. Addict Biol 17 (2):365–377. https://doi.org/10.1111/j.1369-1600.2011.00358.x

26. Chauvet C, Goldberg SR, Jaber M et al (2012) Effects of environmental enrichment on the incubation of cocaine craving. Neuropharmacology 63(4):635–641. https://doi.org/10.1016/j.neuropharm.2012.05.014

27. Johnson AR, Thibeault KC, Lopez AJ et al (2019) Cues play a critical role in estrous cycle-dependent enhancement of cocaine reinforcement. Neuropsychopharmacology 44(7):1189–1197. https://doi.org/10.1038/s41386-019-0320-0

28. Nicolas C, Russell TI, Pierce AF et al (2019) Incubation of cocaine craving after intermittent-access self-administration: sex differences and estrous cycle. Biol Psychiatry 85(11):915–924. https://doi.org/10.1016/j.biopsych.2019.01.015

29. Cannella N, Cosa-Linan A, Roscher M et al (2017) [18F]-Fluorodeoxyglucose-positron emission tomography in rats with prolonged cocaine self-administration suggests potential brain biomarkers for addictive behavior. Front Psychiatry 8:218. https://doi.org/10.3389/fpsyt.2017.00218

30. Mahler SV, Smith RJ, Aston-Jones G (2013) Interactions between VTA orexin and glutamate in cue-induced reinstatement of cocaine seeking in rats. Psychopharmacology 226 (4):687–698. https://doi.org/10.1007/s00213-012-2681-5

31. Bernardi RE, Olevska A, Morella I et al (2019) The inhibition of RasGRF2, but not RasGRF1, alters cocaine reward in mice. J Neurosci 39(32):6325–6338. https://doi.org/10.1523/JNEUROSCI.1120-18.2019

32. Bilbao A, Rieker C, Cannella N et al (2014) CREB activity in dopamine D1 receptor expressing neurons regulates cocaine-induced behavioral effects. Front Behav Neurosci 8:212. https://doi.org/10.3389/fnbeh.2014.00212

33. Xie X, Ramirez DR, Lasseter HC et al (2010) Effects of mGluR1 antagonism in the dorsal hippocampus on drug context-induced reinstatement of cocaine-seeking behavior in rats. Psychopharmacology 208(1):1–11. https://doi.org/10.1007/s00213-009-1700-7

34. Fuchs RA, Evans KA, Ledford CC et al (2005) The role of the dorsomedial prefrontal cortex, basolateral amygdala, and dorsal hippocampus in contextual reinstatement of cocaine seeking in rats. Neuropsychopharmacology 30(2):296–309. https://doi.org/10.1038/sj.npp.1300579

35. McFarland K, Lapish CC, Kalivas PW (2003) Prefrontal glutamate release into the core of the nucleus accumbens mediates cocaine-induced reinstatement of drug-seeking behavior. J Neurosci 23(8):3531–3537

36. Cervo L, Carnovali F, Stark JA et al (2003) Cocaine-seeking behavior in response to drug-associated stimuli in rats: involvement of D3 and D2 dopamine receptors. Neuropsychopharmacology 28(6):1150–1159. https://doi.org/10.1038/sj.npp.1300169

37. Ciccocioppo R, Martin-Fardon R, Weiss F (2004) Stimuli associated with a single cocaine experience elicit long-lasting cocaine-seeking. Nat Neurosci 7 (5):495–496. https://doi.org/10.1038/nn1219

38. Weiss F, Martin-Fardon R, Ciccocioppo R et al (2001) Enduring resistance to extinction of cocaine-seeking behavior induced by drug-related cues. Neuropsychopharmacology 25

(3):361–372. https://doi.org/10.1016/S0893-133X(01)00238-X

39. Backstrom P, Hyytia P (2006) Ionotropic and metabotropic glutamate receptor antagonism attenuates cue-induced cocaine seeking. Neuropsychopharmacology 31(4):778–786. https://doi.org/10.1038/sj.npp.1300845

40. Nugent AL, Anderson EM, Larson EB et al (2017) Incubation of cue-induced reinstatement of cocaine, but not sucrose, seeking in C57BL/6J mice. Pharmacol Biochem Behav 159:12–17. https://doi.org/10.1016/j.pbb.2017.06.017

41. Koob GF (2019) Introduction to addiction : addiction, animal models, and theories. Elsevier, San Deigo

42. Cannella N, Cosa-Linan A, Buchler E et al (2018) In vivo structural imaging in rats reveals neuroanatomical correlates of behavioral sub-dimensions of cocaine addiction. Addict Biol 23(1):182–195. https://doi.org/10.1111/adb.12500

43. Cannella N, Halbout B, Uhrig S et al (2013) The mGluR2/3 agonist LY379268 induced anti-reinstatement effects in rats exhibiting addiction-like behavior. Neuropsychopharmacology 38(10):2048–2056. https://doi.org/10.1038/npp.2013.106

44. Garcia-Rivas V, Fiancette JF, Cannella N et al (2019) Varenicline targets the reinforcing-enhancing effect of nicotine on its associated salient cue during nicotine self-administration in the rat. Front Behav Neurosci 13:159. https://doi.org/10.3389/fnbeh.2019.00159

45. Shen Q, Deng Y, Ciccocioppo R et al (2017) Cebranopadol, a mixed opioid agonist, reduces cocaine self-administration through nociceptin opioid and mu opioid receptors. Front Psych 8:234. https://doi.org/10.3389/fpsyt.2017.00234

46. Cannella N, Oliveira AMM, Hemstedt T et al (2018) Dnmt3a2 in the nucleus accumbens shell is required for reinstatement of cocaine seeking. J Neurosci 38(34):7516–7528. https://doi.org/10.1523/JNEUROSCI.0600-18.2018

47. Rubio FJ, Quintana-Feliciano R, Warren BL et al (2019) Prelimbic cortex is a common brain area activated during cue-induced reinstatement of cocaine and heroin seeking in a polydrug self-administration rat model. Eur J Neurosci 49(2):165–178. https://doi.org/10.1111/ejn.14203

48. Fattore L, Piras G, Corda MG et al (2009) The Roman high- and low-avoidance rat lines differ in the acquisition, maintenance,

extinction, and reinstatement of intravenous cocaine self-administration. Neuropsychopharmacology 34(5):1091–1101. https://doi.org/10.1038/npp.2008.43

49. See RE, Grimm JW, Kruzich PJ et al (1999) The importance of a compound stimulus in conditioned drug-seeking behavior following one week of extinction from self-administered cocaine in rats. Drug Alcohol Depend 57 (1):41–49. https://doi.org/10.1016/s0376-8716(99)00043-5

50. de Wit H, Stewart J (1981) Reinstatement of cocaine-reinforced responding in the rat. Psychopharmacology 75(2):134–143. https://doi.org/10.1007/bf00432175

51. Meil WM, See RE (1996) Conditioned cued recovery of responding following prolonged withdrawal from self-administered cocaine in rats: an animal model of relapse. Behav Pharmacol 7(8):754–763

52. Grimm JW, Hope BT, Wise RA et al (2001) Neuroadaptation. Incubation of cocaine craving after withdrawal. Nature 412 (6843):141–142. https://doi.org/10.1038/35084134

53. Fuchs RA, Tran-Nguyen LT, Specio SE et al (1998) Predictive validity of the extinction/reinstatement model of drug craving. Psychopharmacology 135(2):151–160. https://doi.org/10.1007/s002130050496

54. Weiss F (2010) Advances in animal models of relapse for addiction research. In: Kuhn CM, Koob GF (eds) Advances in the neuroscience of addiction. Frontiers in neuroscience. CRC Press, Boca Raton (FL)

55. See RE (2005) Neural substrates of cocaine-cue associations that trigger relapse. Eur J Pharmacol 526(1–3):140–146. https://doi.org/10.1016/j.ejphar.2005.09.034

56. Shalev U, Grimm JW, Shaham Y (2002) Neurobiology of relapse to heroin and cocaine seeking: a review. Pharmacol Rev 54 (1):1–42. https://doi.org/10.1124/pr.54.1.1

57. Le AD, Harding S, Juzytsch W et al (2005) Role of alpha-2 adrenoceptors in stress-induced reinstatement of alcohol seeking and alcohol self-administration in rats. Psychopharmacology 179(2):366–373. https://doi.org/10.1007/s00213-004-2036-y

58. Cippitelli A, Cannella N, Braconi S et al (2008) Increase of brain endocannabinoid anandamide levels by FAAH inhibition and alcohol abuse behaviours in the rat. Psychopharmacology 198(4):449–460. https://doi.org/10.1007/s00213-008-1104-0

59. Hollander JA, Carelli RM (2007) Cocaine-associated stimuli increase cocaine seeking and activate accumbens core neurons after abstinence. J Neurosci 27(13):3535–3539. https://doi.org/10.1523/JNEUROSCI.3667-06.2007

60. Manuszak M, Harding W, Gadhiya S et al (2018) (−)-Stepholidine reduces cue-induced reinstatement of cocaine seeking and cocaine self-administration in rats. Drug Alcohol Depend 189:49–54. https://doi.org/10.1016/j.drugalcdep.2018.04.030

61. Kruzich PJ (2007) Does response-contingent access to cocaine reinstate previously extinguished cocaine-seeking behavior in C57BL/6J mice? Brain Res 1149:165–171. https://doi.org/10.1016/j.brainres.2007.02.037

62. Saunders BT, Robinson TE (2010) A cocaine cue acts as an incentive stimulus in some but not others: implications for addiction. Biol Psychiatry 67(8):730–736. https://doi.org/10.1016/j.biopsych.2009.11.015

63. Kallupi M, de Guglielmo G, Cannella N et al (2013) Hypothalamic neuropeptide S receptor blockade decreases discriminative cue-induced reinstatement of cocaine seeking in the rat. Psychopharmacology 226 (2):347–355. https://doi.org/10.1007/s00213-012-2910-y

64. Kallupi M, Cannella N, Economidou D et al (2010) Neuropeptide S facilitates cue-induced relapse to cocaine seeking through activation of the hypothalamic hypocretin system. Proc Natl Acad Sci U S A 107 (45):19567–19572. https://doi.org/10.1073/pnas.1004100107

65. Deroche-Gamonet V, Martinez A, Le Moal M et al (2003) Relationships between individual sensitivity to CS- and cocaine-induced reinstatement in the rat. Psychopharmacology 168(1–2):201–207. https://doi.org/10.1007/s00213-002-1306-9

66. Perry CJ, Reed F, Zbukvic IC et al (2016) The metabotropic glutamate 5 receptor is necessary for extinction of cocaine-associated cues. Br J Pharmacol 173(6):1085–1094. https://doi.org/10.1111/bph.13437

67. Shaham Y, Shalev U, Lu L et al (2003) The reinstatement model of drug relapse: history, methodology and major findings. Psychopharmacology 168(1–2):3–20. https://doi.org/10.1007/s00213-002-1224-x

68. Knackstedt LA, Moussawi K, Lalumiere R et al (2010) Extinction training after cocaine self-administration induces glutamatergic plasticity to inhibit cocaine seeking. J

Neurosci 30(23):7984–7992. https://doi.org/10.1523/JNEUROSCI.1244-10.2010

69. Bongiovanni M, See RE (2008) A comparison of the effects of different operant training experiences and dietary restriction on the reinstatement of cocaine-seeking in rats. Pharmacol Biochem Behav 89(2):227–233. https://doi.org/10.1016/j.pbb.2007.12.019

70. Chesworth R, Brown RM, Kim JH et al (2013) The metabotropic glutamate 5 receptor modulates extinction and reinstatement of methamphetamine-seeking in mice. PLoS One 8(7):e68371. https://doi.org/10.1371/journal.pone.0068371

71. Yan Y, Nitta A, Mizoguchi H et al (2006) Relapse of methamphetamine-seeking behavior in C57BL/6J mice demonstrated by a reinstatement procedure involving intravenous self-administration. Behav Brain Res 168(1):137–143. https://doi.org/10.1016/j.bbr.2005.11.030

72. Arguello AA, Richardson BD, Hall JL et al (2017) Role of a lateral orbital frontal cortex-basolateral amygdala circuit in cue-induced cocaine-seeking behavior. Neuropsychopharmacology 42(3):727–735. https://doi.org/10.1038/npp.2016.157

73. Di Ciano P, Robbins TW, Everitt BJ (2008) Differential effects of nucleus accumbens core, shell, or dorsal striatal inactivations on the persistence, reacquisition, or reinstatement of responding for a drug-paired conditioned reinforcer. Neuropsychopharmacology 33(6):1413–1425. https://doi.org/10.1038/sj.npp.1301522

74. Rich MT, Huang YH, Torregrossa MM (2019) Plasticity at thalamo-amygdala synapses regulates cocaine-cue memory formation and extinction. Cell Rep 26(4):1010–1020.e1015. https://doi.org/10.1016/j.celrep.2018.12.105

75. Chen YY, Zhang LB, Li Y et al (2019) Post-retrieval extinction prevents reconsolidation of methamphetamine memory traces and subsequent reinstatement of methamphetamine seeking. Front Mol Neurosci 12:157. https://doi.org/10.3389/fnmol.2019.00157

76. Ahmed SH, Koob GF (1999) Long-lasting increase in the set point for cocaine self-administration after escalation in rats. Psychopharmacology 146(3):303–312. https://doi.org/10.1007/s002130051121

77. Algallal H, Allain F, Ndiaye NA et al (2019) Sex differences in cocaine self-administration behaviour under long access versus intermittent access conditions. Addict Biol 25:

e12809. https://doi.org/10.1111/adb.12809

78. Belin D, Balado E, Piazza PV et al (2009) Pattern of intake and drug craving predict the development of cocaine addiction-like behavior in rats. Biol Psychiatry 65(10):863–868. https://doi.org/10.1016/j.biopsych.2008.05.031

79. Allain F, Samaha AN (2019) Revisiting long-access versus short-access cocaine self-administration in rats: intermittent intake promotes addiction symptoms independent of session length. Addict Biol 24(4):641–651. https://doi.org/10.1111/adb.12629

80. Ciccocioppo R, Martin-Fardon R, Weiss F (2002) Effect of selective blockade of mu (1) or delta opioid receptors on reinstatement of alcohol-seeking behavior by drug-associated stimuli in rats. Neuropsychopharmacology 27(3):391–399. https://doi.org/10.1016/S0893-133X(02)00302-0

81. Weiss F, Maldonado-Vlaar CS, Parsons LH et al (2000) Control of cocaine-seeking behavior by drug-associated stimuli in rats: effects on recovery of extinguished operant-responding and extracellular dopamine levels in amygdala and nucleus accumbens. Proc Natl Acad Sci U S A 97(8):4321–4326. https://doi.org/10.1073/pnas.97.8.4321

82. Cannella N, Economidou D, Kallupi M et al (2009) Persistent increase of alcohol-seeking evoked by neuropeptide S: an effect mediated by the hypothalamic hypocretin system. Neuropsychopharmacology 34(9):2125–2134. https://doi.org/10.1038/npp.2009.37

83. Crombag HS, Bossert JM, Koya E et al (2008) Review. Context-induced relapse to drug seeking: a review. Philos Trans R Soc Lond Ser B Biol Sci 363(1507):3233–3243. https://doi.org/10.1098/rstb.2008.0090

84. Crombag HS, Grimm JW, Shaham Y (2002) Effect of dopamine receptor antagonists on renewal of cocaine seeking by reexposure to drug-associated contextual cues. Neuropsychopharmacology 27(6):1006–1015. https://doi.org/10.1016/S0893-133X(02)00356-1

85. Saunders BT, O'Donnell EG, Aurbach EL et al (2014) A cocaine context renews drug seeking preferentially in a subset of individuals. Neuropsychopharmacology 39(12):2816–2823. https://doi.org/10.1038/npp.2014.131

86. Bossert JM, Liu SY, Lu L et al (2004) A role of ventral tegmental area glutamate in contextual cue-induced relapse to heroin seeking. J

Neurosci 24(47):10726–10730. https://doi.org/10.1523/JNEUROSCI.3207-04.2004

87. Adhikary S, Caprioli D, Venniro M et al (2017) Incubation of extinction responding and cue-induced reinstatement, but not context- or drug priming-induced reinstatement, after withdrawal from methamphetamine. Addict Biol 22(4):977–990. https://doi.org/10.1111/adb.12386

88. Pelloux Y, Hoots JK, Cifani C et al (2018) Context-induced relapse to cocaine seeking after punishment-imposed abstinence is associated with activation of cortical and subcortical brain regions. Addict Biol 23(2):699–712. https://doi.org/10.1111/adb.12527

89. Fuchs RA, Branham RK, See RE (2006) Different neural substrates mediate cocaine seeking after abstinence versus extinction training: a critical role for the dorsolateral caudate-putamen. J Neurosci 26(13):3584–3588. https://doi.org/10.1523/JNEUROSCI.5146-05.2006

90. Bouton ME, Westbrook RF, Corcoran KA et al (2006) Contextual and temporal modulation of extinction: behavioral and biological mechanisms. Biol Psychiatry 60(4):352–360. https://doi.org/10.1016/j.biopsych.2005.12.015

91. Rosas JM, Bouton ME (1997) Additivity of the effects of retention interval and context change on latent inhibition: toward resolution of the context forgetting paradox. J Exp Psychol Anim Behav Process 23 (3):283–294. https://doi.org/10.1037//0097-7403.23.3.283

92. Bouton ME, King DA (1983) Contextual control of the extinction of conditioned fear: tests for the associative value of the context. J Exp Psychol Anim Behav Process 9 (3):248–265

93. Todd TP, Vurbic D, Bouton ME (2014) Mechanisms of renewal after the extinction of discriminated operant behavior. J Exp Psychol Anim Learn Cogn 40(3):355–368. https://doi.org/10.1037/xan0000021

94. Rescorla RA (2008) Within-subject renewal in sign tracking. Q J Exp Psychol (Hove) 61 (12):1793–1802. https://doi.org/10.1080/17470210701790099

95. Cuevas K, Learmonth AE, Rovee-Collier C (2016) A dissociation between recognition and reactivation: the renewal effect at 3 months of age. Dev Psychobiol 58 (2):159–175. https://doi.org/10.1002/dev.21357

96. Yap CS, Richardson R (2007) Extinction in the developing rat: an examination of renewal effects. Dev Psychobiol 49(6):565–575. https://doi.org/10.1002/dev.20244

97. Laborda MA, Witnauer JE, Miller RR (2011) Contrasting AAC and ABC renewal: the role of context associations. Learn Behav 39 (1):46–56. https://doi.org/10.3758/s13420-010-0007-1

98. Schmajuk NA, Larrauri JA, Labar KS (2007) Reinstatement of conditioned fear and the hippocampus: an attentional-associative model. Behav Brain Res 177(2):242–253. https://doi.org/10.1016/j.bbr.2006.11.026

99. Crombag HS, Shaham Y (2002) Renewal of drug seeking by contextual cues after prolonged extinction in rats. Behav Neurosci 116(1):169–173. https://doi.org/10.1037//0735-7044.116.1.169

100. Bouton ME (2002) Context, ambiguity, and unlearning: sources of relapse after behavioral extinction. Biol Psychiatry 52(10):976–986. https://doi.org/10.1016/s0006-3223(02)01546-9

101. Trask S, Thrailkill EA, Bouton ME (2017) Occasion setting, inhibition, and the contextual control of extinction in Pavlovian and instrumental (operant) learning. Behav Process 137:64–72. https://doi.org/10.1016/j.beproc.2016.10.003

102. Bossert JM, Marchant NJ, Calu DJ et al (2013) The reinstatement model of drug relapse: recent neurobiological findings, emerging research topics, and translational research. Psychopharmacology 229 (3):453–476. https://doi.org/10.1007/s00213-013-3120-y

103. Bouton ME, Winterbauer NE, Todd TP (2012) Relapse processes after the extinction of instrumental learning: renewal, resurgence, and reacquisition. Behav Process 90 (1):130–141. https://doi.org/10.1016/j.beproc.2012.03.004

104. Khoo SY, Gibson GD, Prasad AA et al (2017) How contexts promote and prevent relapse to drug seeking. Genes Brain Behav 16 (1):185–204. https://doi.org/10.1111/gbb.12328

105. Perry CJ, Zbukvic I, Kim JH et al (2014) Role of cues and contexts on drug-seeking behaviour. Br J Pharmacol 171(20):4636–4672. https://doi.org/10.1111/bph.12735

106. Sinha R, Fuse T, Aubin LR et al (2000) Psychological stress, drug-related cues and cocaine craving. Psychopharmacology 152 (2):140–148. https://doi.org/10.1007/s002130000499

107. Childress AR, Mozley PD, McElgin W et al (1999) Limbic activation during cue-induced cocaine craving. Am J Psychiatry 156 (1):11–18. https://doi.org/10.1176/ajp.156.1.11

108. Gawin FH, Kleber HD (1986) Abstinence symptomatology and psychiatric diagnosis in cocaine abusers. Clinical observations. Arch Gen Psychiatry 43(2):107–113. https://doi.org/10.1001/archpsyc.1986.01800020013003

109. Reichel CM, Bevins RA (2009) Forced abstinence model of relapse to study pharmacological treatments of substance use disorder. Curr Drug Abuse Rev 2(2):184–194. https://doi.org/10.2174/1874473710902020184

110. Venniro M, Caprioli D, Shaham Y (2016) Animal models of drug relapse and craving: from drug priming-induced reinstatement to incubation of craving after voluntary abstinence. Prog Brain Res 224:25–52. https://doi.org/10.1016/bs.pbr.2015.08.004

111. Luis C, Cannella N, Spanagel R et al (2017) Persistent strengthening of the prefrontal cortex - nucleus accumbens pathway during incubation of cocaine-seeking behavior. Neurobiol Learn Mem 138:281–290. https://doi.org/10.1016/j.nlm.2016.10.003

112. Ben-Shahar O, Sacramento AD, Miller BW et al (2013) Deficits in ventromedial prefrontal cortex group 1 metabotropic glutamate receptor function mediate resistance to extinction during protracted withdrawal from an extensive history of cocaine self-administration. J Neurosci 33(2):495–506a. https://doi.org/10.1523/JNEUROSCI.3710-12.2013

113. Lee BR, Ma YY, Huang YH et al (2013) Maturation of silent synapses in amygdala-accumbens projection contributes to incubation of cocaine craving. Nat Neurosci 16 (11):1644–1651. https://doi.org/10.1038/nn.3533

114. Lu L, Wang X, Wu P et al (2009) Role of ventral tegmental area glial cell line-derived neurotrophic factor in incubation of cocaine craving. Biol Psychiatry 66(2):137–145. https://doi.org/10.1016/j.biopsych.2009.02.009

115. Ma YY, Lee BR, Wang X et al (2014) Bidirectional modulation of incubation of cocaine craving by silent synapse-based remodeling of prefrontal cortex to accumbens projections. Neuron 83(6):1453–1467. https://doi.org/10.1016/j.neuron.2014.08.023

116. Suska A, Lee BR, Huang YH et al (2013) Selective presynaptic enhancement of the prefrontal cortex to nucleus accumbens pathway by cocaine. Proc Natl Acad Sci U S A 110 (2):713–718. https://doi.org/10.1073/pnas.1206287110

117. Zlebnik NE, Carroll ME (2015) Prevention of the incubation of cocaine seeking by aerobic exercise in female rats. Psychopharmacology 232(19):3507–3513. https://doi.org/10.1007/s00213-015-3999-6

118. Shin CB, Serchia MM, Shahin JR et al (2016) Incubation of cocaine-craving relates to glutamate over-flow within ventromedial prefrontal cortex. Neuropharmacology 102:103–110. https://doi.org/10.1016/j.neuropharm.2015.10.038

119. Li X, Caprioli D, Marchant NJ (2015) Recent updates on incubation of drug craving: a mini-review. Addict Biol 20(5):872–876. https://doi.org/10.1111/adb.12205

120. Zlebnik NE, Anker JJ, Gliddon LA et al (2010) Reduction of extinction and reinstatement of cocaine seeking by wheel running in female rats. Psychopharmacology 209 (1):113–125. https://doi.org/10.1007/s00213-010-1776-0

121. Smith MA, Pennock MM, Walker KL et al (2012) Access to a running wheel decreases cocaine-primed and cue-induced reinstatement in male and female rats. Drug Alcohol Depend 121(1–2):54–61. https://doi.org/10.1016/j.drugalcdep.2011.08.006

122. Peterson AB, Abel JM, Lynch WJ (2014) Dose-dependent effects of wheel running on cocaine-seeking and prefrontal cortex Bdnf exon IV expression in rats. Psychopharmacology 231(7):1305–1314. https://doi.org/10.1007/s00213-013-3321-4

123. Lynch WJ, Piehl KB, Acosta G et al (2010) Aerobic exercise attenuates reinstatement of cocaine-seeking behavior and associated neuroadaptations in the prefrontal cortex. Biol Psychiatry 68(8):774–777. https://doi.org/10.1016/j.biopsych.2010.06.022

124. Grimm JW, Lu L, Hayashi T et al (2003) Time-dependent increases in brain-derived neurotrophic factor protein levels within the mesolimbic dopamine system after withdrawal from cocaine: implications for incubation of cocaine craving. J Neurosci 23 (3):742–747

# Chapter 8

# Influence of Social Defeat Stress on the Rewarding Effects of Drugs of Abuse

María Pilar García-Pardo, José Enrique De la Rubia-Ortí, Claudia Calpe-López, M. Ángeles Martínez-Caballero, and María A. Aguilar

## Abstract

Drug addiction is a serious problem in our society. Some individuals develop dependence to different substances very accessible in the market, mainly young people. It is known that different biological and environmental variables facilitate the initiation, maintenance and relapse to drug use. In this sense, social stress is an important factor involved in the development of drug addiction and animal models are an optimal tool to study neurobiological systems associated with stress and addictive disorders. Among the main paradigms of social stress, the social defeat in an agonistic encounter with a conspecific male rodent has a notable ethological validity. Two main procedures, "acute" or "repeated" social defeat, may be distinguished, being the main differences between both the duration and intensity of the social defeat episodes and the evaluation of their short/long-term effects. Indeed, it has been demonstrated that the effects of both types of stress on the self-administration or conditioned place preference induced by different drugs are different. Although acute and repeated social defeat procedures have some limitations, in general terms both paradigms can help us to draw conclusions about the relationship between stress and drug addiction.

**Key words** Acute social defeat, Repeated social defeat, Reward, Drug, Stress, Animal models

## 1  Introduction

Drug addiction is a serious problem in our society. It is defined as a mental disorder according with diagnostic criteria on DSM-5 that involved emotional and behavioral aspects [1, 2]. Some people developed dependence to different drugs and even after long periods of abstinence relapse can appear. It is known that an important variety of addictive substances are available in the market

The original version of this chapter was revised. The correction to this chapter is available at https://doi.org/10.1007/978-1-0716-1748-9_14

María A. Aguilar (ed.), *Methods for Preclinical Research in Addiction*, Neuromethods, vol. 174,
https://doi.org/10.1007/978-1-0716-1748-9_8, © Springer Science+Business Media, LLC, part of Springer Nature 2022,
Corrected Publication 2022

[3]. Substances such as alcohol, cannabis, psychostimulant, or opioid drugs are heavily consumed mainly between young people [4].

Animal models are essential for the identification of factors that induce vulnerability to drug addiction, which is important in order to develop preventive and treatment strategies for this disorder. Some studies have evidenced that drug addiction is associated with biological variables, for example, genetic aspects [5]. Furthermore, several studies show that environmental factors have a profound influence on the acquisition, maintenance, and relapse to drug addiction [6–9]. Between these environmental factors, stress can be determinant not only in the transition from drug use to abuse but also in the development of mental and cognitive disorders related with drug addiction [7, 10]. A wide variety of stressful stimuli have been employed to model the influence of stress on the rewarding effects of drugs of abuse [11]. Some studies used pharmacological stressors (e.g., corticotropin-releasing factor CRF) [12], physical (e.g., intermittent shock, immobilization, or tail pinch) [9, 13, 14], emotional (food deprivation, forced swim, the chronic unpredictable stress) [15–17], or social stress (maternal separation, social isolation, crowding, social defeat) [9, 18–20]. Among these types of stressors used in animal models, social defeat has showed high ethological validity. Fortunately, food deprivation or intermittent shocks are unlikely to be experienced nowadays. However, social adverse events, including bullying, are common in our society.

The association between stress and drug addiction seems to have a neurobiological explanation, since there is a nexus between the brain and the stress system. It is known that under stress the hypothalamus–pituitary–adrenal (HPA) is activated as a central control and regulatory system involved in the stress response together with the sympathetic-adrenal system. Moreover, the mesocorticolimbic dopaminergic system involved in drug reward [21, 22] has a close connection with the brain stress system [11]. In fact, the activation of brain stress system is key to induce the emotional adverse state that characterizes abstinence and leads to relapse to drug-seeking behavior [23]. The common nexus between stress and reward systems is the "extended amygdala," composed by the bed nucleus of the stria terminalis (BNST) and the central nucleus of the amygdala (CeA) and interconnected with the main brain areas of the reward system including the nucleus accumbens, the ventral tegmental area (VTA), and the prefrontal cortex (PFC). Several neurotransmitters involved in stress and reward such as CRF, noradrenaline (NA), and dopamine (DA) interact in the extended amygdala [11]. The PFC has an important role in the stress response, via attenuation of the activity of amygdala that process the emotional stimulus [24]. However, different types of stress can alter these brain regions in a distinctive way. As social stress does not represent an immediate or direct threat for the survival of the organism, this stimulus needs to be processed by the prefrontal cortex (PFC), a rational part of our brain. A recent study demonstrated that the activity of individual neurons of the PFC is altered by social defeat stress [25].

Several animal models have been used to evaluate the rewarding effects of drugs of abuse, although the self-administration (SA) and the conditioned place preference (CPP) are the most commonly employed [26]. These paradigms also have allowed to evaluate the influence of stress in the rewarding properties of drug abuse [20, 27–29]. At the same time, different paradigms have been used to induce social stress although acute or repeated social defeat are the most used. Acute social defeat (ASD) is usually used to evaluate the short-term effects of social stress while the repeated social defeat (RSD) paradigm allow to observe its long-term effects. Both will be described in the next section.

This chapter is aimed to provide an update of studies about the role of social stress on the rewarding effects of drugs of abuse using animal models. First, we describe as social defeat modifies the rewarding properties of the most consumed drugs of abuse (alcohol, psychostimulants, cannabis, and opioids) using the SA and CPP paradigms. Next, we describe the materials and methods used to perform the ASD and RSD procedures in our laboratory. Finally, we draw the final conclusions and propose future lines of research with these paradigms of social stress.

## 2 Social Defeat Modifies the Rewarding Properties of Drugs of Abuse

### 2.1 Influence of Social Stress on the Rewarding Effects of Ethanol

Ethanol is the drug of abuse most typically consumed in our society. The study of the environmental variables that can modify its rewarding properties is essential in order to determine factors that can enhance the vulnerability to the abusive consumption and for the development of new behavioral therapies.

However, the studies about the effects of social stress on the rewarding properties of ethanol are controversial and sometimes depend on the paradigm of reward employed as well as on the strain or age of animals used [30]. In the *SA paradigm* it has been showed that mild social defeat stress is sufficient to increase alcohol consumption in nonpreferring strains of rats [31]. Similarly, exposure to RSD during adolescence increases vulnerability to the rewarding effects of ethanol in mice [32] and moderately stressed mice showed SA during intermittent access to ethanol and escalated intake [33]. Administered during a period of deprivation of alcohol, social defeat caused a smaller increase in alcohol intake but only after a first deprivation and stress cycle [34]. However, other studies reported that exposure to different patterns of social defeat could induce a transient suppression rather than a facilitation of ethanol intake [35, 36]. The interval between social stress and the evaluation of alcohol intake influences the results observed. A decrease of alcohol self-administration was observed 24 h after the previous social stress episode; however, no changes in intake or alcohol reinforcements were observed 4 h after exposure of rats to social defeat [35]. Acute exposure to social defeat decreased alcohol self-administration, reduced rates of responding during extinction, and did not reinstate alcohol seeking [37]. Exposure to a discrete odor cue previously paired with social defeat also decreased alcohol self-administration but induced a modest reinstatement of alcohol seeking [37].

With respect to the results obtained with the *CPP paradigm* most studies support the idea that social stress increased ethanol-induced CPP [38–41]. In the same line we have observed that acute and intermittent/long-term types of social defeat stress (ASD and RSD) reversed the conditioned place aversion induced by high doses of ethanol [42]. In this study, we also evaluated the acute and long-term effects of social defeat on alcohol consumption in the two-bottle choice paradigm. Mice exposed to social defeat showed an increase in alcohol intake in comparison to control nonstressed mice [42].

## 2.2 Influence of Social Stress on the Rewarding Effects of Psychostimulants

Cocaine and "ecstasy" or MDMA (3,4-methylenedioxymethamphetamine) are psychostimulant drugs widely consumed in our society [4]. Different studies have evaluated the influence of social defeat stress on their rewarding properties.

*Using the SA paradigm* Several studies have demonstrated that social defeat increased vulnerability to acquiring and maintaining **cocaine** self-administration and prompts an escalation of cocaine-seeking behavior [43–51], probably the VTA DA neurons and CRF being involved in this effect [52]. Moreover, it seems that behavioral characteristics during social defeat are predictive of later cocaine self-administration [53]. On the other hand, intermittent social defeat did not modify cocaine self-administration in mice [54] or modestly increased it [55]. Individual differences have been observed in the effects of continuous exposure to social defeat; it increased cocaine self-administration and sucrose intake in a subset of animals, but decreased these measures in another subpopulation of mice. Acquisition of cocaine self-administration (0.5 mg/kg per injection) was delayed in adult mice exposed to RSD during adolescence [56].

Social defeat potentiated methamphetamine seeking. In rats trained to acquire the self-administration of methamphetamine and that subsequently underwent extinction of lever pressing, a single social defeat did not reinstate drug seeking; however, a reminder of social defeat followed by a priming injection of methamphetamine potentiated reinstatement over drug-priming alone [57].

No study has evaluated the role of social stress on the rewarding properties of MDMA using the SA paradigm, so more studies are needed in this area.

*Using the CPP paradigm* It has been evidenced that social defeat stress modifies the rewarding properties of cocaine (see the review of Montagud-Romero et al. [30]). Mice exposed to social defeat exhibited an increased place-preference for the cocaine-paired chamber over the control group and the pretreatment with a kappa opioid receptor antagonist blocked this stress-induced potentiation of cocaine-CPP [58]. The effects of social defeat on drug reward are in function of several variables such as the age of the animals when are exposed to stress, the social stress schedule and the dose of drug used [59]. A single exposure to social defeat before the reinstatement test increased the vulnerability to reinstatement induced by a cocaine priming [19] and induces reinstatement of cocaine CPP [60, 61]. Adult mice exposed to ASD showed an increase in the conditioned rewarding effects of cocaine [28] and in the reinstatement of cocaine CPP [62]. Conversely, adolescent mice exposed to ASD display a reduction of the conditioned rewarding effects of cocaine. In fact, CPP was not observed with a low dose of cocaine (1 mg/kg) that is effective to induce CPP in adolescent mice non exposed to social defeat; furthermore, when adolescent mice were conditioned with high doses (25 mg/kg), the CPP extinguished faster in those exposed to social defeat [28].

With respect to the long-term effects of RSD, an enhancement in the rewarding effects of cocaine is observed in both adult and adolescent mice. Three weeks after exposure to RSD in adulthood, mice showed CPP with a low dose of cocaine (1 mg/kg) that is ineffective to induce place conditioning in control mice [63–68] This increase in the rewarding effects of cocaine induced by RSD is mediated by an upregulation of histone acetylation induced by social defeat [63], by DA neurotransmission [64] and is reversed by indomethacin [66], the antagonism of CRF1 receptors [67] and oxytocin [68]. Furthermore, the RSD-induced increase of cocaine CPP is observed regardless of the genotype of mice [69]. Similarly, animals socially defeated during adolescence showed an increase in the conditioned rewarding effects of cocaine in the adulthood [56, 70]. Such increase is expressed as an enhanced vulnerability to the reinstatement of the CPP induced by a low dose of cocaine (1 mg/kg) in mice defeated during adolescence and as a prolonged duration of the CPP induced by 25 mg/kg of cocaine, since defeated mice needed more sessions for the preference to be extinguished [56]. It is especially relevant since adolescent mice experienced social defeat less intensely than adults and showed lower levels of corticosterone.

With respect to **MDMA**, most studies have showed that social defeat modified the conditioned rewarding properties of this drug, although as commented before with cocaine, the results obtained depend on different variables [59]. For example, age is an important factor for the effects of ASD. Exposure to defeat in agonistic encounters with an aggressive conspecific immediately before each session of place conditioning does not induce alterations on the rewarding or other behavioral effects of MDMA (1.25–10 mg/kg) in adolescent mice. However, in young adult mice exposure to the same type of defeat decreases the sensitivity of adult mice to the rewarding effects of MDMA since defeated mice did not show CPP with doses (1.25–10 mg/kg) that are effective to induce CPP in control mice [27]. Changes in the glutamatergic system seem to play a role in the effects of social defeat stress on the rewarding properties of MDMA. In particular, the pretreatment with memantine (NMDA antagonist) to young adult mice exposed to an episode of ASD before each conditioning session with MDMA (1.25 mg/kg) reverse the effects of social defeat since defeated mice treated with memantine showed CPP [76]. Conversely, the pretreatment with 6-cyano-7-nitroquinoxaline-2,3-dione (CNQX) (AMPA antagonist) did not modify the effects of ASD since defeated mice treated with CNQX did not show CPP. The results of this study also indicate that social defeat induces modifications in the brain areas involved in the reward. Social defeat decreased the expression of several subunits of glutamatergic receptors NMDA and AMPA, for example, GluN1 and GluA1 [71]. A recent study also indicates the role of the nitric oxide (NO) pathway in the effects of ASD on the rewarding properties of MDMA. In particular, pretreatment with a low dose of the NO synthase inhibitor 7-nitroindazole (7.25 mg/kg) reversed the effects of ASD since defeated mice show CPP. Moreover, social defeat exposure increased the nitrites in the prefrontal cortex and hippocampus [72].

On the other hand, the long-term effects of social defeat on the conditioned rewarding effects of MDMA are consistent in adolescent and adult mice. Exposure to RSD 3 weeks before place conditioning with MDMA enhanced vulnerability to priming-induced reinstatement in mice conditioned with a low dose of MDMA (1.25 mg/kg) and increased the duration of CPP induced by a high dose (10 mg/kg) [20].

## 2.3 Influence of Social Stress on the Rewarding Effect of Other Drugs of Abuse

Few studies have evaluated the effects of social defeat stress on the rewarding properties of opioid drugs. Rats with a history of intermittent social defeat persisted in self-administering a heroin–cocaine mixture and showed escalated cocaine-taking behavior during the 24-h binge session, although no effects on heroin taking were observed [48]. On the other hand, social defeat prevented the acquisition of morphine CPP, which suggests a decrease in sensitivity to the rewarding effects of opiates [73]. However, social defeat in an agonistic encounter induced reinstatement of CPP in morphine-conditioned animals [14]. These data support the idea that social stress can modulate opiate reward and promote relapse [74].

Studies about the influence of social stress on the rewarding properties of nicotine or cannabinoid drugs are inexistent.

# 3    Main Paradigms Used to Induce Social Defeat Stress

As we have described in the previous sections, social defeat is an animal model of stress exposure with high ethological and ecological validity. However, we used two main variations of the paradigm (acute and repeated social defeat) that differed in function of several factors, mainly the number of exposures to defeat experience, the duration and context of the agonistic encounters and the time elapsed between the stress experience and the exposure to drugs of abuse (*see* Table 1).

## 3.1    Acute Social Defeat

### 3.1.1    Materials

- *Experimental animals.* We use male OF1 mice (Charles River, France) of 21 days or 42 days of age at the arrival to the laboratory (adolescents and young adults, respectively). They are housed in groups of four in plastic cages (25 × 25 × 14.5 cm) for 8 days before the experiments began (*see* **Note 1**).

- *Aggressive Opponents.* We use male OF1 mice of 42 days of age at the arrival to the laboratory. They were individually housed in plastic cages (23 × 13.5 × 13 cm) for a month before experiments to induce aggression [75] (*see* **Note 2**).

**Table 1**
**Differences between acute social defeat (ASD) and repeated social defeat (RSD)**

|  | Objective | Duration and environment | Number of episodes | Evaluation of rewarding properties of drugs | Animals used |
|---|---|---|---|---|---|
| Acute social defeat (ASD) | Evaluate the **short-term** effect of social defeat | The agonistic encounters are performed in a **neutral box** for **10 min.** | **Only** one episode of social defeat. | **Immediately,** before each conditioning or self-administration or reinstatement. | Mice and rats Adults and adolescents |
| Repeated social defeat (RSD) | Evaluate the **long-term** effect of social defeat | The agonistic encounters are performed in the **resident home cage** for **25 min.** Only 5 min of interaction. | **Four** episodes of social defeat each 72 h separated in three phases each of them. | **Three weeks** after the last social defeat. | Mice and rats Adults and adolescents |

- *Cage for the agonistic encounters*: We use a neutral transparent plastic cage ($23 \times 13.5 \times 13$ cm), that is to say, an area that is not the home cage of the experimental or the aggressive animal.

- *Video camera*: The agonistic encounters are videotaped for the behavioral analysis.

- *Computerized program* (Raton time) and computer with screen and keyboard for behavioral analysis.

*3.1.2 Methods*

1. *Conditions of housing*: All mice are housed in a room under constant temperature, a reversed light schedule (white lights on 19:30–07:30 h) (*see* **Note 3**) and food and water freely available, except during behavioral tests.

2. *Induction of aggressiveness in the opponents*: As commented before, to ensure that opponents showed aggressive behaviors, they live isolated (*see* **Note 2**) and are briefly confronted with other isolated mice to instigate threat and attack behaviors.

3. *Screening of aggressive behaviors in opponents*: The opponent animals, which had previous fighting experience, are screened for a high level of aggression (with short latency to show threat and attack behaviors) in a brief encounter with a grouped conspecific (*see* **Note 4**). This screening encounter ends after the presence of threat and the first attack of the opponent and is performed typically 24 h before the agonistic encounter of social defeat with the experimental mouse.

4. *Agonistic encounter with the result of defeat for the experimental mouse*: The agonistic encounter between the experimental mouse and the aggressive opponent (*see* **Note 5**) takes place in a neutral transparent plastic cage (*see* Fig. 1a). First, animals are placed in this cage separated by a transparent plastic barrier during 1 min. Then, this barrier is removed and the physical interaction between them is allowed for 10 min. In response to the aggressive behaviors of the opponent, experimental animals (that are not housed in isolation and have not fighting experience) exhibited avoidance/flee and defensive/submissive behaviors. The criteria used to define an animal as defeated is a specific posture, characterized by an upright position, limp forepaws, upwardly angled head, and retracted ears [76] (*see* **Note 6**). The agonistic encounters are video recorded to subsequently evaluate the behaviors of both animals (*see* **Note 7**).

5. *Behavioral analysis of agonistic encounters*: The behavioral acts and postures showed by mice during the confrontations are recorded, in particular, the frequencies, durations, latencies, and temporal and sequential patterns of the different behavior (ethogram). Submissive and fleeing behaviors of the experimental animals and the aggressive behaviors of the opponents

a) Acute social defeat

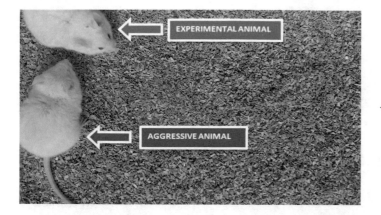

-Neutral area

-10 minutes of free interaction

-Without wire mesh

-Only one episode immediately before each acquisition session or before reinstatement test of the rewarding properties of drugs

b) Repeated social defeat

-Home cage of the resident

-25 minutes of interaction with 3 phases:

    a.- 10 minutes with wire mesh

    b.- 5 minutes of free physical interaction without wire mesh

    c.- 10 minutes with wire mesh

- Four episodes (each 72 hours)

- Three weeks before evaluation of the rewarding properties of drugs

**Fig. 1** Methodology of Acute social defeat (ASD) versus Repeated social defeat (RSD). (**a**) Acute social defeat. (**b**) Repeated social defeat

are evaluated using a custom-developed program that allows estimation of the time engaged in different behaviors (mainly threat, attack, avoidance/flee, and defense/submission) [75] (*see* Fig. 2 and Table 2) (*see* **Note 8**).

6. *Control group without defeat*: We used a control group of mice that do not suffer stress. In this case, the experimental animal is placed in the neutral area with the barrier for 1 min and when the barrier is removed it explore freely the cage for 10 min without any aggressive opponent mouse [27].

7. *Evaluation of corticosterone levels*: Immediately after the first and fourth agonistic encounter, we obtained blood samples from the mice exposed to social defeat (stress group) or

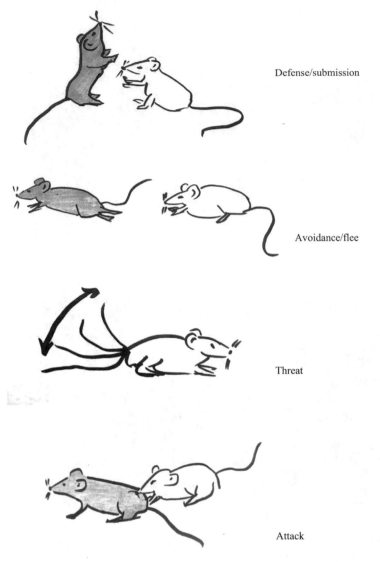

Defense/submission

Avoidance/flee

Threat

Attack

**Fig. 2** Main behavioral categories recorded during episodes of defeat in the intruder mouse (in brown) and the aggressive opponent mouse

exploration of a new cage without opponent (control group) for corticosterone determination (*see* **Note 9**). Blood sampling is performed using the tail-nick procedure. In this procedure, the animal is wrapped in a cloth and a 2-mm incision is made at the end of the tail artery. Then, the tail is massaged to facilitate blood collection that is performed in an ice-cold Microvette CB 300 capillary tube (Sarstedt, Nümbrecht, Germany). We collect 50 μL of blood. Tubes with blood samples are kept on ice until they are placed in a centrifuge apparatus. Centrifugation (5 min, 5000 × *g*) separates plasma from whole blood. Plasma is transferred to sterile, 2 mL microcentrifuge tubes,

**Table 2**
**Main behavioral category, elements and description**

| Behavioral category | Elements | Description |
|---|---|---|
| Body care | Abbreviated groom | A single rapid wipe of the head or snout using the forepaws |
| | Self-groom | Licking of the fur on the flanks or abdomen or preening of the tail |
| | Wash | Forepaws licked and then stroked repeatedly over the head |
| | Shake | A brief, mild quiver of the body |
| | Scratch | Hindlimb used to scratch at the flanks |
| Digging | Dig | Forepaws used to direct sawdust to the rear of the animal |
| | Kick dig | Hindpaws used to kick sawdust backward |
| | Push dig | Forepaws used to push sawdust forward |
| Non-social exploration | Explore | Walking or running around the cage, not directed toward partner |
| | Rear | Bipedal posture, front part of the body and forepaws raised |
| | Supported rear | Rear with forepaws resting on the walls of the cage |
| | Scan | Side-to-side movement of the head, attention not directed to partner |
| Explore from a distance | Approach | Ambulation and attention toward the partner |
| | Attend | Attention directed toward the partner from a distance |
| | Circle | Cycles of approaching and leaving with no intervening activity |
| | Head orient | Head turned toward the partner |
| | Stretched attention | As in attend, except that the body and head neck craned forward |
| Social investigation | Crawl over | Both forepaws placed on the partner |
| | Crawl under | Head and anterior part of the body pushed underneath the partner |
| | Follow | Moving in close proximity to the partner as it walks around the cage |
| | Groom | Grooming the body of the partner using the mouth |
| | Walk around | Walking around the partner, in close proximity |
| | Investigate | Sniffing the body or tail of the partner |
| | Sniff | Sniffing the anogenital region of the partner |
| | Nose sniff | Sniffing the head or snout of the partner |
| | Push past | The animals moving in opposite directions come into lateral contact |
| Threat | Aggressive groom | Vigorous grooming of the immobile partner from a lateral position, using the teeth and the forepaws |
| | Sideways offensive | Tripedal posture oriented toward the partner forepaws nearest the partner being raised, eyes slitted and ears flattened |
| | Upright offensive | Bipedal posture oriented toward the partner, characterized by slitting of the eyes and flattening of the ears |
| | Tail rattle | Rapid lateral quivering or thrashing of the tail |
| Attack | Charge | Running rapidly toward the partner |
| | Lunge | body, as if to bite the partner, but failing to make contact with teeth |
| | Attack | Biting the partner |
| | Chase | Pursuing a fleeing partner |

(continued)

**Table 2**
**(continued)**

| Behavioral category | Elements | Description |
|---|---|---|
| Avoidance/flee | Evade | Move anterior part of the body and/or head away from the partner |
| | flinch | Head rapidly retracted from the partner |
| | Retreat | Running away from approaching partner |
| | Ricochet | Fast undirected movement, with jumping from the walls of the cage |
| | Wheel | Turns as opponent approaches |
| | Startle | Sudden vertical movement in response to approach of partner |
| | Jump | Animal jumps into the air, all four feet leaving the substrate |
| | Leave | Ambulation away from the partner |
| | Wall clutch | Ventral surface of the body pressed against the cage wall, with forelimbs widely splayed |
| Defense/ submission | Upright defensive | Bipedal postures, usually oriented toward the partner accompanied by widening of the eyes and raising of the ears |
| | Upright submissive | An extreme form of "upright defensive" with the head pushed far backward and with the forelimbs held rigid and widely splayed |
| | Sideways defensive | Tripedal postures, one forepaw raised from the ground, accompanied by widening of the eyes and raising of the ears |
| Sexual | Attempted mount | Attempts to dorsally mount the partner, a motion which is incomplete |
| | Mount | Climbing dorsally on partner, palpating the flanks, and pelvis thrusts |
| Immobility | Squat | Complete immobility, no movements of any part of the body |
| | Cringe | As squat, but the body is pushed against the cage wall, animal pulls itself away from the partner and may make quivering motions |

which are stored at $-80\ °C$ until determination of corticosterone. To perform the assay of corticosterone levels we use a corticosterone EIA kit (Catalog No. ADI-900-097, 96 Well kit; Enzo Life Sciences, Taper SA, Madrid, Spain). According to the manufacturer's instructions, plasma samples are diluted, in a proportion of ~1:40, in the Steroid Displacement Reagent mix provided with the kit. Corticosterone levels in diluted plasma are analysed using an iMark microplate reader and Microplate Manager 6.2. software (Bio-Rad, Madrid, Spain). The optical density was read at 405 nm, with 590 nm correction.

8. *Effects of acute social defeat on the rewarding effects of drugs of abuse*: In this paradigm of acute social defeat, as its name indicated, the experimental animal suffers a short and punctual experience of social defeat (10 min). When we used this paradigm in order to evaluate the effects of social defeat on the acquisition of the conditioned place preference induced by drugs of abuse, such as alcohol, cocaine or MDMA [20, 27,

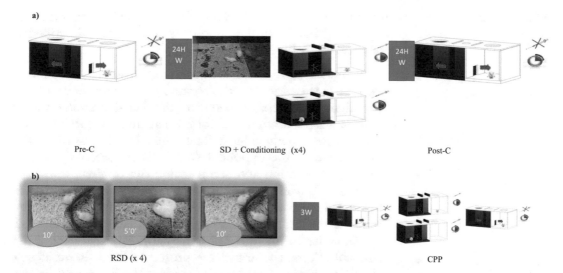

**Fig. 3** Schedule of defeat episodes and CPP procedure. (**a**) Acute social defeat. (**b**) Repeated social defeat

28, 59, 71, 72] experimental mice suffer an experience of social defeat immediately before each conditioning session with the drug (thus, the agonistic encounter is performed only on 4 consecutive days or in alternative days, depending the drug used). Few hours after (24 or 48 h in function of the drug used) the effects of social defeat on the acquisition of place conditioning is evaluated (*see* Fig. 3a). On the other hand, to test the effects of social defeat on the reinstatement of CPP, experimental mouse only performed one agonistic encounter of 10 or 15 min with result of defeat, immediately or 30 min, before the reinstatement test [14, 19, 61].

## 3.2 Repeated Social Defeat

### 3.2.1 Materials

*Experimental animals*: We use male mice of the OF1 or C57BL/6 strain (Charles River, France). See details about the age and housing of mice in Subheading 3.1.

*Aggressive Opponents*: We use male OF1 mice of 42 days of age at the arrival to the laboratory. They were individually housed in plastic cages (23 × 32 × 20 cm) for a month before experiments to induce aggression [74]. Please note that in this case the home-cage of the aggressive opponents is longer because we used a resident–intruder model to induce social defeat.

*Wire mesh barriers*: to separate the experimental and opponent mice during the first and last 10 min of social defeat encounters.

*Video camera, computer and computerized program* (see details in the Subheading 3.1).

Conditions of housing, induction of aggressiveness and screening of aggressive behavior in the opponents are the same described in the Subheading 3.1.

1. *Repeated intermittent social defeat*: In this case, the experimental mouse is defeated in the context of a "intruder–resident" paradigm of aggression, based on the fact that an adult male rodent will establish a territory when given sufficient living space. The experimental animal (intruder) is placed in the home cage of the opponent (resident) mouse. As a consequence of isolation and territoriality, the resident will show offensive aggression versus the unfamiliar male mouse intruding in its home cage. In response to the offensive attacks by the resident, the intruder will show defensive/submissive behavior.

   In order to minimize physical harm but maintaining the stressful effects, the intruder is protected from the resident's attack by a wire mesh barrier during the main part of the encounter (*see* Fig. 1b) and physically exposed to the resident only for a brief time. In particular, each episode of social defeat (25 min) consisted of three phases.

   (a) *First phase*: Initially, the experimental animal is introduced in the home cage of resident aggressive opponent for 10 min but both animals are separated by the wire mesh protecting the intruder from the attack (bites) of the resident animal. However, social interaction between both animals and species-typical threats from the aggressive resident, as provocation and instigation, are allowed [20, 59, 77].

   (b) *Second phase*: The wire mesh is removed and the direct confrontation between animals is allowed for 5 min. We consider that the experimental (intruder) was defeated because of the adoption of the upright submissive position (as described in the Subheading 3.1) for 5 s [75, 78]. This posture appears normally after 3–5 attacks by the resident.

   (c) *Third phase*: In this last phase, both animals are separated again with the wire mesh for 10 min and the intruder animal is exposed to provocation and threat behaviors from the resident animal.

   Mice are exposed to several episodes of social defeat, usually one episode per day for 10 consecutive days (chronic social defeat) or, as we performed in our laboratory, one episode each 72 h until a total of 4 episodes (intermittent RSD). An important question is that in each aggressive episode the experimental animal is exposed to a different aggressive animal. However, when we confront experimental C57BL/6 mice with OF1 residents, we always use the same opponent in order to reduce the aggressive contacts received by the small-size experimental animal. The first and fourth defeat episodes are video recorded to subsequently evaluate the offensive behaviors (threat and attack) of the resident and the defensive/submissive and avoidance/flee behaviors of the experimental animal.

2. *Behavioral analysis of agonistic encounters: see* Subheading 3.1.

3. *Control group without defeat*: We used a control group of mice that do not suffer stress. In this case, the experimental animal is placed in a cage (equal to the home-cage of the resident) without any mouse during 25 min, with a wire mesh wall on the first and last 10 min [20, 79].

4. *Evaluation of corticosterone levels: see* Subheading 3.1.

5. *Effects of repeated social defeat on the rewarding effects of drugs of abuse*: In our laboratory, the main objective of the RSD paradigm is to evaluate the long-term effects of the exposure to an intermittent situation of social stress on the acquisition, extinction and reinstatement of the CPP induced by different drugs of abuse [20, 59, 71]. Furthermore, in some studies we have evaluated the long-term effects of RSD on other behaviors (anxiety) or learning and memory processes.

In particular, the experimental animals are exposed to four episodes of social defeat that lasted 25 min each, separated by 72 h [20, 32, 80, 81]. To model exposure to social stress during adolescence animals are exposed to RSD on PND 29, 32, 35, and 38, while to model stress in young adult mice they experience social defeat on PND 47, 50, 53, and 56. In both cases, 3 weeks after suffering the last episode of defeat stress, animals undergo the place conditioning procedure with cocaine, MDMA, or alcohol [20, 42, 79] (*see* Fig. 3b).

# 4  Notes

1. To reduce the stress levels of mice in response to experimental manipulations, they are handled for 5 min/day on each of the 3 days before initiation of the behavioral experiments.

2. Mice of the OF1 strain are more aggressive than mice of other strains (e.g., Balb/c or C57). In addition, a month of isolation heightens aggression in OF1 mice [75].

3. The light schedule was reversed in order to facilitate the performance of the experimental procedures in the dark phase of the cycle, because mice are more active on night.

4. Mice confronted with aggressive opponents in the screening of aggressive behaviors are not other finality (usually they are animals employed in previous studies of our laboratory).

5. Usually, experimental adult animals are exposed to a conspecific of equal age and body weight. However, when adolescent mice are employed as experimental animals, the size of the opponent is higher.

6. Defeated mice always exhibit an extreme form of "upright submissive" behavior [75] (*see* Fig. 2). Furthermore, usually we performed subsequently a more profound analysis of the behavior of experimental and opponent mice (see the list of behaviors and postures in Table 2). A detailed description of all elements can be found in Brain et al. [82].

7. Agonistic encounters (social defeat) take place in a different room of that used for place conditioning or other behavioral experiments.

8. Usually only the behavioral analysis of the first and fourth agonistic encounter is performed.

9. Usually, a separate set of mice divided into two groups (social defeat and control) is used only to evaluate corticosterone levels. In other cases, the same mice that are exposed to CPP after social defeat or exploration are used (in this case, blood samples are obtained immediately after the place conditioning session).

## 5    Advantages, Main Applications, and Limitations of Social Defeat

Social stress is highly prevalent in our society. An elevated number of people suffer this type of stress in multiple situations affecting their mental and physical health. As we commented in the introduction of this chapter situations such as bullying and mobbing are common in different social interactions. The study of the consequences, prevention, treatment or variables associated with these social adverse situations are necessary in order to understand the neurobiological processes that are involved in the social defeat situations.

However, the induction of social stress in humans is not ethical and the use of animal models help us to progress on this research area. Acute/repeated social defeat paradigms allow us to draw conclusions about the effects of social stress. Moreover, the simulation in the laboratory of social stress using defeat paradigms have high ecological and ethological validity. In human social stress situations, the individuals also suffer a subordination by others that are hierarchically in a superior social scale (e.g., the boss at work or the leader in a group in high school). So, the use of defeat animal models is a great advantage to study aspects of social defeat that in other cases would be impossible. For example, the study of neuroanatomical structures and neurobiological pathways involved in the stress response is determinant in order to develop effective treatments for the consequences of stress-related disorders. Moreover, defeat paradigms allow us to understand the main variables that influence the acute/long-term effects of social defeat (duration, intensity, frequency, intermittency) in order to develop preventive strategies to decrease the negative consequences of social stress. Although we used mice, the ASD and RSD paradigms can be performed in different strain of rats [83, 84].

As we have commented around this chapter, drug addiction is a chronic affective-cognitive disorder for which currently there is no cure. Thus, the study of variables involved in the initiation and maintenance of this disorder and in the vulnerability to relapse after withdrawal periods is essential. In agreement with clinical observations, animal models of social defeat have demonstrated an intimate relationship between social stress experiences and enhanced vulnerability to drug addiction. Several models of social defeat modify the rewarding properties of different drugs of abuse and a clear long-term increase in the sensitivity to psychostimulant drugs has been observed in defeated animals [29, 30, 59, 85]. As commented before defeat paradigms allow us draw conclusions about the consequences of different types of stress, different drugs of abuse, strain, sex, and age of animals. Moreover, these paradigms have allowed advances in the knowledge of the brain changes involved in the effects of social stress on drug addiction vulnerability. In fact, studies in our laboratory have demonstrated that NMDA and AMPA glutamatergic receptors are involved in the effects of social defeat stress [71, 86] and that the NMDA antagonist memantine reversed the effects of social defeat on MDMA [71] and cocaine [86] reward. Social defeat also increased the level of nitrites in several brain areas and the inhibition of NO synthesis reversed the effects of defeat on MDMA CPP [72]. These results suggested that the manipulation of glutamatergic system and/or NO pathway could be a useful therapeutic tool for the treatment of stressed individuals dependent on psychostimulant drugs. Other neurobiological systems have been involved in the effects of social defeat. Dopaminergic receptors [64, 65], the oxytocin system [68], BDNF [65, 87], and inflammatory signaling [66, 69, 88] are involved in the long-term effects of RSD on cocaine reward. These studies may contribute to developing pharmacological treatments for stress and addiction-related disorders.

A disadvantage of social defeat stress is that this paradigm is not free from ethical concerns and the difficulty to distinguish between the effects of physical and emotional stress. This is more important when experimental animals are confronted to more aggressive conspecifics that repeatedly bites them, for example, adolescent mice defeated by adult opponents. In fact, physical interaction between the animals during the agonistic encounter is reduced to a minimum, until the experimental mouse displays the behavioral posture of defeat. However, the main limitation of social defeat paradigms is the difficulty to translate the results observed in animals to the clinical settings. Animal studies involve relatively brief exposure to the aggressive opponent that are often not representative of natural conditions or the duration of social interactions. In laboratory animal studies all the variables and conditions are under control, thus, the situation is more artificial. However, in humans other noncontrolled variables can affect the stress response. For example, personality traits can modify the physiological, cognitive, and behavioral responses to stress. Some of these aspects are being studied in new research projects (see the next section).

## 6 Conclusions and Future Perspective

As described in this chapter social stress can increase the initiation and maintenance of drug abuse and the vulnerability to relapse. In this sense, animal models help us to draw these conclusions. Social stress is an environmental factor that modifies our brain and in consequence our behavior. Moreover, the neurobiological systems involved in social stress are intimately associated with the reward brain pathway and with the structures related with the development of drug addiction [11]. Studies with different animal models of reward evidence that social defeat modifies the rewarding properties of different drugs of abuse including ethanol, opioid or psychostimulant drugs. However, the different paradigms of social defeat (single, intermittent, or continuous) and the time in which their effects on drug reward are evaluated (short- or long-term after social defeat) may influence the results observed [20, 27, 28, 65]. Moreover, other factors, such as the type of drug used or even the dose, can also modify the effects of social defeat stress [59]. Finally, individual variables such as sex [89], age [59], and/or behavioral traits of the experimental animals [79] have a profound influence in the effects of social defeat. For example, mice with a more active coping strategy during the episodes of defeat (with lower defensive behaviors) and mice with lower novelty-seeking profile are resilient to the potentiating effects of social defeat on cocaine reward (they do not develop CPP with low doses of cocaine while defeat mice usually do it).

Animal models of acute/repeated social defeat stress have demonstrated to be an essential tool to the progress gained regarding knowledge of the underlying mechanisms of stress and addiction. The use of these models has the advantage of greater control of experimental variables, has provided much valuable information and possesses high predictive value. On the other hand, the main drawback to animal studies is that any model reproduces totally all aspects that composed human stress and addictive disorders. Future lines of research should be proposed to enlarge the knowledge about the neurobiological substrates underlying the interaction between these disorders. As commented before, behavioral traits associated with resilience to the effects of RSD on cocaine reward have been identified [79]. Future studies using other types of social defeat (such as ASD), other drugs of abuse (e.g., MDMA or ethanol) and animals of different sex or age (e.g. female or adolescent mice) can be important in order to draw conclusion about the role of these factors in the effects of social stress.

On the other hand, although it seems that stress is a negative event, some studies have indicated that a short and punctual stress in early life can cause the individuals to be more resistant to overcome future adverse stress situations, a phenomenon named as "inoculation of stress" [90, 91]. In this sense, more studies are necessary to demonstrate the effects of inoculation stress (using a single social defeat or other stressful events) on the subsequent effects of social defeat on the rewarding properties of drugs of abuse in other periods of life.

Other novel line of research of our laboratory is the study of the influence of social stress in individuals with neurodevelopment disorders, in particular, with attention deficit hyperactivity disorder. In children and adolescents with previous problems the effect of social defeat on the posterior rewarding properties of the drugs of abuse may be different. Moreover, sometimes these patients are treated with drugs to reduce the symptoms of these neurodevelopment disorders. For example, methylphenidate (an amphetamine derivate) is used for the attention deficit hyperactivity disorder and will be interesting to study the effect of this treatment in early age on the posterior response to the drugs of abuse in subjects that suffer social stress.

Finally, future works should study the neurobehavioral substrates of resilience after social stress in order to develop interventions that increase resilience to develop drug abuse and addiction in individuals at a high risk of suffering from stress.

# References

1. Koob GF, Volkow ND (2016) Neurobiology of addiction: a neurocircuitry analysis. Lancet Psychiatry 3(8):760–773. S2215-0366(16) 00104-8 [pii]

2. American Psychiatric Association (2013) Diagnostic and statistical manual of mental disorders, 5th. Edition (DSM-5). American Psychiatric Association, Washington, DC

3. UNODC (2019) World drug report 2019. https://wdr.unodc.org/wdr2019/prelaunch/pre-launchpresentation_WDR_2019.pdf

4. EMCDDA (2019) European drug report 2019: trends and developments. https://www.emcdda.europa.eu/system/files/

publications/11364/20191724_TDAT19001ENN_PDF.pdf

5. Meng W, Sjoholm LK, Kononenko O, Tay N, Zhang D, Sarkisyan D et al (2019) Genotype-dependent epigenetic regulation of DLGAP2 in alcohol use and dependence. Mol Psychiatry. https://doi.org/10.1038/s41380-019-0588-9

6. Koob GF (2010) The role of CRF and CRF-related peptides in the dark side of addiction. Brain Res 1314:3–14. https://doi.org/10.1016/j.brainres.2009.11.008

7. Sinha R (2008) Chronic stress, drug use, and vulnerability to addiction. Ann N Y Acad Sci

1141:105–130. https://doi.org/10.1196/annals.1441.030

8. Sinha R, Shaham Y, Heilig M (2011) Translational and reverse translational research on the role of stress in drug craving and relapse. Psychopharmacology 218(1):69–82. https://doi.org/10.1007/s00213-011-2263-y

9. Logrip ML, Zorrilla EP, Koob GF (2012) Stress modulation of drug self-administration: implications for addiction comorbidity with post-traumatic stress disorder. Neuropharmacology 62(2):552–564. https://doi.org/10.1016/j.neuropharm.2011.07.007

10. Gould TJ (2010) Addiction and cognition. Add Sci Clin Pract 5(2):4–14. https://www.ncbi.nlm.nih.gov/pubmed/22002448

11. Rodríguez-Arias M, García-Pardo MP, Montagud-Romero S, Miñarro J, Aguilar MA (2013) The role of stress in psychostimulant addiction: treatment approaches based on animal models. In: Drug use and abuse: signs/symptoms, physical and psychological effects and intervention approaches, pp 153–220

12. Buffalari DM, Baldwin CK, Feltenstein MW, See RE (2012) Corticotrophin releasing factor (CRF) induced reinstatement of cocaine seeking in male and female rats. Physiol Behav 105(2):209–214. https://doi.org/10.1016/j.physbeh.2011.08.020

13. Shalev U, Erb S, Shaham Y (2010) Role of CRF and other neuropeptides in stress-induced reinstatement of drug seeking. Brain Res 1314:15–28. https://doi.org/10.1016/j.brainres.2009.07.028

14. Ribeiro Do Couto B, Aguilar MA, Manzanedo C, Rodríguez-Arias M, Armario A, Minarro J (2006) Social stress is as effective as physical stress in reinstating morphine-induced place preference in mice. Psychopharmacology 185(4):459–470. https://doi.org/10.1007/s00213-006-0345-z

15. Shalev U, Marinelli M, Baumann MH, Piazza PV, Shaham Y (2003) The role of corticosterone in food deprivation-induced reinstatement of cocaine seeking in the rat. Psychopharmacology 168(1–2):170–176. https://doi.org/10.1007/s00213-002-1200-5

16. Mantsch JR, Weyer A, Vranjkovic O, Beyer CE, Baker DA, Caretta H (2010) Involvement of noradrenergic neurotransmission in the stress- but not cocaine-induced reinstatement of extinguished cocaine-induced conditioned place preference in mice: role for beta-2-adrenergic receptors. Neuropsychopharmacology 35(11):2165–2178. https://doi.org/10.1038/npp.2010.86

17. Miller LL, Ward SJ, Dykstra LA (2008) Chronic unpredictable stress enhances cocaine-conditioned place preference in type 1 cannabinoid receptor knockout mice. Behav Pharmacol 19(5–6):575–581. https://doi.org/10.1097/FBP.0b013e32830ded11

18. Ding YJ, Kang L, Li BM, Ma L (2005) Enhanced cocaine self-administration in adult rats with adolescent isolation experience. Pharmacol Biochem Behav 82:673–677. https://doi.org/10.1016/j.pbb.2005.11.007

19. Ribeiro Do Couto B, Aguilar MA, Lluch J, Rodríguez-Arias M, Minarro J (2009) Social experiences affect reinstatement of cocaine-induced place preference in mice. Psychopharmacology 207(3):485–498. https://doi.org/10.1007/s00213-009-1678-1

20. García-Pardo MP, Blanco-Gandía MC, Valiente-Lluch M, Rodríguez-Arias M, Miñarro J, Aguilar MA (2015) Long-term effects of repeated social stress on the conditioned place preference induced by MDMA in mice. Prog Neuropsychopharmacol Biol Psychiatry 63:98–109. https://doi.org/10.1016/j.pnpbp.2015.06.006

21. Dalley JW, Everitt BJ (2009) Dopamine receptors in the learning, memory and drug reward circuitry. Semin Cell Dev Biol 20(4):403–410. https://doi.org/10.1016/j.semcdb.2009.01.002

22. Wise RA (2009) Ventral tegmental glutamate: a role in stress-, cue-, and cocaine-induced reinstatement of cocaine-seeking. Neuropharmacology 56(Suppl 1):174–176. https://doi.org/10.1016/j.neuropharm.2008.06.008

23. Koob GF (2009) Dynamics of neuronal circuits in addiction: reward, antireward, and emotional memory. Pharmacopsychiatry 42(Suppl 1):32. https://doi.org/10.1055/s-0029-1216356

24. Rosenkranz JA, Moore H, Grace AA (2003) The prefrontal cortex regulates lateral amygdala neuronal plasticity and responses to previously conditioned stimuli. J Neurosci 23(35):11054–11064. https://doi.org/10.1523/JNEUROSCI.23-35-11054.2003

25. Barthas F, Hu MY, Siniscalchi MJ, Ali F, Mineur YS, Picciotto MR, Kwan AC (2020) Cumulative effects of social stress on reward-guided actions and prefrontal cortical activity. Biol Psychiatry. S0006-3223(20)30097-4 [pii]

26. García Pardo MP, Roger Sánchez C, De la Rubia Ortí JE, Aguilar Calpe MA (2017) Animal models of drug addiction. Adicciones 29(4):278–292. https://doi.org/10.20882/adicciones.862

27. García-Pardo MP, Rodríguez-Arias M, Maldonado C, Manzanedo C, Miñarro J, Aguilar MA (2014) Effects of acute social stress on the conditioned place preference induced by MDMA in adolescent and adult mice. Behav Pharmacol 25(5–6):532–546. https://doi.org/10.1097/FBP.0000000000000065

28. Montagud-Romero S, Aguilar MA, Maldonado C, Manzanedo C, Miñarro J, Rodríguez-Arias M (2015) Acute social defeat stress increases the conditioned rewarding effects of cocaine in adult but not in adolescent mice. Pharmacol Biochem Behav 135:1–12. https://doi.org/10.1016/j.pbb.2015.05.008

29. Miczek KA, Yap JJ, Covington HE (2008) Social stress, therapeutics and drug abuse: preclinical models of escalated and depressed intake. Pharmacol Ther 120(2):102–128. https://doi.org/10.1016/j.pharmthera.2008.07.006

30. Montagud-Romero S, Blanco-Gandía MC, Reguilón MD, Ferrer-Pérez C, Ballestín R, Miñarro J, Rodríguez-Arias M (2018) Social defeat stress: mechanisms underlying the increase in rewarding effects of drugs of abuse. Eur J Neurosci 48(9):2948–2970. https://doi.org/10.1111/ejn.14127

31. Caldwell EE, Riccio DC (2010) Alcohol self-administration in rats: modulation by temporal parameters related to repeated mild social defeat stress. Alcohol (Fayetteville, NY) 44(3):265–274. https://doi.org/10.1016/j.alcohol.2010.02.012

32. Rodríguez-Arias M, Navarrete F, Blanco-Gandia MC, Arenas MC, Bartoll-Andrés A, Aguilar MA, Rubio G, Miñarro J, Manzanares J (2016) Social defeat in adolescent mice increases vulnerability to alcohol consumption. Addict Biol 21(1):87–97. https://doi.org/10.1111/adb.12184

33. Norman KJ, Seiden JA, Klickstein JA, Han X, Hwa LS, DeBold JF, Miczek KA (2015) Social stress and escalated drug self-administration in mice I. alcohol and corticosterone. Psychopharmacology 232(6):991–1001. https://doi.org/10.1007/s00213-014-3733-9

34. Funk D, Vohra S, Le AD (2004) Influence of stressors on the rewarding effects of alcohol in wistar rats: studies with alcohol deprivation and place conditioning. Psychopharmacology 176(1):82–87. https://doi.org/10.1007/s00213-004-1859-x

35. van Erp AM, Miczek KA (2001) Persistent suppression of ethanol self-administration by brief social stress in rats and increased startle response as index of withdrawal. Physiol Behav 73(3):301–311. S0031-9384(01)00458-9 [pii]

36. van Erp AM, Tachi N, Miczek KA (2001) Short or continuous social stress: suppression of continuously available ethanol intake in subordinate rats. Behav Pharmacol 12(5):335–342. https://doi.org/10.1097/00008877-200109000-00004

37. Funk D, Harding S, Juzytsch W, Le AD (2005) Effects of unconditioned and conditioned social defeat on alcohol self-administration and reinstatement of alcohol seeking in rats. Psychopharmacology 183(3):341–349. https://doi.org/10.1007/s00213-005-0194-1

38. Bahi A (2013) Increased anxiety, voluntary alcohol consumption and ethanol-induced place preference in mice following chronic psychosocial stress. Stress (Amsterdam, Netherlands) 16(4):441–451. https://doi.org/10.3109/10253890.2012.754419

39. Karlsson C, Schank JR, Rehman F, Stojakovic A, Bjork K, Barbier E et al (2017) Proinflammatory signaling regulates voluntary alcohol intake and stress-induced consumption after exposure to social defeat stress in mice. Addict Biol 22(5):1279–1288. https://doi.org/10.1111/adb.12416

40. Macedo GC, Morita GM, Domingues LP, Favoretto CA, Suchecki D, Quadros IMH (2018) Consequences of continuous social defeat stress on anxiety- and depressive-like behaviors and ethanol reward in mice. Horm Behav 97:154–161. S0018-506X(17)30138-1 [pii]

41. Bahi A, Dreyer JL (2020) Environmental enrichment decreases chronic psychosocial stress-impaired extinction and reinstatement of ethanol conditioned place preference in C57BL/6 male mice. Psychopharmacology 237(3):707–721. https://doi.org/10.1007/s00213-019-05408-8

42. García-Pardo MP, Roger-Sánchez C, Rodríguez-Arias M, Miñarro J, Aguilar MA (2016) Effects of social stress on ethanol responsivity in adult mice. Neuropsychiatry 6. https://doi.org/10.4172/Neuropsychiatry.1000146

43. Covington HE, Kikusui T, Goodhue J, Nikulina EM, Hammer RP, Miczek KA (2005) Brief social defeat stress: long lasting effects on cocaine taking during a binge and zif268 mRNA expression in the amygdala and prefrontal cortex. Neuropsychopharmacology 30(2):310–321

44. Covington HE, Miczek KA (2005) Intense cocaine self-administration after episodic social defeat stress, but not after aggressive behavior: dissociation from corticosterone activation. Psychopharmacology 183(3):331–340.

https://doi.org/10.1007/s00213-005-0190-5

45. Burke AR, Miczek KA (2015) Escalation of cocaine self-administration in adulthood after social defeat of adolescent rats: role of social experience and adaptive coping behavior. Psychopharmacology 232(16):3067–3079. https://doi.org/10.1007/s00213-015-3947-5

46. Boyson CO, Miguel TT, Quadros IM, Debold JF, Miczek KA (2011) Prevention of social stress-escalated cocaine self-administration by CRF-R1 antagonist in the rat VTA. Psychopharmacology 218(1):257–269. https://doi.org/10.1007/s00213-011-2266-8

47. Boyson CO, Holly EN, Shimamoto A, Albrechet-Souza L, Weiner LA, DeBold JF, Miczek KA (2014) Social stress and CRF-dopamine interactions in the VTA: role in long-term escalation of cocaine self-administration. J Neurosci 34 (19):6659–6667. https://doi.org/10.1523/JNEUROSCI.3942-13.2014

48. Cruz FC, Quadros IM, Hogenelst K, Planeta CS, Miczek KA (2011) Social defeat stress in rats: escalation of cocaine and "speedball" binge self-administration, but not heroin. Psychopharmacology 215(1):165–175. https://doi.org/10.1007/s00213-010-2139-6

49. Han X, Albrechet-Souza L, Doyle MR, Shimamoto A, DeBold JF, Miczek KA (2015) Social stress and escalated drug self-administration in mice II. cocaine and dopamine in the nucleus accumbens. Psychopharmacology 232(6):1003–1010. https://doi.org/10.1007/s00213-014-3734-8

50. Miczek KA, Nikulina EM, Shimamoto A, Covington HE (2011) Escalated or suppressed cocaine reward, tegmental BDNF, and accumbal dopamine caused by episodic versus continuous social stress in rats. J Neurosci 31 (27):9848–9857. https://doi.org/10.1523/JNEUROSCI.0637-11.2011

51. Yap JJ, Chartoff EH, Holly EN, Potter DN, Carlezon WA, Miczek KA (2015) Social defeat stress-induced sensitization and escalated cocaine self-administration: the role of ERK signaling in the rat ventral tegmental area. Psychopharmacology 232(9):1555–1569. https://doi.org/10.1007/s00213-014-3796-7

52. Holly EN, Boyson CO, Montagud-Romero S, Stein DJ, Gobrogge KL, DeBold JF, Miczek KA (2016) Episodic social stress-escalated cocaine self-administration: role of phasic and tonic corticotropin releasing factor in the anterior and posterior ventral tegmental area. J

Neurosci 36(14):4093–4105. https://doi.org/10.1523/JNEUROSCI.2232-15.2016

53. Boyson CO, Holly EN, Burke AR, Montagud-Romero S, DeBold JF, Miczek KA (2016) Maladaptive choices by defeated rats: link between rapid approach to social threat and escalated cocaine self-administration. Psychopharmacology 233(17):3173–3186. https://doi.org/10.1007/s00213-016-4363-1

54. Yap JJ, Miczek KA (2007) Social defeat stress, sensitization, and intravenous cocaine self-administration in mice. Psychopharmacology 192(2):261–273. https://doi.org/10.1007/s00213-007-0712-4

55. Arena DT, Covington HE, DeBold JF, Miczek KA (2019) Persistent increase of I.V. cocaine self-administration in a subgroup of C57BL/6J male mice after social defeat stress. Psychopharmacology 236(7):2027–2037. https://doi.org/10.1007/s00213-019-05191-6

56. Rodríguez-Arias M, Montagud-Romero S, Rubio-Araiz A, Aguilar MA, Martín-García E, Cabrera R, Maldonado R, Porcu F, Colado MI, Miñarro J (2017) Effects of repeated social defeat on adolescent mice on cocaine-induced CPP and self-administration in adulthood: integrity of the blood-brain barrier. Addict Biol 22(1):129–141. https://doi.org/10.1111/adb.12301

57. Blouin AM, Pisupati S, Hoffer CG, Hafenbreidel M, Jamieson SE, Rumbaugh G, Miller CA (2019) Social stress-potentiated methamphetamine seeking. Addict Biol 24 (5):958–968. https://doi.org/10.1111/adb.12666

58. McLaughlin JP, Land BB, Li S, Pintar JE, Chavkin C (2006) Prior activation of kappa opioid receptors by U50,488 mimics repeated forced swim stress to potentiate cocaine place preference conditioning. Neuropsychopharmacology 31(4):787–794

59. García-Pardo MP, de la Rubia JE, Aguilar MA (2017) The influence of social stress on the reinforcing effect of ecstasy under the conditioned place preference paradigm: the role played by age, dose and type of stress. Rev Neurol 65(10):469–476

60. Guerrero-Bautista R, Do Couto BR, Hidalgo JM, Cárceles-Moreno FJ, Molina G, Laorden ML, Núñez C, Milanés MV (2019) Modulation of stress- and cocaine prime-induced reinstatement of conditioned place preference after memory extinction through dopamine D3 receptor. Prog Neuro-Psychopharmacol Biol Psychiatry 92:308–320. https://doi.org/10.1016/j.pnpbp.2019.01.017

61. Titomanlio F, Manzanedo C, Rodríguez-Arias M, Mattioli L, Perfumi M, Minarro J,

Aguilar MA (2013) Rhodiola rosea impairs acquisition and expression of conditioned place preference induced by cocaine. Evid Based Complement Alternat Med 2013:697632. https://doi.org/10.1155/2013/697632

62. Reguilon MD, Montagud-Romero S, Ferrer-Perez C, Roger-Sanchez C, Aguilar MA, Minarro J, Rodríguez-Arias M (2017) Dopamine D2 receptors mediate the increase in reinstatement of the conditioned rewarding effects of cocaine induced by acute social defeat. Eur J Pharmacol 799:48–57. S0014-2999(17) 30048-1 [pii]

63. Montagud-Romero S, Montesinos J, Pascual M, Aguilar MA, Roger-Sanchez C, Guerri C, Miñarro J, Rodríguez-Arias M (2016) Up-regulation of histone acetylation induced by social defeat mediates the conditioned rewarding effects of cocaine. Prog Neuro-Psychopharmacol Biol Psychiatry 70:39–48. https://doi.org/10.1016/j.pnpbp.2016.04.016

64. Montagud-Romero S, Reguilon MD, Roger-Sanchez C, Pascual M, Aguilar MA, Guerri C, Miñarro J, Rodríguez-Arias M (2016) Role of dopamine neurotransmission in the long-term effects of repeated social defeat on the conditioned rewarding effects of cocaine. Prog Neuro-Psychopharmacol Biol Psychiatry 71:144–154. https://doi.org/10.1016/j.pnpbp.2016.07.008

65. Montagud-Romero S, Nuñez C, Blanco-Gandia MC, Martínez-Laorden E, Aguilar MA, Navarro-Zaragoza J, Almela P, Milanés MV, Laorden ML, Miñarro J, Rodríguez-Arias M (2017) Repeated social defeat and the rewarding effects of cocaine in adult and adolescent mice: dopamine transcription factors, proBDNF signaling pathways, and the TrkB receptor in the mesolimbic system. Psychopharmacology 234(13):2063–2075. https://doi.org/10.1007/s00213-017-4612-y

66. Ferrer-Pérez C, Martinez TE, Montagud-Romero S, Ballestín R, Reguilón MD, Miñarro J, Rodríguez-Arias M (2018) Indomethacin blocks the increased conditioned rewarding effects of cocaine induced by repeated social defeat. PLoS One 13(12): e0209291. https://doi.org/10.1371/journal.pone.0209291

67. Ferrer-Pérez C, Reguilón MD, Manzanedo C, Aguilar MA, Miñarro J, Rodríguez-Arias M (2018) Antagonism of corticotropin-releasing factor CRF1 receptors blocks the enhanced response to cocaine after social stress. Eur J Pharmacol 823:87–95. https://doi.org/10.1016/j.ejphar.2018.01.052

68. Ferrer-Pérez C, Castro-Zavala A, Luján MÁ, Filarowska J, Ballestín R, Miñarro J, Valverde O, Rodríguez-Arias M (2019) Oxytocin prevents the increase of cocaine-related responses produced by social defeat. Neuropharmacology 146:50–64. https://doi.org/10.1016/j.neuropharm.2018.11.011

69. Montagud-Romero S, Montesinos J, Pavón FJ, Blanco-Gandia MC, Ballestín R, Rodríguez de Fonseca F, Miñarro J, Guerri C, Rodríguez-Arias M (2020) Social defeat-induced increase in the conditioned rewarding effects of cocaine: role of CX3CL1. Prog Neuro-Psychopharmacol Biol Psychiatry 96:109753. https://doi.org/10.1016/j.pnpbp.2019.109753

70. Rodríguez-Arias M, Montagud-Romero S, Guardia Carrión AM, Ferrer-Pérez C, Pérez-Villalba A, Marco E, López Gallardo M, Viveros MP, Miñarro J (2018) Social stress during adolescence activates long-term microglia inflammation insult in reward processing nuclei. PLoS One 13(10):e0206421. https://doi.org/10.1371/journal.pone.0206421

71. García-Pardo MP, Miñarro J, Llansola M, Felipo V, Aguilar MA (2019) Role of NMDA and AMPA glutamatergic receptors in the effects of social defeat on the rewarding properties of MDMA in mice. Eur J Neurosci 50 (3):2623–2634. https://doi.org/10.1111/ejn.14190

72. García-Pardo MP, Llansola M, Felipo V, De la Rubia Ortí JE, Aguilar MA (2020) Blockade of nitric oxide signalling promotes resilience to the effects of social defeat stress on the conditioned rewarding properties of MDMA in mice. Nitric Oxide 98:29–32. https://doi.org/10.1016/j.niox.2020.03.001

73. Coventry TL, D'Aquila PS, Brain P, Willner P (1997) Social influences on morphine conditioned place preference. Behav Pharmacol 8(6–7):575–584. https://doi.org/10.1097/00008877-199711000-00015

74. Tomek SE, Stegmann GM, Olive MF (2019) Effects of heroin on rat prosocial behavior. Addict Biol 24(4):676–684. https://doi.org/10.1111/adb.12633

75. Rodríguez-Arias M, Minarro J, Aguilar MA, Pinazo J, Simon VM (1998) Effects of risperidone and SCH 23390 on isolation-induced aggression in male mice. Eur Neuropsychopharmacol 8(2):95–103. S0924-977X(97) 00051-5 [pii]

76. Miczek KA, Thompson ML, Shuster L (1982) Opioid-like analgesia in defeated mice. Science (New York, NY) 215(4539):1520–1522. https://doi.org/10.1126/science.7199758

77. Covington HE, Miczek KA (2001) Repeated social-defeat stress, cocaine or morphine. effects on behavioral sensitization and intravenous cocaine self-administration "binges". Psychopharmacology 158(4):388–398. https://doi.org/10.1007/s002130100858

78. Miczek KA, Weerts EM, Tornatzky W, DeBold JF, Vatne TM (1992) Alcohol and "bursts" of aggressive behavior: ethological analysis of individual differences in rats. Psychopharmacology 107(4):551–563. https://doi.org/10.1007/BF02245270

79. Calpe-Lopez C, Garcia-Pardo MP, Martinez-Caballero MA, Santos-Ortiz A, Aguilar MA (2020) Behavioral traits associated with resilience to the effects of repeated social defeat on cocaine-induced conditioned place preference in mice. Front Behav Neurosci 13:278. https://doi.org/10.3389/fnbeh.2019.00278

80. Tornatzky W, Miczek KA (1993) Long-term impairment of autonomic circadian rhythms after brief intermittent social stress. Physiol Behav 53(5):983–993. 0031-9384(93)90278-N [pii]

81. Rodríguez-Arias M, Montagud Romero S, Rubio A, Aguilar M, Martin-Garcia E, Cabrera R et al (2015) Effects of repeated social defeat on adolescent mice on cocaine-induced CPP and self-administration in adulthood: integrity of the blood-brain barrier. Addict Biol 22:129. https://doi.org/10.1111/adb.12301

82. Brain PF, McAllister KH, Wamsley SV (1989) Drug effects on social behavior. Methods in ethopharmacology. In: Boulton AA, Baker GB, Greeshaw AJ (eds) Neuromethods, vol 13. The Human Press, Clifton, NJ, pp 687–739

83. Holly EN, DeBold JF, Miczek KA (2015) Increased mesocorticolimbic dopamine during acute and repeated social defeat stress: modulation by corticotropin releasing factor receptors in the ventral tegmental area. Psychopharmacology 232(24):4469–4479. https://doi.org/10.1007/s00213-015-4082-z

84. Leonard MZ, DeBold JF, Miczek KA (2017) Escalated cocaine "binges" in rats: enduring effects of social defeat stress or intra-VTA CRF. Psychopharmacology 234 (18):2823–2836. https://doi.org/10.1007/s00213-017-4677-7

85. Vasconcelos M, Stein DJ, de Almeida RM (2015) Social defeat protocol and relevant biomarkers, implications for stress response physiology, drug abuse, mood disorders and individual stress vulnerability: a systematic review of the last decade. Trends Psychiatry Psychother 37(2):51–66. https://doi.org/10.1590/2237-6089-2014-0034

86. García-Pardo MP, Calpe-López C, Miñarro J, Aguilar MA (2019) Role of N-methyl-D-aspartate receptors in the long-term effects of repeated social defeat stress on the rewarding and psychomotor properties of cocaine in mice. Behav Brain Res 361:95–103. https://doi.org/10.1016/j.bbr.2018.12.025

87. Wang J, Bastle RM, Bass CE, Hammer RP, Neisewander JL, Nikulina EM (2016) Overexpression of BDNF in the ventral tegmental area enhances binge cocaine self-administration in rats exposed to repeated social defeat. Neuropharmacology 109:121–130. https://doi.org/10.1016/j.neuropharm.2016.04.045

88. Montesinos J, Castilla-Ortega E, Sanchez-Marin L, Montagud-Romero S, Araos P, Pedraz M et al (2020) Cocaine-induced changes in CX3CL1 and inflammatory signaling pathways in the hippocampus: association with IL1beta. Neuropharmacology 162:107840. S0028-3908(19)30406-X [pii]

89. Shimamoto A (2018) Social defeat stress, sex, and addiction-like behaviors. Int Rev Neurobiol 140:271–313. S0074-7742(18)30038-2 [pii]

90. Ashokan A, Sivasubramanian M, Mitra R (2016) Seeding stress resilience through inoculation. Neural Plasticity 2016:4928081. https://doi.org/10.1155/2016/4928081

91. Brockhurst J, Cheleuitte-Nieves C, Buckmaster CL, Schatzberg AF, Lyons DM (2015) Stress inoculation modeled in mice. Transl Psychiatry 5:e537. https://doi.org/10.1038/tp.2015.34

# Chapter 9

# Environmental Enrichment and the Effects on Drug Abuse Vulnerability: The Last Ten Years

## Dustin J. Stairs, Taena Hanson, and Kendall Kellerman

## Abstract

Environmental enrichment in rats during development has been shown to reliably alter the sensitivity to various drugs of abuse. The current chapter will attempt to summarize new research that has come out investigating the effects of environmental enrichment on different drugs of abuse in the last 10 years. At the same time the chapter also aims to give a detailed description of the methods we employ in the environmental enrichment model used in our laboratory. We focused on details in the different the use of the environmental conditions that are often not reported in peer-reviewed publications. The review of studies conducted in the past decade indicated that studies continue to confirm that animals raised in enriched conditions show a protective effect in the behavioral sensitivity to different drugs of abuse relative to the impoverished condition counterparts. Studies focusing on changes in the nervous system indicate changes in mesolimbic structures as a result of enrichment, areas like the NAcc, VTA, and the bed nucleus of the stria terminalis. There are also alterations in the HPA-axis in enriched vs. impoverished animals that plays a role in differences seen in stimulant self-administration. Future studies using the environmental enrichment model should focus on understudied drugs like cannabinoids, opiates, and nicotine.

Key words Environmental enrichment, Rodents, Isolation, Novelty, Drug abuse, Mesocorticolimbic reward system, Drug self-administration, Conditioned place preference, Locomotor activity, Dopamine, Glutamate, Corticosterone

## 1 Introduction

It has been about 10 years since the last comprehensive review of the effect of environmental enrichment on drug abuse vulnerability was written [1]. In the previous review we attempted to give a complete review of the work dealing with the effects of environmental enrichment on both behavioral and neurochemical effects of abused drugs and how this manipulation alters vulnerability in rodents that are exposed to these differential conditions. This paper, although 10 years old, is still cited numerous times each year in the environmental enrichment literature. While as the first author of the paper, I am very happy it is still being cited, there has been a considerable amount of new research conducted in the last

María A. Aguilar (ed.), *Methods for Preclinical Research in Addiction*, Neuromethods, vol. 174,
https://doi.org/10.1007/978-1-0716-1748-9_9, © Springer Science+Business Media, LLC, part of Springer Nature 2022

10 years looking at the effects of environmental enrichment. The current chapter is put together to serve as an update to the previous review. We have tried to gather all the new empirical papers investigating the effects of environmental enrichment since 2009 and create a comprehensive review of the research completed in the last 10 years.

Over the last decade, the bulk of the research that has been published investigating the effects of enrichment on drug abuse have focused on the psychomotor stimulants; amphetamines, cocaine and nicotine, while there have only been three papers investigating the effects of opiates and a handful of papers looking at the effects of ethanol. Given the papers that were found, we have decided to organize this review by drugs investigated and the behavioral assay used. Also given the methods focus of this chapter we will start with a detailed discussion of the environmental enrichment procedures and materials need and a discussion of why we use the specific social control group that we do in our laboratory.

## 2    General Enrichment Procedures

One of the difficult aspects of studying the effects of environmental enrichment on drugs of abuse is that there is variability in how labs define enrichment [2] as well as debate about what is considered an appropriate control group [3]. Given the variety of enrichment procedures that can be used, our laboratory has purposefully tried to mimic the enrichment manipulations developed in Dr. Michael Bardo's laboratory at the University of Kentucky. We have mimicked this manipulation in order to improve the comparison of results across different laboratories. In the typical Bardo environmental enrichment experiment, rats (often, male Sprague-Dawley rats) are received from a commercial vendor at approximately 21 days of age and housed in one of three conditions. In the enriched condition (EC) rats are socially housed in large stainless steel cages in groups of 12. The rats also have 14 novel objects placed throughout the cage with seven of these objects changed daily and the remaining toys being rearranged. The novel configuration of toys is done daily throughout the duration of the experiment (Fig. 1). In the isolated condition (IC) rats are housed in hanging stainless steel cages with no novel objects nor social cohorts (Fig. 2). The third condition is the social condition (SC) in which the rats are pair-housed in standard National Institute of Health (NIH) caging with no novel objects (Fig. 3). After being placed into one of these three conditions the animals are allowed to develop, typically for 30 days, until approximately postnatal Day 51. At this point behavioral procedures typically begin, with animals remaining in these conditions throughout the duration of the experiments. While this is only one enrichment

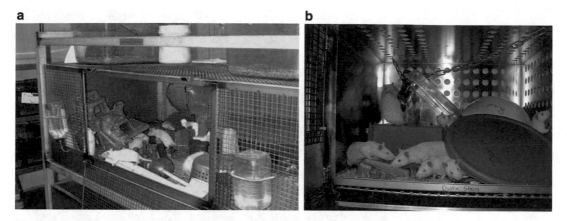

**Fig. 1** Environmental enrichment conditions (EC). (**a**) Enrichment caging used in Bardo laboratory. (**b**) Enrichment caging used in Stairs laboratory. Male rats are typically housed in groups of 12 with the novel objects being change and reconfigured daily in large metal cages containing a solid stainless-steel floor with shaved aspen bedding. Rats are typically group fed with two water bottles available ab libitum

**Fig. 2** Isolated condition (IC). Male rats are housed individually in hanging stainless steel cages (17 × 24 × 20 cm) with a wire mesh floor and front wall and solid top and side walls. A food basket hangs inside the cage with a water bottle available ab libitum

procedure that can be employed, many of the trainees that have come through the Bardo laboratory have chosen to mimic this enrichment manipulation. Throughout this review I will only refer to EC, SC, and IC rats in experiments that employ this form of environmental enrichment. We will not use these abbreviations

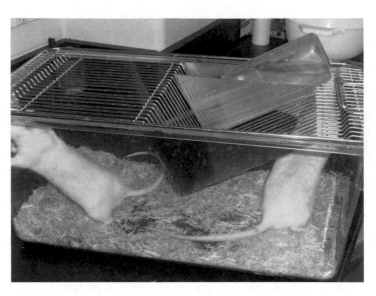

**Fig. 3** Social condition (SC). Male rats are pair-housed in shoebox cages (48 × 27 × 20 cm) with shaved aspen bedding. Rats are fed in pairs with a water bottle available ab libitum

when describing studies that employ an alternative enrichment procedure.

## 3   Environmental Enrichment Procedures

### 3.1   Materials

#### 3.1.1   Enrichment Caging

For the enrichment caging a larger stainless-steel cage with solid (no wire grid) flooring is used. This cage needs to be large enough to house up to 12 rats while still meeting NIH standards for space per rat in addition to supporting the 14 novel objects or toys. While in Michael Bardo's laboratory we used custom-made large stainless-steel caging (60 × 120 × 45 cm, Fig. 1a) that also had framing with wheels in which the cages rested. This allowed for the heaver cages to be wheeled down to be cleaned. The cost of having these custom-made could be in the price range of $2–3000 per cage. At Creighton University I was able to fine old stainless steel ferret cages in a warehouse that were no longer in use (62 × 62 × 42 cm, Fig. 1b). Both of these designs are large enough to accommodate 12 rats with 14 novel objects. The added height allows toys to be hung from the ceiling, which the rats are often found climbing up into. In terms of the larger enrichment cages you will need to coordinate how the cages will be cleaned with your animal resource facility staff. Given the size of the cages they often have to be cleaned through a separate mechanism than standard plastic shoebox caging and you have to concern yourself with how they will be transported to be cleaned.

### 3.1.2 Toy-Change

For the enrichment condition you will also need to purchase a number of hard plastic toys and objects. We often purchase these objects/toys at basic dollar general type stores or colleagues will donate their children's old toys. A factor to keep in mind with the toys is they need to be a hard enough plastic to go through the cage washing system (withstand heat and not melt) in your animal resource facilities if you plan to reuse them. Once the toys are acquired we group them into four different groups that are given a color name (red, green, blue, orange) and marked with a Sharpie. Then each day we do a "toy-change"; this requires seven toys from two different color groups to be placed and arranged in the cage (always hanging at least two toys). On the next day's toy-change, the seven toys from one of the color groups are removed and placed to the side, the remaining seven toys from the other color group stay in the cage but are rearranged. Then seven new toys are added from another color group. On third day of toy-change, the color group that has been in for 2 days are removed and set to the side and the fourth and final color group toys are added to the remaining seven toys that have been rearranged. The toy-change continues in this daily manner until all permutations of color group combinations have been done, at which point the toys are washed and the process starts over with 14 fresh toys. Doing this toy-change procedure allows for the EC rats to have a novel configuration in their environment everyday throughout the duration of the experiment.

### 3.1.3 Feeding

Our enriched rats are always group fed whether they are on ad libitum access or food restricted. Ad libitum access to food is requires their food hopper to be topped off daily and they always have two full water bottles available at all times. When we do have to food restrict our rats we typically try to keep our rats at ~85% of free-feeding weight. We typically achieve this by first changing the bedding of EC cage, because they tend to nest their food. Then we give 15 g of food per rat ($15 \times 12 = 180$ g) that is placed in a pile on the front of the cage. Each animal is tail marked with a number using a Sharpie and animals are weighed daily to monitor their weight. After the animals' weight comes down we often increase the daily food allotment to 18 g of food per rat. At this daily amount all animals tend to increase their weight while in a level of food deprivation. One thing of note is you have to keep up on maintaining the tail marks, particularly while they are young, to make sure you can tell the animals apart.

### 3.1.4 Special Considerations with EC Rats

Once you have the large caging and toys most other things are fairly standard in terms of investigating the abuse potential of abused drugs, with one important exception; running intravenous drug self-administration. Since the EC rats are group-housed you have to worry about an animal's social cohorts chewing and opening up their indwelling jugular vein catheter. In order to overcome this limitation, we make custom-made catheters for both our EC and SC rats. We use a facial jugular vein catheter system where the port for the catheter exits out the top of the animals' head via a dental acrylic head mount. With this system we make a modified catheter for our group-housed rats (EC and SC rats) which has a plastic thread round the port so a stainless steel "stand-off" can be screwed over the port to the catheter. This stand-off prevents the other animals from opening an animal's catheter while they are in the home cage. Without these modified catheters you cannot maintain patent catheters in group-housed rats. This also explains why some researchers that investigate enrichment effects on drug abuse do not look at active drug self-administration while the animals are in the enriched condition ("treatment" approach described below). Both the group-housed catheter and the catheter used in IC rats can be seen in Fig. 4.

## 4   Impoverished Condition Procedures

### 4.1   Materials

#### 4.1.1   Impoverished Caging

We use the single hanging stainless steel caging (*see* Fig. 2). This caging used to be the standard rat caging up until the 2000s in which there were changes to NIH standard in housing of rodents. Our institutions IACUC has to get special approval for our use of these cages, which is not necessarily an issue for the individual investigator. Although, the investigator needs to clarify in the IACUC protocol that this is an important experimental manipulation in the experiment and needs to be done for the validity of the animal model. One issue the researcher may face when dealing with this type of caging is how to purchase this sort of caging, as it is not the standard type of caging produced today. I would recommend talking with those in charge of animal facilities to see if old caging is being stored somewhere on campus. These cages can also be found for sale online in places like Ebay. Besides the caging you will also want to have stainless steel food bins that can be hung from the wall to place the animals food in. These work better than placing the animal's food on the floor as the food placed on the floor tends to fall through the wire mesh flooring where the animals cannot reach. Due to this loss of some food falling down below when IC rats are on food deprivation, we tend to give an extra 2 g of food compared to what the EC and SC rats receive (20 g vs. 18 g). For the most part, IC rats do not tend to show any health issues being in these cages for an extended period of time. Although being in this condition does seem to create a behavioral phenotype in which the IC rats will be more skittish in terms of handling by the researchers, especially early on.

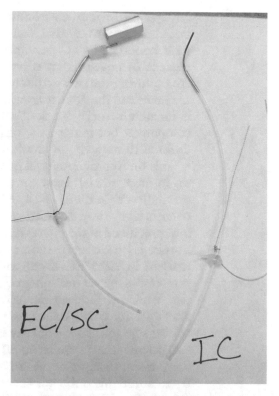

**Fig. 4** Custom-made catheters for the group-housed (EC and SC) and individually housed (IC) rats. The EC/SC catheters have a plastic threading glued onto the stainless-steel catheter port. This threading allows for the stainless-steel stand-off (pictured at the top) to be screwed over the catheter opening after the opening is closed with a small pieced of closed off tubing

## 5 Social Control Condition

Before I begin discussion of the effects of enrichment on various drugs of abuse I would like to touch on the question of what is the appropriate control to the enriched group. This is a question of considerable debate in the field and of journal editors [20], and in my opinion a question without a clear right answer. The discussion of the various arguments of what is the appropriate control is beyond the scope of this review, but what we would like to layout is how we see the manipulation of enrichment in the current paradigm and why we have settled on using the pair-housed group (SC rats) in our laboratory. The enrichment model developed in the Bardo lab was really developed to investigate the ability of novelty exposure to "create" rats with different behavioral phenotypes that alters the rat's vulnerability to drugs of abuse. The manipulation of novelty exposure is on a continuum, the EC and IC conditions are set-up to capture the extremes of the novelty spectrums. The EC rats are exposed to high levels of novelty both in terms of the physical environment (daily novel object arrangement) but also the novelty that comes from the social interaction with their cohort. The IC condition is trying to capture the virtual absence of novelty, that is no physical or social novelty. If these are capturing the two extremes the control group is somewhere in the middle of the novelty continuum. Given that the rodent species is innately a social species, we elect to keep the social novelty component but remove the physical novelty. Once you make the decision to remove the physical novelty you still have to consider what is the appropriate level of social novelty. From my time training in the Bardo lab I saw the use of different social controls, some researchers used the same large EC cage without toys and 12 social cohorts all the way down to just pair-housed rats with no toys. My interpretations of the effects of those different groups is that social-novelty is on a continuation as well, that is, SC groups where the rats are in large EC cages with no toys but 12 social cohorts "creates" a behavioral phenotype that will be closer to the EC rats, while pair-housed rats tend to have a behavioral phenotype closer to IC rats. Given this potential variability depending on cohort size and space we really have settled on using a pair-housed social group that is similar to NIH standard housing and allows for comparison to labs that do not do enrichment research. We do not try to argue this is the perfect control, but given the group design nature of enrichment research the experimenter has to decide on a control group that allows for control of the manipulation of interest but is also financially feasible to carry out.

## 6    Enrichment and Amphetamines

### 6.1 Locomotor Effects
#### D-Amphetamine

The bulk of the studies investigating the effects of environmental enrichment on amphetamine locomotor effects have come out of Dr. Mary Cain's laboratory at Kansas State University. For instance, Gill, Arnold, and Cain [4] looked at the ability of the metabolic glutamate receptor 5 (mGluR5) antagonist MTEP to alter the locomotor effects of both acute and repeated D-amphetamine injections in EC, SC, and IC rats. The authors found that MTEP pretreatments significantly decreased the effects of acute low dose of amphetamine but only in the EC and SC rats, not in the IC rats. Additionally, the authors found that MTEP also attenuated the heightened locomotor sensitization seen in IC rats following repeated amphetamine pretreatments. The results of this study indicated the role of the glutamate systems in the differential sensitivity to the locomotor effects of D-amphetamine in EC, SC, and IC rats. The next paper to further investigate the role of glutamate system in the locomotor effects of D-amphetamine was done by Arndt, Arnold, and Cain [5]. In this study the authors investigated the role of the metabolic glutamate receptor 2 and 3 receptor (mGluR$_{2/3}$) by administering the mGluR$_{2/3}$ agonist LY-379268 following injections with acute or repeated D-amphetamine. Results from this study found that LY-379268 decreased the acute locomotor effects of D-amphetamine but only in IC and SC rats; there was no significant decrease in D-amphetamine induced locomotor behavior in EC rats. The authors also found that the low dose of LY-379268 decreased the locomotor effects of repeated D-amphetamine pretreatments again only in IC and SC rats, although the high dose decreased this effect in all three groups of rats. The results from this study strengthen the argument that the glutamatergic system is altered by environmental enrichment and that both the mGluR$_{2/3}$ and the mGluR5 receptors are in part mediating the differential sensitivity to the locomotor stimulant effects of D-amphetamine in EC, SC, and IC rats.

Another study put out by the Cain group [6] aimed to elucidate the potential interacting effects of environmental enrichment and D-amphetamine exposure on the mesolimbic dopamine pathway. They did so by looking at *c-fos* expression in the nucleus accumbens (NAcc), medial prefrontal cortex (mPFC), and different nuclei in the amygdala in rats following either saline or D-amphetamine injections and locomotor screens. This study found that following D-amphetamine exposure (1.0 mg/kg) IC rats had a significantly higher number of *c-fos* stained neurons compared to their saline control and also compared to D-amphetamine EC rats in the NAcc. In the cingulate and prelimbic cortices the number of *c-fos* stained neurons were significantly higher in the saline-treated EC rats compared to the SC saline-treated rats; there were no effects of D-amphetamine in these two regions. There were no other significant effects seen in any other brain areas tested. These results indicate that environmental enrichment alters prelimbic areas and the NAcc which can alter the sensitivity to D-amphetamine drugs.

Besides looking at brain changes as a result of enrichment and D-amphetamine exposure, the Cain group recently published a study trying to tease apart the role of enrichment versus the role of the response to novelty in some of the D-amphetamine locomotor differences seen in EC, SC, and IC rats [7]. To test this the authors exposed EC, SC, and IC rats to an inescapable novel environment while they measured locomotor behavior. This showed that after 30 days in the differential environments IC rats demonstrated significant increase in their response to the novel environment compared to their baseline measure prior to enrichment exposure and significantly higher than the response seen in both EC and SC rats. When animals were then exposed to repeated D-amphetamine injections (0.5 mg/kg) and later tested in the novel environment, only the SC rats showed significant effects of D-amphetamine compared to saline controls. This experiment indicates that response to novelty in differentially raised rats is an important component to the D-amphetamine locomotor effects seen in EC, SC, and IC rats. In a second experiment the authors showed that this differential locomotor response to a novel environment in EC and IC rats is also seen in animals that are not exposed to environmental enrichment until adulthood (PND 60). The last study we found that was published looking at the locomotor effects of D-amphetamine in differentially reared rats was conducted by our laboratory [8]. In this study we wanted to determine whether environmental enrichment could block the ability of adolescent nicotine pretreatments to increases sensitivity to the locomotor effects of D-amphetamine in adulthood. In this study we pretreated EC and IC rats with nicotine (0.4 mg/kg given PND 28–34) while they were developing in the conditions, then after a 30 day washout period in their conditions we tested the animals' locomotor sensitivity to an acute challenge to D-amphetamine. We found that only nicotine-treated IC rats showed a significant cross sensitization to a low dose of D-amphetamine (0.5 mg/kg), although when a higher dose was tested (1.0 mg/kg) only nicotine-treated EC rats showed a significant cross-sensitization to D-amphetamine. This paper indicated that environmental enrichment can protect against the ability of adolescent nicotine exposure to increase the sensitivity to low doses of D-amphetamine in adulthood.

## 6.2 Locomotor Effects MDMA and Methylphenidate

While the majority of locomotor studies have used D-amphetamine a couple of studies have investigated the locomotor effects of MDMA or methylphenidate (MPD) in EC and IC rats. A study by Wooters et al. [9] found that there were no differences in the acute locomotor response to MPD, but EC rats had a decreased locomotor response to repeated pretreatments of the low dose of MPD (3 mg/kg) compared to IC rats. This study also showed the basal levels of the dopamine transporter protein (DAT) function was blunted in EC rats compared to IC rats in the PFC. Furthermore, it was found that an acute dose of MPD decreased DAT function in the PFC only in EC rats while repeated dosing decreased DAT binding in the PFC in EC, SC, and IC rats. Although when the striatum was investigated, repeated MPD decreased DAT binding only in EC rats not SC or IC rats. These results extended the literature on environmental enrichment by showing that this manipulation can also decrease the locomotor sensitivity to repeated MPD and that enrichment and MPD exposure interact to alter DAT function in key mesolimbic structures involved in drugs of abuse. In a second study looking at the interaction between enrichment and repeated MPD on locomotor behavior, when the MPD exposure was done via an osmotic mini-pump, there was no differential effects of MPD in enriched, social or isolated rats. Although the authors did find regardless of MPD exposure, isolated rats had significantly more locomotor behavior compared to enriched rats [10]. It appears the route of administration may be an important factor to consider with looking at the effects of repeated MPD on the locomotor response. Finally, a study looking at the effects of differential housing and repeated MDMA [11], they found that there was no effect on locomotor behavior after acute injections of MDMA (5 mg/kg) but by the second day of injections animals that were socially housed had a significantly greater locomotor response than compared to the isolated rats, this persisted for the 5 days of injections. In this study following MDMA pretreatments the authors also looked at sensitivity to cocaine conditioned placed preference (CPP). The authors found that MDMA exposure increased the level of cocaine CPP and that the differential environments altered this effect depending on whether the animals had social interaction versus physical enrichment [10].

## 6.3 Conditioned Place Preference (CPP) with Amphetamines

While the previous literature tends to consistently show that enrichment decreases the locomotor effects of amphetamines, the effects of enrichment on amphetamine CPP in the past have shown either the opposite effect or no effect of enrichment [1]. The research done in the last 10 years seem to have continued to support this lack of an effect of enrichment. For instance, Hofford et al. [12] found the EC and IC rats displayed similar levels of methamphetamine (METH) induced CPP with both 0.3 and 1.0 mg/kg doses. This lack of an effect of enrichment is similar to results shown in mice which also indicated a lack of differences in METH-induced CPP in EC or IC mice [13]. Thiriet et al. [12] also showed the enrichment did not block the neurotoxic effects of METH on striatal dopamine content when compared to SC mice. A study by Yates et al. [14] again found that both adolescent and adult SC and IC rats did not differ in their ability to acquire a D-amphetamine-induced CPP response. Although this study did find that only adolescent IC rats displayed a social interaction-induced CPP response. Finally, these authors also found that when given a choice between a compartment that had been paired with D-amphetamine versus a compartment that had been paired with social interaction, adolescent IC rats choose the compartment with social interaction over the D-amphetamine compartment, whereas SC rats chose the D-amphetamine compartment over the social interaction chamber [13]. This study indicates that IC rats, when given a choice between drug-paired stimuli vs. social-paired stimuli, will choose the social-paired stimuli; this indicates that the social isolation component of the IC condition is an important motivational manipulation.

## 6.4 D-Amphetamine Self-Administration

While results with amphetamine CPP have continued to show a lack of an effect of environmental enrichment, studies looking at the effect of enrichment on self-administration consistently show enrichment-induced differences with D-amphetamine and methamphetamine self-administration. In a study by Arndt et al. [15], they found under baseline conditions IC rats self-administered higher levels of a low dose of D-amphetamine under both an fixed-ratio (FR) or a progressive-ratio (PR) compared to EC rats. At a higher dose of D-amphetamine the effect of enrichment was lost when behavior was maintained under an FR, but IC rats continued to display higher levels of intake on the PR schedule with the higher dose of D-amphetamine compared to both EC and SC rats [14]. Also this study found that there was a differential role of the MGluR5 receptor in D-amphetamine self-administration in EC and IC rats in that pretreatments with the antagonist MTEP had a greater decreasing effect in IC rats when the amphetamine dose was low, whereas MTEP had greater decreasing effects in EC rats when the D-amphetamine dose was low.

Not only does there seem to be a role for the MGluR5 receptor in the differential effects of enrichment on D-amphetamine, there also appears to be difference in EC and IC rats in how the stress axis plays a role in D-amphetamine differences [16]. In this study we measured basal levels of corticosterone in EC, SC, and IC rats and found that IC rats showed a much higher corticosterone response following the first blood draw. IC rats also showed a more rapid rise in D-amphetamine-induced corticosterone response. Finally, in this study we found the pretreatments with a glucocorticoid receptor antagonist (RU486) IC rats had a blunted antagonistic effect on D-amphetamine self-administration (0.03 mg/kg/infusion) compared to EC rats. These results indicated that the differential environments alter the hypothalamic–adrenal stress (HPA) axis and how D-amphetamine interacts with that system.

Besides investigating the potential neural mechanism for how environmental enrichment alters D-amphetamine self-administration, two studies have looked at the ability of environmental enrichment to protect against the deleterious effects of prenatal and adolescent drug exposure for increasing abuse liability for D-amphetamine in adulthood [17, 18]. In the study from our laboratory we looked at the ability of environmental enrichment to block adolescent nicotine-induced increases in D-amphetamine self-administration in adulthood. We found that IC rats treated with nicotine in adolescence took significantly more D-amphetamine under both an FR and PR schedule compared to their saline-treated counterparts and environmental enrichment reversed this effect [17]. While our study investigated the effects of adolescent nicotine exposure, Wang et al. [18] found that rats exposed to ethanol prenatally self-administered more D-amphetamine in adulthood and that raising rats in enriched environments after birth reversed this effect relative to rats raised in impoverished environments.

A handful of studies have been published looking at the effects of enrichment on METH self-administration. One study found that

IC rats displayed greater level of acquisition of METH self-administration (0.03 mg/kg/infusion) under an FR schedule but there were no differences between EC and IC rats at higher doses or when METH was self-administered under a PR schedule of reinforcement [19]. These authors also found the EC rats showed a faster rate of extinction of METH self-administration and IC rats showed a greater level of cue-induced METH-seeking responses compared to EC rats [19]. A later study by the Bardo laboratory, did not find any differences between EC and IC rats in METH acquisition using an autoshaping procedure [12]. Also, when Hofford et al. [11] tested a complete dose effect curve they found no significant differences between EC and IC rats at any METH dose tested. Although when they extinguished METH self-administration, they again found the EC rats extinguished at a faster rate compared to IC rats and that IC rats displayed a higher number of cue-induced METH-seeking responses compared to EC rats. While the bulk of research with EC and IC rats have looked at either D-amphetamine or METH, one study found the environmental enrichment decreased levels of MPD self-administration under both FR and PR schedules but this effect was specific to low unit doses of MPD [20].

While the previous studies just discussed looked at the effects of enrichment using a more "classic" design (i.e., start environmental enrichment at PND 21 and continues throughout the study) Marcello Solinas's laboratory used a different procedure. They have been looking at the ability for environmental enrichment to alter sensitivity to drugs of abuse but do not apply the enrichment manipulation until after drug self-administration has been established. In addition, they start the enrichment during an abstinence phase where the animals are not allowed access to drug and remain in their home cages in either enriched or impoverished conditions [3]. Using this "treatment" form of enrichment exposure Sikora et al. [21] found that enriched rats had less METH-seeking responses after the forced abstinence period compared to social controls.

## 6.5 METH Drug Discrimination

While considerable research has been done looking at amphetamine self-administration or even CPP, only one study has investigated the effects of enrichment on METH drug discrimination. In this study we determined the effect of environmental enrichment on METH drug discrimination using a two-lever-operant procedure. In this study we found that EC and IC rats did not differ in acquisition of the discrimination between METH and saline, but we found the IC rats showed greater generalization of a moderate dose of METH to the 1.0 mg/kg training dose of METH compared to EC rats. We also found the EC rats had a greater sensitivity to the antagonist effects of the D1 dopamine receptor antagonist SCH23390 on the training dose compared to IC rats. This experiment extended our understanding of the effects of environmental enrichment on METH to not only the drug's rewarding effects but also its discriminative stimulus effects and indicated the D1 dopamine receptor may be meditating the enrichment differences.

## 7    Enrichment and Cocaine

### 7.1    Conditioned Place Preference (CPP) with Cocaine

Another psychomotor stimulant that has received considerable attention in the last 10 years when looking at the effects of environmental enrichment is cocaine. A study done by Green et al. [22] found that EC rats displayed a significantly greater cocaine-induced CPP response compared to IC rats following conditioning with 10 mg/kg of cocaine. Although interestingly, they found that IC rats had a trend toward a higher cocaine-induced CPP compared to SC control animals. In this study the investigators also investigated whether the behavioral phenotype of EC rats was in part due to decreases in cAMP response element binding protein (CREB) in the NAcc. The authors found that there were decreased levels of phosphorylated CREB in the NAcc of EC rats compared to IC rats [22]. They also found that if they decreased CREB activity in the NAcc of regularly housed rats, the rats displayed a behavioral phenotype similar to EC rats. The neural data from this study indicates that decreases in CREB activity in the NAcc may in part mediate some of the changes in drug sensitivity seen between EC and IC rats. Another study using mice found that mice raised in an enriched condition from weaning had a decreased cocaine-induced CPP response relative to social controls [23]. Interestingly, this protective effect was lost if enriched mice were switched to living in the social condition during adulthood and this switch actually lead to a heightened cocaine-induced CPP response compared to mice that lived in the social condition throughout the entire study. The authors found that this loss of the protective effect of being switched out of the enriched condition was associated with increases in levels of corticotropin releasing factor in the bed nucleus of the stria terminalis and increases in phosphorylated CREB in the NAcc [23]. These two studies seem to indicate the roll of CREB activity in the NAcc in some of the effects of enrichment on cocaine CPP, although there is a discrepancy between the two studies on the effects of enrichment on cocaine-induced CPP. Green et al. [22] found EC rats had a greater response compared to IC rats while Nader et al. [23] found a decreased response in enriched mice compared to social controls. The discrepancy in CPP may be due to the difference of rat versus mouse or the differences in complexities of the enriched conditions used in the two different studies.

Finally, a study looking at the effect of enrichment on cocaine-induced CPP used the "treatment" enrichment manipulation, where they first ran conditioning trials for the cocaine CPP and then put the mice in either enriched or social conditions for 30 days before testing for a cocaine CPP response [24]. Using this procedure, they found the enriched mice did not show a cocaine-induced CPP response while the social controls did. This protective effect of enrichment on cocaine-induced CPP was related to decreased cFos expression in various brain areas while social control mice showed an increase in cFos expression in the same brain structures [24].

## 7.2 Cocaine Self-Administration

A considerable number of studies have looked at the effects of environmental enrichment on different aspects of cocaine self-administration, including: acquisition, maintenance, escalation, extinction, and reinstatement, as well as, cocaine incubation effects. A study by Puhl et al. [25] found that nonenriched rats tended to self-administer higher levels of cocaine during acquisition on an FR10 and FR 20 schedule of reinforcement compared to enriched rats. The authors also found the nonenriched rats had significantly higher breakpoints on a PR schedule of reinforcement compared to enriched rats. In a study looking at the effects of enrichment on the escalation of cocaine self-administration following exposure to extended cocaine access, it was found that during short access sessions of cocaine, EC rats had slower rates of acquisition compared to IC rats for both a low (0.1 mg/kg/infusion) and a high (0.5 mg/kg/infusion) dose of cocaine [26]. With exposure to 6 h sessions for the lower dose of cocaine only IC rats showed an escalation of intake while EC rats did not, but when the higher dose of cocaine was tested in the 6 h sessions both EC and IC rats showed an escalation of intake [26].

While the previous two studies employed the more "classic" enrichment manipulation the majority of the studies investigating the effects of enrichment on cocaine self-administration have used the "treatment" manipulation of enrichment where cocaine self-administration is established first then the enrichment manipulation is implemented. For instance, in one study animals were first allowed to stabilize their levels of cocaine self-administration, before being given 10 days off from self-administration, at which point that were placed into either enriched or control housing conditions. After the 10 days off, the animals were given 10 extinction sessions in a different context before being tested for context dependent reinstatement [27]. In this study, Ranaldi et al. [27] found that rats that were placed in the enriched conditions had less resistance to extinction and lower levels of responding on the previously active lever during the context-dependent reinstatement sessions. This study indicates that even just 10 days of exposure to enrichment can speed extinction and decrease risks of relapse. Two studies by the Solinas's group used the "treatment" enrichment procedure in which the rats were exposed to enrichment conditions during a period of forced abstinence after cocaine self-administration (0.6 mg/kg/infusion) was established. In the first study, Chauvet et al. [28] found the enriched rats had faster rates of extinction relative to social controls and that enriched rats showed a decrease in cocaine-seeking responses following either a cue-induced or stress-induced prime compared to social controls. Although, when an experimentally-delivered cocaine pretreatment was used as a prime, there were no effects of enrichment of cocaine-seeking behavior [28]. In a second follow up study the authors again used the "treatment" environmental procedure but looked at the effect of enrichment to block the ability of forced abstinence to lead to an incubation of cocaine craving or seeking behavior [29]. In this study, the authors looked at the ability of enrichment during cocaine abstinence to decrease the incubation effect of forced abstinence following 1, 30, or 60 days of abstinence. The authors found that after 30 and 60 days of abstinence enrichment blocked the abstinence-induced increase in cocaine seeking behavior. They also found that if the housing was reversed the protective effect of enrichment was lost [29]. Again, these studies indicated the environmental enrichment does not have to start prior to administration of the drug but can be used as a treatment to decrease vulnerability to relapse.

A series of studies out of the Neisewander laboratory at Arizona State University replicated the protective effects of environmental enrichment when used as a "treatment" to block the cocaine incubation effect during abstinence [30–33]. For example, Thiel et al. [33] found that rats placed into EC conditions following acquisition of cocaine self-administration showed a significant reduction in the number of cocaine-seeking responses following a cue prime after 21 days of abstinence compared to IC and SC rats. Although, the protective effects of the EC condition was not seen when cocaine-seeking behavior was induced by a cocaine injection [33]. A follow up study looked at the relationship between this protective effect of enrichment and changes in cFos levels in different brain structures [32]. In particular, EC rats showed decreased cFos expression in the infralimbic and anterior cingulate cortices, NAcc core and shell and the bed nucleus of the stria terminalis, and ventral tegmental area compared to IC and SC rats. These results indicate that exposure to enrichment can alter these brain structures and these structures play a role in the decreased vulnerability to the reinforcing effects of cocaine. These neural effects are also consistent with the cFos expression seen in mice following cocaine CPP [24]. In another study by Theil et al. [30] using the "treatment" enrichment approach, EC rats were found to have faster rates of extinction and a decrease in the level of cocaine-seeking behavior following a cue prime compared to SC rats. The effect on cocaine-seeking behavior was lost when the animals were removed from the EC condition.

Finally, two different studies have investigated the role of the HPA stress axis in the effects of enrichment on cocaine self-administration. First, in a study by Hofford, Prendergast, and Bardo [34], pretreatments with the glucocorticoid antagonist RU486 prior to a cocaine self-administration session found that EC rats self-administered significantly less cocaine after antagonist pretreatments when the dose of cocaine was low compared to IC rats. Also, pretreatment with the glucocorticoid agonist corticosterone resulted in a decrease in cocaine intake in ECs but an increase in cocaine intake in IC rats. Although, in this study the authors did not find any differences between EC and IC rats in the level of glucocorticoid (GR) receptors in various brain areas, indicating enrichment may alter the sensitivity of the GR receptor not the number of GR receptors [34]. Hofford, Prendergast, and Bardo [35] attempted to determine if environmental enrichment would alter the ability of a modified single prolonged stressor (modSPS) to decrease cocaine self-administration. The authors overall found little effect of enrichment on the effects of a modSPS on cocaine self-administration, although they did find that IC rats showed a steeper or faster acquisition of cocaine self-administration compared to EC and SC rats.

## 8    Enrichment and Nicotine

### 8.1    Nicotine Locomotor Effects

At the time of the Stairs and Bardo [1] review very little research on the effects of environmental enrichment had been done, over the last 10 years there have been a number of studies that have begun to elucidate the effects of enrichment on various nicotine effects. We previously published a review of the effects of enrichment on nicotine effects [36] but will give a brief review of the behavioral effects in EC and IC rats, including new studies that have come out since the last review. In a study by Coolon and Cain [37] they found that following repeated pretreatments of nicotine, EC rats displayed less locomotor sensitization compared to IC and SC rats. This study also found that the nicotine antagonist mecamylamine was able to block the expression of conditioned hyper-locomotion but only in EC and SC rats, with IC rats still displaying the conditioned locomotion effects [37]. A study from our lab found that EC rats did not show a nicotine sensitization when tested with a low dose of nicotine in adulthood following repeated nicotine pretreatments of nicotine in adolescence, while IC rats did [8]. This protective effect was only seen with low doses of nicotine as both EC and IC rats show a sensitized locomotor response when tested with a higher dose of nicotine. This same protective effect of enrichment for nicotine sensitization was found when we looked at the effects of repeated nicotine pretreatments to lead to locomotor cross-sensitization to D-amphetamine [8]. The protective effect of environmental enrichment to block nicotine locomotor sensitization was also found to occur in female rats [38]. This was an important finding given that all the previous nicotine locomotor studies had used male rats. There was one study that failed to replicate the protective effects of enrichment on nicotine locomotor sensitization: Gomez et al. [39] found no differences in EC and IC rats in nicotine locomotor sensitization. This incongruent effect may be due to differences in the pretreatment regimen of nicotine compared to the previous studies [8, 37, 38].

**8.2 Nicotine and CPP**

Only two studies have investigated the effects of environmental enrichment on a nicotine-induced CPP response [40, 41]. Ewin, Kangiser, and Stairs [40] found that only EC rats displayed a CPP response to nicotine at three different doses tested while IC rats failed to show a nicotine CPP response at any dose tested. EC rats also showed extinction of the nicotine CPP response and showed a nicotine-primed reinstatement of the CPP response following extinction while IC rats showed no signs of reinstatement following nicotine primes. This apparent increase in sensitivity in EC rats to the rewarding effects of nicotine was not replicated in a subsequent study: Nawaz et al. [41] actually found that rats exposed to enrichment actually showed a decreased in nicotine CPP response compared to rats not in enriched conditions. The discrepancy between these two studies may be due to a number of different procedural differences, including differences in enrichment conditions, differences in CPP apparatus, and number of conditioning trials. Regardless of the discrepancy further studies will need to be done to clarify the effects of enrichment on the rewarding effects of nicotine.

**8.3 Nicotine Self-Administration, Consumption, and Drug Discrimination**

Similar to the limited number of studies looking at the effects of enrichment and nicotine CPP, there are only a few studies looking at the effects of enrichment on nicotine self-administration or drinking behavior [21, 39, 42]. Gomez et al. [39] looked at whether EC and IC rats differed in the level of i.v. nicotine self-administration at one dose (0.03 mg/kg/infusion) for 21 days. The authors found that IC rats acquired nicotine self-administered while EC rats never did acquire nicotine self-administration [39]. While this study indicated enrichment has protective effects on the reinforcing effects of nicotine, some caution should be used in the interpretation of these results as the level of nicotine intake, even in IC rats, was generally low. A second study looked at the effects of enrichment on nicotine drinking behavior [42]. In this study the authors found that rats raised in enriched conditions consumed significantly less nicotine in a two-bottle free-choice procedure compared to rats raised in a social control condition. The results from this study do strengthen the argument that enrichment may decrease the intake of nicotine as the effect has now been shown in both the i.v. self-administration paradigm and a drinking consumption paradigm. The final study that investigated the effect of enrichment on the reinforcing effects of nicotine utilized the "treatment" procedure of enrichment [21]. In this study the authors first established i.v. nicotine self-administration in rats, then put the rats in either an enriched condition or a control social condition while they had forced abstinence from nicotine for 3–4 weeks, then the animals were tested for the level of nicotine-seeking behavior when placed back in the operant chambers. The authors found the enriched rats showed a significantly lower number of nicotine-seeking responses compared to social control animals [21].

Only one study from our laboratory has looked at the effects of enrichment on the discriminative stimulus effects of nicotine. In this study we found that EC rats were less sensitive to the discriminative stimulus effects of nicotine as they showed less generalization of lower doses of nicotine to substitute for the training dose of nicotine compared to IC rats [43]. We also found that EC rats were more sensitive to the ability of mecamylamine pretreatments to block the discriminative stimuli effects of the training dose of nicotine compared to IC rats. We also believe the protective effects of enrichment may be mediated through differences in density of nicotinic acetylcholine receptors in the ventral tegmental area based on quantitative autoradiography data from the rats brains [43].

## 9 Enrichment and Opiates

Three studies have investigated the effects of environmental enrichment on sensitivity to the behavioral effects of opiates [21, 44, 45]. The first study done by Galaj, Manuszak, and Ranaldi [44] in one experiment used the "treatment" enrichment approach where they first established i.v. heroin self-administration (0.05 mg/kg/infusion) then put the rats into either an enriched or nonenriched condition. Fifteen days later the animals were tested for extinction of heroin self-administration across 15 sessions. One day after the last extinction session the rats were tested for levels of heroin-seeking with a cue-induced reinstatement session. The results from this experiment found there were no differences between enriched and nonenriched in rates of extinction, but enriched rats did show a significant reduction in the level of heroin-seeking responses following a heroin-cue prime compared to nonenriched controls. In a second experiment, following heroin conditioning in a CPP apparatus rats were placed into either enriched or nonenriched conditions for 30 days before they were tested for the appearance of a CPP response [44]. The authors found that the enriched rats did not display significant levels of a heroin-induced

CPP response while the nonenriched animals did. These experiments indicate that enrichment using the "treatment" approach can block the rewarding effects of heroin and decrease the relapse to the reinforcing effects of heroin. A second study again using the "treatment" enrichment manipulation replicated that enrichment can decrease the level of heroin-seeking after a period of abstinence [21]. Finally, a study that used the more "classic" enrichment manipulation found that that EC rats self-administered significantly less of the short-acting opiate agonist remifentanil compared to both IC and SC rats [45]. The authors found that when they tested a low dose of remifentanil (1 mg/kg/infusion) EC rats did not show significant levels of drug intake while IC and SC rats did: the entire dose effect curve was flattened in EC rats relative to IC and SC rats [45].

## 10    Enrichment and Ethanol

### 10.1   Ethanol and CPP

The final drug we will discuss that has been investigated for effects of environmental enrichment is ethanol. A study conducted by de Carvalho et al. [46] looked at the effects of enrichment on female spontaneously hypertensive rats (SHR), which are a validated rodent model of attention deficit hyperactivity disorder, and sensitivity to ethanol-induced CPP. The authors found that SHR rats raised in enriched conditions did not acquire a CPP response to either doses of ethanol tested (0.5 or 1.2 mg/kg) and actually showed a conditioned place aversion to the high dose of ethanol whereas the social control group showed a significant ethanol CPP response [46]. In a study using the "treatment" enrichment manipulation in mice, researchers investigated the ability of enrichment to alter the extinction and reinstatement of an ethanol-induced CPP response [47]. After establishing ethanol CPP the mice were placed into either an enriched condition or maintained in their standard condition. Extinction sessions were then conducted across 8 days followed by a reinstatement session where the mice were injected with a priming dose of ethanol. Using this procedure, the authors found that there were no differences between enriched and standard-caged mice, although they did find that only standard-caged mice showed a reinstated CPP response following an ethanol prime while enriched mice did not [47].

### 10.2   Ethanol and Drinking Behavior

While environmental enrichment appears to have consistent protective effects against the rewarding effect of ethanol when using the CPP procedure, enrichment also appears to have consistent protective effects on ethanol drinking behavior [46, 48, 49]. For instance, not only did de Carvalho et al. [46] find that female SHR rats raised in enriched conditions had decreased ethanol CPP response, but they also found that enriched SHR rats also drank significantly less ethanol in both a free- and forced-bottle-choice procedures compared to standard caged SHR rats. Using an operant self-administration drinking procedure Deehan et al. [49] found that alcohol-preferring (P) rats raised in EC conditions showed a decreased preference for ethanol over water compared to P rats in IC conditions. The P rats in the EC conditions also had a significantly lower breakpoint under a PR schedule of reinforcement compared to IC rats [49]. Results from this study indicated that enrichment decreased the reinforcing effects of ethanol and resulted in an increase in choice of a non–drug reinforcer. Finally, a study by Chappell et al. [48] found that group-housed rats had significantly lower levels of ethanol drinking in a two-bottle choice procedure compared to standard-caged rats. The last study indicates that the social component that is part of the enriched condition is an important factor in the protective effects of enrichment in decreasing ethanol consumption.

## 11  Conclusions and Future Directions

The current review of the literature on the effects of enrichment on drug abuse vulnerability in the last 10 years has only strengthened the finding that, in general, environmental enrichment tends to be protective against numerous drugs of abuse and the various behavioral effects of these compounds. The protective effects are generally consistent even if different enrichment procedures are employed ("classic" vs. "treatment"). While for the most part enrichment seems to be protective against vulnerability to different drugs of abuse, there are still some areas of inconsistency. For instance, results from CPP often find that EC rats are more sensitive to the rewarding effects of the psychomotor stimulants, but when cocaine is used there are some studies which find EC animals had decreased CPP response. Regardless of the inconsistencies in CPP, when self-administration behavior is investigated enriched animals consistently show a decrease in the sensitivity to the reinforcing effects of various drugs of abuse. This protective effect tends to be more pronounced when the unit dose of the drug is low relative to higher unit doses. In terms of some of the neural data that we discussed, data across the different drugs of abuse seem to indicate changes in mesolimbic structures as a result of enrichment, in particular areas like the NAcc, VTA, and the bed nucleus of the stria terminalis. There also appears to be alterations in the HPA axis of enriched vs. impoverished animals and this seems to play a role in differences seen in stimulant self-administration, although what changes are taking place need to be better clarified with future studies.

While our understanding of the effects of enrichment on drug abuse vulnerability have grown considerably over the last 10 years, there are still some areas that need to be better studied. For example, I believe more research could be done looking at the effect of enrichment on nicotine and opiate self-administration. For these two compounds, more work needs to be done looking at the effects of enrichment on acquisition, maintenance, extinction and reinstatement under a wider range of doses. Also, one area that has not been investigated much at all is looking at the effects of enrichment on sensitivity to cannabinoids. Only one study has been done which just looked at gene expression of endocannabinoids in different brain areas of enriched- and standard-caged mice [50]. As improvements are being made in the development of viable rodent models to investigate the behavioral effects of cannabinoids [51, 52] this may open up the opportunity to look at the effects of enrichment on this commonly used drug of abuse. Given this area and the other unanswered questions of enrichment effects on drugs of abuse we look forward to seeing what the next 10 years of research discovers.

## References

1. Stairs DJ, Bardo MT (2009) Neurobehavioral effects of environmental enrichment and drug abuse vulnerability. Pharmacol Biochem Behav 92(3):377–382

2. Simpson J, Kelly JP (2011) The impact of environmental enrichment in laboratory rats--behavioural and neurochemical aspects. Behav Brain Res 222(1):246–264

3. Solinas M et al (2010) Prevention and treatment of drug addiction by environmental enrichment. Prog Neurobiol 92(4):572–592

4. Gill MJ, Arnold JC, Cain ME (2012) Impact of mGluR5 during amphetamine-induced hyperactivity and conditioned hyperactivity in differentially reared rats. Psychopharmacology 221(2):227–237

5. Arndt DL, Arnold JC, Cain ME (2014) The effects of mGluR2/3 activation on acute and repeated amphetamine-induced locomotor activity in differentially reared male rats. Exp Clin Psychopharmacol 22(3):257–265

6. Gill MJ, Weiss ML, Cain ME (2014) Effects of differential rearing on amphetamine-induced c-fos expression in rats. Drug Alcohol Depend 145:231–234

7. Garcia EJ et al (2017) Differential housing and novelty response: protection and risk from locomotor sensitization. Pharmacol Biochem Behav 154:20–30

8. Adams E et al (2013) Effects of environmental enrichment on nicotine-induced sensitization and cross-sensitization to D-amphetamine in rats. Drug Alcohol Depend 129(3):247–253

9. Wooters TE et al (2011) Effect of environmental enrichment on methylphenidate-induced locomotion and dopamine transporter dynamics. Behav Brain Res 219(1):98–107

10. Gill KE et al (2013) The effects of rearing environment and chronic methylphenidate administration on behavior and dopamine receptors in adolescent rats. Brain Res 1527:67–78

11. Starosciak AK et al (2012) Differential alteration of the effects of MDMA (ecstasy) on locomotor activity and cocaine conditioned place preference in male adolescent rats by social and environmental enrichment. Psychopharmacology 224(1):101–108

12. Hofford RS et al (2014) Environmental enrichment reduces methamphetamine cue-induced reinstatement but does not alter methamphetamine reward or VMAT2 function. Behav Brain Res 270:151–158

13. Thiriet N et al (2011) Environmental enrichment does not reduce the rewarding and neurotoxic effects of methamphetamine. Neurotox Res 19(1):172–182

14. Yates JR et al (2013) Concurrent choice for social interaction and amphetamine using conditioned place preference in rats: effects of age and housing condition. Drug Alcohol Depend 129(3):240–246

15. Arndt DL et al (2015) Environmental condition alters amphetamine self-administration: role of the MGluR(5) receptor and schedule of reinforcement. Psychopharmacology 232(20):3741–3752

16. Stairs DJ, Prendergast MA, Bardo MT (2011) Environmental-induced differences in corticosterone and glucocorticoid receptor blockade of amphetamine self-administration in rats. Psychopharmacology 218(1):293–301

17. Stairs DJ et al (2017) Effects of environmental enrichment on d-amphetamine self-administration following nicotine exposure. Exp Clin Psychopharmacol 25(5):393–401

18. Wang R et al (2018) Environmental enrichment reverses increased addiction risk caused by prenatal ethanol exposure. Drug Alcohol Depend 191:343–347

19. Lu X et al (2012) The effects of rearing condition on methamphetamine self-administration and cue-induced drug seeking. Drug Alcohol Depend 124(3):288–298

20. Alvers KM et al (2012) Environmental enrichment during development decreases intravenous self-administration of methylphenidate at low unit doses in rats. Behav Pharmacol 23(7):650–657

21. Sikora M et al (2018) Generalization of effects of environmental enrichment on seeking for different classes of drugs of abuse. Behav Brain Res 341:109–113

22. Green TA et al (2010) Environmental enrichment produces a behavioral phenotype mediated by low cyclic adenosine monophosphate response element binding (CREB) activity in the nucleus accumbens. Biol Psychiatry 67(1):28–35

23. Nader J et al (2012) Loss of environmental enrichment increases vulnerability to cocaine addiction. Neuropsychopharmacology 37(7):1579–1587

24. Chauvet C et al (2011) Brain regions associated with the reversal of cocaine conditioned place preference by environmental enrichment. Neuroscience 184:88–96

25. Puhl MD et al (2012) Environmental enrichment protects against the acquisition of cocaine self-administration in adult male rats, but does not eliminate avoidance of a drug-associated saccharin cue. Behav Pharmacol 23(1):43–53

26. Gipson CD et al (2011) Effect of environmental enrichment on escalation of cocaine self-administration in rats. Psychopharmacology 214(2):557–566

27. Ranaldi R et al (2011) Environmental enrichment, administered after establishment of cocaine self-administration, reduces lever pressing in extinction and during a cocaine context renewal test. Behav Pharmacol 22(4):347–353

28. Chauvet C et al (2009) Environmental enrichment reduces cocaine seeking and reinstatement induced by cues and stress but not by cocaine. Neuropsychopharmacology 34(13):2767–2778

29. Chauvet C et al (2012) Effects of environmental enrichment on the incubation of cocaine craving. Neuropharmacology 63(4):635–641

30. Thiel KJ et al (2011) The interactive effects of environmental enrichment and extinction interventions in attenuating cue-elicited cocaine-seeking behavior in rats. Pharmacol Biochem Behav 97(3):595–602

31. Thiel KJ et al (2012) Environmental enrichment counters cocaine abstinence-induced stress and brain reactivity to cocaine cues but fails to prevent the incubation effect. Addict Biol 17(2):365–377

32. Thiel KJ et al (2010) Environmental living conditions introduced during forced abstinence alter cocaine-seeking behavior and Fos protein expression. Neuroscience 171(4):1187–1196

33. Thiel KJ et al (2009) Anti-craving effects of environmental enrichment. Int J Neuropsychopharmacol 12(9):1151–1156

34. Hofford RS, Prendergast MA, Bardo MT (2015) Pharmacological manipulation of glucocorticoid receptors differentially affects cocaine self-administration in environmentally enriched and isolated rats. Behav Brain Res 283:196–202

35. Hofford RS, Prendergast MA, Bardo MT (2018) Modified single prolonged stress reduces cocaine self-administration during acquisition regardless of rearing environment. Behav Brain Res 338:143–152

36. Stairs DJ et al (2016) Environmental enrichment and nicotine addiction. In: Preedy VR (ed) Neuropathology of drug addictions and substance misuse; Volume 1: Common substances of abuse/tobacco, alcohol, cannabinoids and opioids. Academic Press, London

37. Coolon RA, Cain ME (2009) Effects of mecamylamine on nicotine-induced conditioned hyperactivity and sensitization in differentially reared rats. Pharmacol Biochem Behav 93(1):59–66

38. Hamilton KR et al (2014) Environmental enrichment attenuates nicotine behavioral sensitization in male and female rats. Exp Clin Psychopharmacol 22(4):356–363

39. Gomez AM et al (2015) Effects of environmental enrichment on ERK1/2 phosphorylation in the rat prefrontal cortex following nicotine-induced sensitization or nicotine self-administration. Eur J Neurosci 41(1):109–119

40. Ewin SE, Kangiser MM, Stairs DJ (2015) The effects of environmental enrichment on

nicotine condition place preference in male rats. Exp Clin Psychopharmacol 23 (5):387–394

41. Nawaz A et al (2017) Enriched environment palliates nicotine-induced addiction and associated neurobehavioral deficits in rats. Pak J Pharm Sci 30(6 Suppl):2375–2381

42. Venebra-Munoz A et al (2014) Enriched environment attenuates nicotine self-administration and induces changes in Delta-FosB expression in the rat prefrontal cortex and nucleus accumbens. Neuroreport 25 (9):688–692

43. Bockman CS et al (2018) Nicotine drug discrimination and nicotinic acetylcholine receptors in differentially reared rats. Psychopharmacology 235(5):1415–1426

44. Galaj E, Manuszak M, Ranaldi R (2016) Environmental enrichment as a potential intervention for heroin seeking. Drug Alcohol Depend 163:195–201

45. Hofford RS et al (2017) Effects of environmental enrichment on self-administration of the short-acting opioid remifentanil in male rats. Psychopharmacology 234 (23–24):3499–3506

46. de Carvalho CR et al (2010) Environmental enrichment reduces the impact of novelty and motivational properties of ethanol in spontaneously hypertensive rats. Behav Brain Res 208(1):231–236

47. Li X et al (2015) Environmental enrichment blocks reinstatement of ethanol-induced conditioned place preference in mice. Neurosci Lett 599:92–96

48. Chappell AM et al (2013) Adolescent rearing conditions influence the relationship between initial anxiety-like behavior and ethanol drinking in male Long Evans rats. Alcohol Clin Exp Res 37(Suppl 1):E394–E403

49. Deehan GA Jr et al (2011) Differential rearing conditions and alcohol-preferring rats: consumption of and operant responding for ethanol. Behav Neurosci 125(2):184–193

50. El Rawas R et al (2011) Early exposure to environmental enrichment alters the expression of genes of the endocannabinoid system. Brain Res 1390:80–89

51. Javadi-Paydar M et al (2019) Effects of nicotine and THC vapor inhalation administered by an electronic nicotine delivery system (ENDS) in male rats. Drug Alcohol Depend 198:54–62

52. Wakeford AGP et al (2017) The effects of cannabidiol (CBD) on Delta(9)-tetrahydrocannabinol (THC) self-administration in male and female Long-Evans rats. Exp Clin Psychopharmacol 25 (4):242–248

# Part III

# Consequences of Chronic Drug Consumption and Treatment of Addiction

# Chapter 10

# Investigation of Individual Differences in Stress Susceptibility and Drug-Seeking in an Animal Model of SUD/PTSD Comorbidity

**Courtney Wilkinson, Harrison Blount, Lori Knackstedt, and Marek Schwendt**

## Abstract

Differential physical, psychological, and treatment outcomes for individuals with comorbid substance use and posttraumatic stress disorder (SUD/PTSD), in comparison for either disorder alone, necessitate the advancement of preclinical translational models for this dual diagnosis. This chapter details a novel animal model of comorbid SUD/PTSD that uses predator scent stress to precipitate heterogenous responses to trauma in a rodent population. Rodents are classified as stress-Susceptible, -Resilient, or Intermediate according to anxiety-like behavior in the elevated plus maze and the acoustic startle response task. Following classification, rats undergo drug-self administration, extinction, and reinstatement. The methods described in this chapter detail a preclinical model of comorbid SUD/PTSD that can be used to identify targets for treatment translatable to the human population.

**Keywords** Substance use disorder, Posttraumatic stress disorder, Comorbidity, Animal models of substance abuse

## 1 Introduction

### 1.1 SUD and PTSD Comorbidity

Substance use disorder (SUD) is frequently comorbid with post-traumatic stress disorder (PTSD) [1–3]. There are currently few effective treatments for either disorder [4, 5], and treatments prescribed specifically for PTSD or substance abuse alone are complicated by the presence of the other. Compared to individuals diagnosed with SUD alone, those diagnosed with comorbid SUD/PTSD report poorer social functioning and overall general health, an increase in cardiovascular and neurological symptoms, an increase in treatment attrition, poorer overall treatment outcomes, and higher suicide risk [1, 3, 6]; clearly highlighting a need for targeted treatments specific to this comorbid pathology. Due to the common comorbidity of PTSD and substance abuse observed in

María A. Aguilar (ed.), *Methods for Preclinical Research in Addiction*, Neuromethods, vol. 174,
https://doi.org/10.1007/978-1-0716-1748-9_10, © Springer Science+Business Media, LLC, part of Springer Nature 2022

the human population, the unique pathology specific to concurrent diagnosis, and the treatment complications that arise from this comorbidity, a preclinical model encompassing both psychiatric conditions is necessary to increase treatment translatability. Current models of SUD or PTSD alone have presented challenges to be met. Accounting for these challenges, this comorbid SUD/PTSD model combines preexisting SUD and PTSD models with the highest translational value.

## 1.2 Methods to Study PTSD-Like Symptoms in Experimental Animals

Many animal models attempt to capture features of PTSD and it is imperative these models accurately reflect the human experience as closely as possible. Some established animal models of PTSD ignore temporal aspects of this disorder and its heterogeneous nature. In humans, a PTSD diagnosis requires symptoms persist for longer than 1 month. However, early models of this disorder analyzed behavior and/or neurobiological alterations only 24–48 h following stress exposure, completely discounting temporal aspects of the diagnosis. Further, epidemiological studies report an 80% lifetime prevalence of experiencing a DSM-V classified traumatic event in the Western world. Of this 80%, 15–25% will be diagnosed with PTSD [2]. Early models of PTSD ignored these differential characteristics and compared all stress-exposed rodents to nonexposed controls. For models of PTSD to reach high translational value, we recommend that (1) animals are exposed to a potentially life threatening stressor (trauma), (2) acute exposure to the traumatic stimulus results in PTSD-like symptoms, (3) increased intensity of exposure increases severity of symptoms, (4) PTSD-like symptoms persist or worsen over time, and (5) result in differential responses [7]. Following these guidelines for back-translation, in our lab, exposure to predator stress yields all of the aforementioned criteria.

Predator stress is initiated by exposure to a predator or predator-related scent without physical contact. Predator stress has ethological validity and results in behavioral and neurobiological effects consistent with PTSD. Clinical PTSD research and animal research are in agreement with the percentage of heterogeneity of responses to trauma; increases in avoidance behavior, negative mood states, hypervigilance, and cognitive impairments; decreases in basal glucocorticoid levels and thymus and adrenal weights; increases in dexamethasone suppression of the HPA axis; and structural alterations in the hippocampus and amygdala [8–19]. In our lab, predator scent stress in the form of the fox pheromone, 2,5-dihydro-2,4,5-trimethylthiazoline (here on referred to as TMT) is used. After a single, 10-min exposure, this model results in differential long-term fear and anxiety responses persisting up to 6 weeks, with approximately 14–25% of rats expressing stress-susceptibility and an equal number exhibiting control-like "Resilient" responses [20–22].

### 1.3 Methods to Study SUD-Like Symptoms in Experimental Animals

Models of substance use disorder are under similar scrutiny. While many behavioral assays using "noncontingent" drug administration have utility in identifying basic pharmacological or behavioral mechanisms of substances of abuse, investigations of SUD must also identify the mechanisms underlying voluntary drug consumption and drug-seeking. To reach high translational value, models are recommended to employ response-contingent drug-self administration with sophisticated schedules of reinforcement and to assess behaviors similar to those observed in humans (abstinence, craving, relapse, etc.) [23]. In drug self-administration, rats are trained to self-administer drugs in standard operant chambers with two levers (one active, one inactive) or nose poke ports. Responses on the active lever or port result in drug delivery, while responses on the inactive lever or port control for nonspecific drug effects. Different schedules of reinforcement can be employed; such schedules prescribe the response requirements to receive drug delivery. The fixed-ratio 1 (FR-1) schedule is commonly used, wherein one lever press delivers drug. Self-administration sessions can vary in length depending on the aspect of drug reinforcement that is of interest. Common procedures include short-access (ShA; 1–2-h sessions), long-access (LgA; 6–9-h sessions) or a combination of both. Other self-administration models, such as progressive ratio or behavioral economic demand models, increase the response requirement within or across sessions [23]. Our lab utilizes a combination of short and long access self-administration to assess escalation of cocaine intake.

### 1.4 Novel Animal Model of SUD/PTSD Comorbidity

Our model of comorbid SUD/PTSD combines the predator scent stress (PSS) model of PTSD, with the drug self-administration, extinction, and reinstatement model of SUD. Beginning with a single, 10-min exposure to TMT, rats (we used Sprague-Dawley male rats) are tested 7 days later in two assays for anxiety-like behavior, the elevated plus maze (EPM) and acoustic startle response (ASR). Rodents have innate preference for dark, closed spaces, while at the same time displaying novelty-induced exploratory behavior. Therefore, the amount of time spent in the open and closed arms of the EPM, as well as the number of entries into each arm, is recorded and used to assess anxiety-like behavior. Rats exposed to loud acoustic stimuli that is repeated intermittently will habituate over time. However, stress can result in hypervigilance, or a decrease in habituation to stimuli. We first verify an overall increase in anxiety-like behavior in PSS-exposed rats, comparing time spent in the open arms (OA) of the EPM and percent habituation of the ASR between PSS-exposed and Control rats. Upon finding significant reduction in time spent in the OA and habituation of the startle response in the PSS group, a median split analysis of EPM and ASR data is conducted in order to segregate TMT-exposed rats into phenotypes. While we have also found

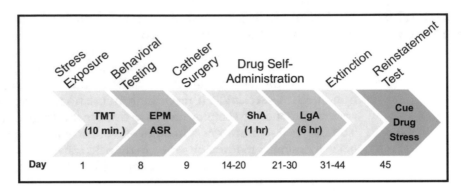

**Fig. 1** Comorbid SUD/PTSD model timeline. Rats are exposed to TMT on Day 1. Seven days later, rats are tested for anxiety and hypervigilance using EPM and ASR, respectively. One day following behavioral testing, rats are surgically implanted with jugular catheters and recover from surgery for a minimum of 5 days. On Day 14, rats self-administer the drug of study on an FR1 during Short Access (ShA; 1 h) sessions for 7 days, then self-administer the drug on an FR1 during Long Access (LgA; 6 h) sessions for 10 days. Rats undergo extinction training until extinction criteria are met (e.g., less than 15 presses per day on the previously active lever) or for a maximum of 10 days, followed by reinstatement testing (cue, drug, or stress)

that TMT decreases OA entries [21], we use only one EPM measure for segregation—time spent in the OA. Rats below the median time spent in the OA and above the median percent habituation of the ASR are classified as stress-Susceptible, rats above the median time spent in the OA and below the median percent habituation of the ASR as classified as stress-Resilient. The remainder of rats scoring as more anxious only on one of the two tests, are classified as Intermediate. These methods were adapted from those used by the Cohen group, who using the PSS of cat urine, have found similar ratios of Susceptible/Resilient rats using EPM and ASR [24–26]. Our attempt at using cat urine from a vivarium colony of cats produced an overall reduction in habituation of the ASR, but did not yield differences in EPM behavior compared to unstressed Controls [27], leading to our use of TMT.

Following segregation into stress-Susceptible and -Resilient phenotypes, rats are implanted with indwelling jugular catheters for self-administration of intravenous cocaine. Rats are first trained to self-administer cocaine on an FR-1 schedule in 1-h "short-access" sessions for 7 days, followed by 10 days of "long-access" (6 h/day) self-administration. Responses on the active lever during self-administration result in drug delivery and the presentation of discrete drug-associated cues (illumination of stimulus light, and a tone presentation). Rats then move into the extinction phase, during which presses on the previously active lever no longer delivers drug or drug-paired cues and the response is gradually extinguished. Once extinction criteria are met (e.g., less than 15 presses per day on the previously active lever), cue-primed reinstatement testing is conducted. During such a test, presses on the active lever (or port) once again yield drug-paired cues, but not drug itself. A significant increase in responding during the test compared to responding during the last 2–3 days of extinction training is termed a "reinstatement of the drug-seeking response" and is considered to be homologous to drug relapse [28]. *See* Fig. 1 for a timeline of this model.

This novel SUD/PTSD model provides a platform on which investigators may assess differential responses to stressors, and their subsequent effects on substance use to identify neurobiological signatures of risk and resilience to SUD/PTSD comorbidity. Thus far, our model identified significant differences in ShA self-administration of Resilient rats versus Control and stress-Susceptible rats, and increased responding during cue induced-reinstatement for stress-Susceptible rats compared to Control and Resilient [21]. We have also found a resistance to extinction in Susceptible rats [21]. Results of this model further demonstrate the differential effects of SUD and PTSD and the importance of addressing these disorders concurrently.

## 2 Materials

### 2.1 Equipment

Recommended equipment and instruments used in this protocol are itemized below; however, these can be substituted with items of comparable parameters or capability.

1. **Behavioral activity monitoring system:** To conduct the proposed experiments, hardware and software for tracking and analysis of behavior is necessary. This includes high-resolution cameras and wiring for remote animal monitoring, and a PC with EthoVision XT (Noldus Leesburg, VA), or a comparable software.

2. **Predator scent exposure chambers:** Cylindrical clear plexiglass chambers with a lid and stainless-steel mesh floor (Bio Bubble Pets, Boca Raton, FL, USA). Dimensions: $L \times W \times H$: $47 \times 47 \times 47$ cm. Such a chamber permits the exposure to PSS without coming in physical contact with the scent, in a medium-sized, inescapable chamber. *See* Fig. 2 for example.

3. **Elevated Plus Maze (EPM), the rat version:** Standard, commercially available elevated plus maze apparatus (e.g., # ENV-560; Med Associates, St. Albans, VT, USA), or a custom-built apparatus can be used. The EPM utilized in this protocol consisted of two opposing open arms and two opposing closed arms ($L \times W \times H$: $51 \times 10 \times 40.5$ cm), a center zone ($L \times W$: $10 \times 10$ cm), mounted approximately 50 cm above the floor. *See* Fig. 3 for example. *See* **Note 1** regarding the optimal illumination conditions.

**Fig. 2** Predator scent exposure chambers (47 × 47 × 47 cm; Bio Bubble Pets, Boca Raton, FL, USA). The small plastic cup located below the mesh flooring holds the TMT saturated filter paper, allowing for exposure to PSS without direct physical contact

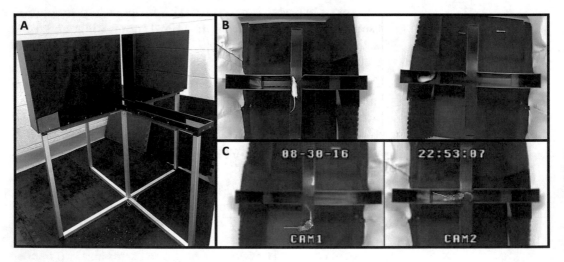

**Fig. 3** Elevated Plus Maze (**a**) mounted approximately 50 cm above the floor with two opposing open arms and two opposing closed arms ($L \times W \times H$: 51 × 10 × 40.5 cm) and a center zone ($L \times W$: 10 × 10 cm). (**b**) Cameras are mounted above platforms and curtains are placed between mazes to minimize subject interaction. (**c**) Behavior tracking software records amount of time spent in the open and closed arms and the number of arm entries for later analysis

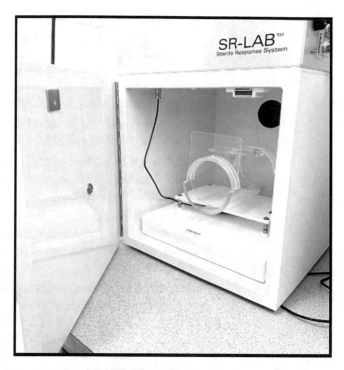

**Fig. 4** Acoustic Startle Response Chambers (51 × 55 × 31 cm; SR-LAB™ Startle System, San Diego Instruments, San Diego, CA, USA) equipped with two speakers. The rat is placed in a holding tube during the session, and a transducer system to transmit startle responses to Advanced Startle software (San Diego Instruments)

4. **Acoustic Startle Response (ASR) chambers:** (SR-LAB™ Startle System, San Diego Instruments, San Diego, CA, USA). ASR chambers (51 × 55 × 31 cm) should be equipped with two speakers, a holding tube with Plexiglass partitions, and a transducer system to transmit startle responses to Advanced Startle software (San Diego Instruments). *See* **Note 2** for the optimal settings and optional adjustments. *See* Fig. 4 for example.

5. **Locomotor activity chambers:** Standard locomotor activity chambers ($L \times W \times H$: 40 × 44 × 37 cm; Photobeam Activity System (PAS)-Open Field, San Diego Instruments) equipped with **infrared photobeam sensors** were modified for the light-dark test. An opaque black plexiglass insert ($L \times W \times H$: 20 × 44 × 37 cm) with an opening was used to darken one side of the chamber while allowing for free movement between light and dark sides. Illumination of the light compartment was adjusted to ~300 lux.

6. **Standard operant chambers, the rat version:** (30 × 24 × 30 cm; Med Associates, St. Albans, VT, USA): Used for drug self-administration. Equip standard operant

chambers with two levers, a stimulus light above each retractable lever, a house light, and a tone generator (78 dB, 4.5 kHz). The operant chamber should be located inside a sound attenuating cubicle with a fan for airflow and to block ambient noise. Operant responses in the chamber will be recorded using MedPC V Software.

7. **Syringe pump:** Single speed or variable speed syringe pump for i.v. drug delivery (PHM-100 series, Med Associates)

8. **Flexible tubing:** Medical- or laboratory-grade Tygon® and/or Silastic® tubing of varied diameter for drug delivery lines and i.v. catheter preparation.

9. **Swivel and arm** (Instech, Plymouth Meeting, PA) to carry the Tygon® tubing from the syringe to the catheter, and metal spring leash (Instech) to protect the tubing that is inside the operant chamber.

10. **Catheter for i.v. drug delivery, the rat version:** Catheters can be custom-prepared or obtained from various commercial vendors (e.g., Instech, or SAI infusion technologies, Lake Villa, IL, USA). Catheters used in this protocol are connected to a cannula (Plastics One, Roanoke, VA, USA) secured by a flexible rubber harness (Instech) worn by the rat for the duration of self-administration.

**2.2 Reagents and Solutions**

1. **2,5-dihydro-2,4,5-trimethylthiazoline (TMT)**, purity >97%: (BioSRQ, Sarasota, FL, USA) a fox pheromone. *See* **Note 3** on alternative predator scents.

2. **Airx X44 ACE Disinfectant Cleaner:** Odor-neutralizing disinfectant for cleaning the ASR equipment between sessions.

3. **70% EtOH:** for cleaning EPM equipment between sessions.

4. **Ketamine:** (87.5 mg/kg, i.p.) for sedation prior to cannula surgery.

5. **Xylazine**: (5 mg/kg, i.p.) for sedation prior to cannula surgery.

6. **Ketorolac**: (3 mg/kg, i.p.) analgesic for postoperative care.

7. **Heparinized saline:** (100 mg/mL) to flush catheters and maintain catheter patency.

8. **Methohexital sodium:** (10 mg/mL) to verify catheter patency.

9. **Drug for self-administration:** In this protocol, cocaine is used as an example. Cocaine hydrochloride (NIDA Controlled Substances Program Research Triangle Institute, NC, USA) is dissolved in sterile 0.9% physiological saline to a final concentration of 4 mg/mL (0.35 mg/infusion). In general, final drug concentrations need to be optimized by the experimenter, as

they are experiment-specific, and should be adjusted according to the weight and sex of animals, and the drug being studied.

### 2.3  Software

1. **MedPC V Software:** (Med Associates) to program the schedule of reinforcement and collect data.

2. **Ethovision XT:** (Noldus) to record and analyze behavior fear and anxiety-related behavior (freezing, activity in the EPM). *See* Fig. 3 for tracking software example.

3. **Advanced Startle Software:** (San Diego Instruments) to record and analyze startle behavior.

4. **Photobeam Activity System (PAS) Software:** (San Diego Instruments) to record and analyze behavioral activity during the light–dark box test.

## 3  Methods

### 3.1  Subjects

1. House rats individually on a reversed 12-h light–dark cycle (lights off at 7 am) in a temperature- and humidity-controlled vivarium. *See* **Note 4** for discussion of subject sex.

2. Maintain rats in the vivarium for 1 week prior to the beginning of experiment, with ad libitum food and water.

3. Three days prior to the beginning of experiment, all study personnel should handle rats for 1 min each, to reduce nonspecific handling stress during the PSS exposure and subsequent anxiety assessments. Initiate food restriction (to approximately 85% of their normal caloric intake; ~20 g of standard chow/day for adult male rats). *See* **Note 5** for discussion of nonspecific stress and personnel assigned to the study.

4. As HPA axis activity (and adrenal glucocorticoid hormone release) show a pronounced circadian rhythm, conduct all relevant behavioral procedures at the same time of the day. In this protocol, behavioral procedures were conducted within the first 5 h of the dark (active) phase.

### 3.2  Predator Scent Stress Procedure

1. Thoroughly clean the exposure chamber prior to the first animal and between subjects with 70% EtOH and let air dry.

2. Prepare small strips or squares of laboratory filter paper (approximately 5 × 5 cm).

3. Saturate a piece of filter paper with 3 μL of undiluted TMT solution and place in the holding cup beneath the mesh-wire flooring of the exposure chamber. Use a new, freshly TMT-saturated filter paper for each rat. Be sure to carefully dispose of each strip, as the scent is strong, and it is not advised

to have the subjects experience the smell outside of the exposure context.

4. Place rat in the exposure chamber, close the lid, and remotely monitor/record behavioral activity (freezing, darting, rearing) for 10 min using a camera and Ethovision XT software. Videos can also be recorded for later analysis through Ethovision or by a blinded observer.

5. Following exposure, immediately remove rat from the exposure chambers and return to the home cage. Repeat this procedure for each rat individually.

6. Maintain rats undisturbed in the vivarium for 1 week. Continue food restriction.

**3.3 Behavioral Assessment of Anxiety-Like Behavior**

One week following exposure, test rats for anxiety-like behavior and hypervigilance using EPM and ASR, respectively. Conduct EPM test first to minimize any carry-over stress effects of the ASR test.

*3.3.1 Elevated Plus Maze (EPM)*

*See* **Note 6** regarding the critical EPM variables that need to be controlled.

1. Clean EPM thoroughly in the beginning and between sessions with 70% EtOH. Let it air dry.

2. Place rat in the center of the cross zone of the open and closed arms.

3. Allow rat to freely explore the maze during a 5-min session.

4. Remotely monitor/record behavioral activity using a camera and Ethovision XT software to quantify: time spent in open arms (OA), time spent in closed arms (CA), number of entries into open arms and overall locomotor activity.

5. Remove rat back to home cage to be then be placed into an ASR chamber.

*3.3.2 Light-Dark Box as an EPM Alternative*

*See* **Note 7** for using Light-Dark Box test in lieu of EPM for behavioral phenotyping.

1. Thoroughly clean apparatus in the beginning and between rats with 70% EtOH. Let air dry.

2. Place rat in the center of the light side of the box and allow to explore the apparatus during a 5-min session. Remotely monitor/record behavioral activity using a camera and Ethovision XT software to measure: latency to enter dark compartment, time spent in dark compartment, and time spent in lit compartment.

3. Return rat to home cage. Repeat for each additional rat.

**3.3.3 Acoustic Startle Response (ASR)**

*See* **Note 8** regarding the settings and calibration of the ASR chambers.

1. Thoroughly clean each holding tube with RX44 ACE Disinfectant Cleaner prior to each session.

2. Program session using Advanced Startle Software to deliver a background noise (68 dB) with 30 startle trials of 110 dB white noise for 40 ms followed by a 30–45 s inter-trial interval.

3. Place rat into the holding tube with plexiglass partitions within the startle chamber.

4. Initiate 30-min session using Advance Startle Software. All experimenters should exit the testing area during the session to exclude confounding external stimuli.

5. Record the amplitude of the startle response to determine its habituation during repeated startle trails.

6. Carefully remove each rat from the holding tube and place back into the home cage.

**3.4 Behavioral Phenotyping**

Classify rats as stress-Susceptible (PTSD-like), Intermediate, or Resilient according to a median split analysis. *See* **Note 9** for strengths/weaknesses of this method and alternative methods for classification.

1. Remove outliers from data (>2 SD from the mean).

2. Identify median of the amount of time spent in the open arms of EPM and percent habituation of the startle response, dividing the mean of the last six ASR trials by the mean of the first six trials.

3. Classify rats with scores below the median EPM and above the median ASR as stress-Susceptible. Classify rats with scores above the median EPM and below the median ASR as Resilient. Classify rats above the median on both EPM and ASR or below the median on both EPM and ASR as Intermediate.

**3.5 Drug Self-Administration, Extinction, and Reinstatement Procedures**

One day following behavioral testing, surgically implant rats with indwelling jugular catheters for i.v. drug self-administration. *See* **Note 10** regarding the standard procedures for rodent aseptic surgery. Detailed protocols for jugular catheter implantation can also be found here [29, 30].

*3.5.1 Surgical Implantation of Intravenous Catheters*

1. Anesthetize rat using a mixture of ketamine (87.5 mg/kg, i.p.) and xylazine (5 mg/kg, i.p.). Before proceeding with surgery. During surgery assess level of anesthesia by pedal reflex (firm toe pinch) and adjust anesthetic delivery as appropriate to maintain surgical plane.

2. Before making any incisions, shave an appropriate surgical margin on the back and around the neck of the animal and sterilize it with betadine.

3. Using a surgical scalpel, make two incisions, first one in the upper part of the back of the animal (3–4 cm wide mid-scapular incision) and second, on the front neck area making a diagonal cut from the right clavicle going upward to the animals jaw (2 cm wide).

4. Find the jugular vein located superficially under the skin of the neck, remove the connective and adipose tissue and elevate the vein using plastic bar.

5. After making a small incision into the vein, insert catheter into the jugular vein all the way to the catheter bulb. Secure catheter in the vein with suture knots that travel around the vein and are located above and below the insertion point.

6. Create a subcutaneous passageway from the back incision to the neck incision area and pull the catheter tubing all the way through using a hemostat. Attach the catheter tubing to the stainless-steel cannula held within a harness (Instech).

7. Test the catheter to make sure there are no leaks by pushing a small amount of sterile saline (0.1 mL) into the vein. Any potential leaks indicate that the vein may have been pierced during surgery, or that the catheter tubing placement needs adjusting.

8. Close the skin incision with sterile surgical staples.

9. Allow a minimum of 5 days for the recovery from jugular catheter surgery prior to self-administration.

10. For the first 3 days of the recovery period, inject rat with ketorolac (3 mg/kg, i.p.) to provide analgesia.

11. Flush catheters daily with 0.1 mL of 100 U/mL of heparinized saline to maintain catheter patency.

*3.5.2 Self-Administration, Extinction, and Cue Reinstatement Procedures*

Self-administration procedures can vary significantly, with respect to drug doses, schedules of reinforcement, the duration of daily sessions, the number of daily sessions, and other factors. *See* **Note 11**. Detailed below is the step-by-step procedure for short and long access cocaine self-administration using an FR-1 schedule of reinforcement as typically utilized in this PTSD/SUD animal model [21].

1. Prior to each daily self-administration session flush catheters with 0.9% sterile saline.

2. Program MedPC V Software to a fixed ratio 1 (FR-1) schedule of reinforcement (1 lever press = 1 infusion) for the session. Program responses on the active lever during self-

administration to result in drug delivery and a light and tone presentation followed by a 20 s time out, during which time lever presses are recorded but do not deliver drug or cues. Program responses on the inactive lever to result in no consequences (no light, tone, or drug delivery).

3. Fill syringes with the drug of study, connect to the Tygon tubing, and press syringe until it fills the tubing entirely and all bubbles are removed. Insert syringe into syringe pump and secure.

4. Remove cannula cap from rat, hold finger over opening to prevent air from entering cannula, and connect tubing for drug delivery from the syringe pump to cannula.

5. Place rat in operant chamber for the duration of the session (1 h for ShA or 6 h for LgA).

6. After session has completed, remove rat from chamber, disconnect drug delivery tubing, and fill cannula with 0.2 mL of heparinized saline (100 mg/mL) to maintain patency. Replace cannula cap. Hold finger over opening between transitions.

7. Return rat to home cage.

8. Self-administration sessions should be run daily (or 6 days a week).

9. During self-administration, periodically check the catheter patency by intravenous administration of methohexital sodium (10 mg/mL). This should be done when signs of a loss of patency are observed, such as leaking around either incision or an increase in inactive or active lever presses that exceeds 130% of the previous day's presses.

10. During extinction training, use a Med PC program (Med Associates) that does not deliver cues or drug upon presses on the active lever. Place rat in the operant chamber and initiate 2 h extinction session. All lever presses will be recorded and responses on the active lever will decrease over the course of ~10–14 days when drug delivery ceases.

11. Once predetermined extinction criteria are met, conduct the cue-induced reinstatement test.

12. For cue-induced reinstatement, program active lever responses to in a presentation of cues associated with drug delivery (light and tone), but not drug delivery. *See* **Note 12** for alternative reinstatement methods.

13. Place rat into chamber for reinstatement for 1-h reinstatement test.

14. Compare reinstatement responses to responses recorded during extinction (2–3 days prior to reinstatement test).

## 4  Conclusion

Well-documented differences in social, psychological, and physical health and treatment outcomes between individuals diagnosed with SUD/PTSD and those diagnosed with either disorder alone necessitate a model encompassing this duality. The novel technique outlined in this chapter allows for the investigation of mechanisms driving these comorbid-specific health and treatment outcomes. By incorporating highly validated models of SUD and PTSD, this dual model evaluates anxiety and hypervigilance in parallel to reliably produce differential responses in percentages comparable to those observed in humans. Following trauma exposure, approximately 15–25% of humans are diagnosed with PTSD [2]. Similarly, following predator scent stress exposure in this model, PTSD-like behavior (as classified by high anxiety and high hypervigilance during EPM/ASR testing) is observed in 14–25% of rodents [20, 21]. Further, these percentages are reliable and have been reproduced in approximately 1200 rats [20, 21].

Using this protocol, our lab has shown phenotypic-specific behaviors and gene expression. Stress-Susceptible rats display increased fear responses during context reexposure and increased active lever pressing during extinction sessions and cue-primed reinstatement testing [21]. Resilient rats showed an enhanced expression of mGlu5 in the medial prefrontal cortex and enhanced expression of both CB1 and mGlu5 in the amygdala [20–22]. Additionally, cue-primed reinstatement of cocaine-seeking was *only* attenuated in stress-Susceptible rats after three treatments were administered (ceftriaxone, CDPPB, and fear extinction training), where as these treatments administered individually in separate rodent models of PTSD or CUD are effective [21]. This finding further validates the discrepancy in treatment outcomes observed in individuals diagnosed with a SUD/PTSD comorbidity in comparison to individuals diagnosed with either disorder alone. In sum, this method has high translational value and therefore utility in identifying underlying mechanisms driving the SUD/PTSD comorbidity and in identifying targets for future treatments.

## 5  Notes

1. Optimal illumination conditions for EPM: utilize dim white light (illumination ≤ 50 lux) in the testing room to decrease rodents' aversion to brightly lit environments, while still providing enough illumination to allow for video capture.

2. Optimal settings and optional adjustments for the ASR testing: for 30-min sessions, set background noise to 68 dB with

30 intermittent (30–45 s inter-trial intervals) startle trials set to 110 dB of white noise.

3. Alternative predator scents: many laboratories use alternate odors including canine and feline odors, such as bobcat urine [31], soiled cat litter or sand [32–34], and cat collars previously worn for 4 weeks [35].

4. It is well known that both PTSD and SUD are sexually dimorphic. Women are twice as likely to be diagnosed with PTSD than men [36]. Clinical data show salient differences in substance use, functional impairments from SUD, treatment outcomes, risk factors for relapse, and relapse to substance use between sexes [37, 38]. Further, rates of comorbid SUD/PTSD diagnoses are higher in females and sex is correlated with treatment course [39]. Thus far, our lab has only assessed comorbid SUD/PTSD in male rats. The differential effects between males and females in clinical data highlight the need for investigations of female rats in this model.

5. Nonspecific stress throughout the study should be minimized by prior handling and assigning specific personnel to interact with animals for the duration of the study.

6. Rat behavior in the EPM is very sensitive to a number of variables present in the environment. This includes illumination (*see* **Note 1**), scent of another animal (thorough cleaning of the apparatus between animals is required), the presence with another animal (use separate rooms or room dividers for each EPM apparatus), and the presence of an investigator (remote monitoring is strongly advised).

7. Results from our lab have identified comparable responses in EPM and Light-Dark Box tests [20], representing similar measures of the same anxiety-like behavior. Therefore, in lieu of EPM, amount of time spent in light side of the light–dark box can be used for behavioral phenotyping.

8. Two parameters need to be adjusted to get optimum and valid results from the SR-LAB System during the ASR testing: sound levels and response levels. Sound levels are calibrated (using a standard sound meter, like the one provided by the manufacturer) to assure that the analog input setting provides the expected dB level (120 dB, is typically a starting point for calibration). Response needs to be "standardized" to assure that all test stations are set to the same baseline allowing data collection from multiple test chambers. Detailed calibration methods are provided in the manufacturer's manual (San Diego Instruments).

9. Alternative statistical analyses for behavioral phenotyping: while a median split analysis allows us to identify a subpopulation of rats most affected by the predator scent exposure which

provides translational value, the use of a median-split analysis comes with limitations. Rats with behavioral scores immediately above or below the median are included, which may decrease classification accuracy. However, the addition of multiple behavioral tasks increases the probability of detecting "true" stress-Susceptible phenotypes. Additionally, PTSD symptoms in humans exist on a spectrum and are not accurately represented by behavioral categories. Therefore, alternative phenotypic analyses should consider regression and cluster analyses to assess changes in drug self-administration and drug-seeking that fall along a continuum of behavioral responses. For other approaches to classification of posttrauma phenotypes in animals see [24] and [40].

10. The surgical bench, the surgical instruments, and the catheters are sterilized prior to surgery using standard aseptic techniques. Surgical instruments and catheters should be sterilized by steam autoclaving. If applicable, catheters can be sterilized using ethylene oxide to prevent their damage. Cleaning and glass bead sterilization of the surgical instruments between animals is strongly recommended.

11. Besides the parameters described here, self-administration can be more broadly used to assess consumption, behavioral economic demand, incentive-motivation, sensitivity to aversive stimuli/punishment, concurrent choice, and other aspects of SUD [23].

12. Other methods of reinstatement for this model could include drug-primed and stress-induced reinstatement. Additionally, context reexposure stress could be used as a reinstatement procedure [23].

## Acknowledgments

This research was supported by the University of Florida Center for OCD, Anxiety, and Related Disorders (COARD).

## References

1. Simpson TL, Lehavot K, Petrakis IL (2017) No wrong doors: findings from a critical review of behavioral randomized clinical trials for individuals with co-occurring alcohol/drug problems and posttraumatic stress disorder. Alcohol Clin Exp Res 41(4):681–702

2. Kessler RC, Berglund P, Demler O, Jin R, Merikangas KR, Walters EE (2005) Lifetime prevalence and age-of-onset distributions of DSM-IV disorders in the National Comorbidity Survey Replication. Arch Gen Psychiatry 62(6):593–602

3. Ouimette P, Goodwin E, Brown PJ (2006) Health and well being of substance use disorder patients with and without posttraumatic stress disorder. Addict Behav 31 (8):1415–1423

4. O'Connor KM, Paauw DS (2016) Pharmacologic therapy. Med Clin N Am 100(4):i

5. Department of Defense (2017) The Management of Posttraumatic Stress DisorderWork GroupVA/DoD clinical practice guideline for the management of posttraumatic stress disorder and acute stress disorderversion 3.0

6. Ronzitti S, Loree AM, Potenza MN et al (2019) Gender differences in suicide and self-directed violence risk among veterans with post-traumatic stress and substance use disorders. Womens Health Issues 29(Suppl 1): S94–S102

7. Siegmund A, Wotjak CT (2006) Toward an animal model of posttraumatic stress disorder. Ann N Y Acad Sci 1071:324–334

8. Daskalakis NP, Cohen H, Cai G, Buxbaum JD, Yehuda R (2014) Expression profiling associates blood and brain glucocorticoid receptor signaling with trauma-related individual differences in both sexes. Proc Natl Acad Sci U S A 111(37):13529–13534

9. Elharrar E, Warhaftig G, Issler O et al (2013) Overexpression of corticotropin-releasing factor receptor type 2 in the bed nucleus of stria terminalis improves posttraumatic stress disorder-like symptoms in a model of incubation of fear. Biol Psychiatry 74(11):827–836

10. Park CR, Zoladz PR, Conrad CD, Fleshner M, Diamond DM (2008) Acute predator stress impairs the consolidation and retrieval of hippocampus-dependent memory in male and female rats. Learn Mem 15(4):271–280

11. Goswami S, Rodríguez-Sierra O, Cascardi M, Paré D (2013) Animal models of post-traumatic stress disorder: face validity. Front Neurosci 7:89

12. Goswami S, Samuel S, Sierra OR, Cascardi M, Paré D (2012) A rat model of post-traumatic stress disorder reproduces the hippocampal deficits seen in the human syndrome. Front Behav Neurosci 6:26

13. Goswami S, Cascardi M, Rodríguez-Sierra OE, Duvarci S, Paré D (2010) Impact of predatory threat on fear extinction in Lewis rats. Learn Mem 17(10):494–501

14. Gilbertson MW, Williston SK, Paulus LA et al (2007) Configural cue performance in identical twins discordant for posttraumatic stress disorder: theoretical implications for the role of hippocampal function. Biol Psychiatry 62 (5):513–520

15. Boscarino JA (2004) Posttraumatic stress disorder and physical illness: results from clinical and epidemiologic studies. Ann N Y Acad Sci 1032:141–153

16. Zoladz PR, Diamond DM (2013) Current status on behavioral and biological markers of

PTSD: a search for clarity in a conflicting literature. Neurosci Biobehav Rev 37(5):860–895

17. Sternberg EM, Glowa JR, Smith MA et al (1992) Corticotropin releasing hormone related behavioral and neuroendocrine responses to stress in Lewis and Fischer rats. Brain Res 570(1–2):54–60

18. Schelling G, Roozendaal B, Krauseneck T, Schmoelz M, DE Quervain D, Briegel J (2006) Efficacy of hydrocortisone in preventing posttraumatic stress disorder following critical illness and major surgery. Ann N Y Acad Sci 1071:46–53

19. Golier JA, Schmeidler J, Legge J, Yehuda R (2006) Enhanced cortisol suppression to dexamethasone associated with Gulf War deployment. Psychoneuroendocrinology 31 (10):1181–1189

20. Shallcross J, Hámor P, Bechard AR, Romano M, Knackstedt L, Schwendt M (2019) The divergent effects of CDPPB and cannabidiol on fear extinction and anxiety in a predator scent stress model of PTSD in rats. Front Behav Neurosci 13:91

21. Schwendt M, Shallcross J, Hadad NA et al (2018) A novel rat model of comorbid PTSD and addiction reveals intersections between stress susceptibility and enhanced cocaine seeking with a role for mGlu5 receptors. Transl Psychiatry 8(1):209

22. Shallcross J, Wu L, Knackstedt LA, Schwendt M (2020) Increased mGlu5 mRNA expression in BLA glutamate neurons facilitates resilience to the long-term effects of a single predator scent stress exposure. BioRxiv

23. Smith MA (2020) Nonhuman animal models of substance use disorders: translational value and utility to basic science. Drug Alcohol Depend 206:107733

24. Cohen H, Matar MA, Buskila D, Kaplan Z, Zohar J (2008) Early post-stressor intervention with high-dose corticosterone attenuates posttraumatic stress response in an animal model of posttraumatic stress disorder. Biol Psychiatry 64(8):708–717

25. Cohen H, Zohar J (2004) An animal model of posttraumatic stress disorder: the use of cut-off behavioral criteria. Ann N Y Acad Sci 1032:167–178

26. Cohen H, Kozlovsky N, Alona C, Matar MA, Joseph Z (2012) Animal model for PTSD: from clinical concept to translational research. Neuropharmacology 62(2):715–724

27. Hadad NA, Wu L, Hiller H, Krause EG, Schwendt M, Knackstedt LA (2016) Conditioned stress prevents cue-primed

cocaine reinstatement only in stress-responsive rats. Stress 19(4):406–418

28. Epstein DH, Preston KL, Stewart J, Shaham Y (2006) Toward a model of drug relapse: an assessment of the validity of the reinstatement procedure. Psychopharmacology 189(1):1–16

29. Sedighim S, Tieu L, George O (2020) Intravenous jugular catheterization for rats. Protocols. io

30. Thomsen M, Caine SB (2005) Chronic intravenous drug self-administration in rats and mice. Curr Protoc Neurosci Chapter 9:Unit 9.20

31. Albrechet-Souza L, Gilpin NW (2019) The predator odor avoidance model of post-traumatic stress disorder in rats. Behav Pharmacol 30(2 and 3-Spec Issue):105–114

32. Zohar J, Matar MA, Ifergane G, Kaplan Z, Cohen H (2008) Brief post-stressor treatment with pregabalin in an animal model for PTSD: short-term anxiolytic effects without long-term anxiogenic effect. Eur Neuropsychopharmacol 18(9):653–666

33. Cohen H, Liu T, Kozlovsky N, Kaplan Z, Zohar J, Mathé AA (2012) The neuropeptide Y (NPY)-ergic system is associated with behavioral resilience to stress exposure in an animal model of post-traumatic stress disorder. Neuropsychopharmacology 37(2):350–363

34. Dremencov E, Lapshin M, Komelkova M et al (2019) Chronic predator scent stress alters serotonin and dopamine levels in the rat thalamus and hypothalamus, respectively. Gen Physiol Biophys 38(2):187–190

35. Mackenzie L, Nalivaiko E, Beig MI, Day TA, Walker FR (2010) Ability of predator odour exposure to elicit conditioned versus sensitised post traumatic stress disorder-like behaviours, and forebrain deltaFosB expression, in rats. Neuroscience 169(2):733–742

36. Alexander KS, Nalloor R, Bunting KM, Vazdarjanova A (2019) Investigating individual pre-trauma susceptibility to a PTSD-like phenotype in animals. Front Syst Neurosci 13:85

37. Walitzer KS, Dearing RL (2006) Gender differences in alcohol and substance use relapse. Clin Psychol Rev 26(2):128–148

38. McHugh RK, Votaw VR, Sugarman DE, Greenfield SF (2018) Sex and gender differences in substance use disorders. Clin Psychol Rev 66:12–23

39. Torchalla I, Strehlau V, Li K, Aube Linden I, Noel F, Krausz M (2014) Posttraumatic stress disorder and substance use disorder comorbidity in homeless adults: prevalence, correlates, and sex differences. Psychol Addict Behav 28(2):443–452

40. Cohen H, Zohar J, Matar MA, Kaplan Z, Geva AB (2005) Unsupervised fuzzy clustering analysis supports behavioral cutoff criteria in an animal model of posttraumatic stress disorder. Biol Psychiatry 58(8):640–650

# Chapter 11

# Working and Reference Memory Impairments Induced by Passive Chronic Cocaine Administration in Mice

M. Carmen Mañas-Padilla, Fabiola Ávila-Gámiz, Sara Gil-Rodríguez, Lourdes Sánchez-Salido, Luis J. Santín, and Estela Castilla-Ortega

## Abstract

Cognitive performance, including working and reference memory functions, is frequently impaired in chronic cocaine users compared with healthy controls and this deficit predicts a worse therapeutic outcome. In clinical samples, it is difficult to elucidate whether the cognitive decline entails a trait present prior to the use of cocaine (i.e., a drug vulnerability factor) or if it is a consequence of the neuroplastic actions of the drug. Basic research in animal models has demonstrated that notable neurocognitive impairment is induced in naïve animals by repeated drug exposure. In the "passive" model of chronic cocaine administration, the animal is briefly administered a daily dose of cocaine by the experimenter and then returned to its home-cage; this process is repeated for several days. Although it is simple, this protocol robustly induces cognitive impairments in rodents that persist several weeks after withdrawal from the drug. In this chapter, we describe the procedure for passive chronic cocaine administration in mice, as well as three different and complementary tasks to evaluate the impact of cocaine on memory: (1) a continuous spontaneous alternation task to evaluate spatial working memory; (2) a novel object recognition task to evaluate reference memory for objects; and (3) a novel place recognition task to evaluate reference memory for places. These tasks are well established paradigms to evaluate memory in rodents, and they are advantageous because they may be performed and analyzed in the absence of costly or sophisticated equipment. The methods are described in sufficient detail so they may be performed by experimenters unfamiliar with cocaine administration or animal behavior research.

Key words Cocaine withdrawal, Cognitive impairment, Home-cage, Continuous spontaneous alternation, Y-maze, Novel object recognition, Novel place recognition, Observational analysis

---

## 1 Introduction

### 1.1 Drug-Induced Cognitive Impairment in Cocaine Use Disorders

Cognitive processes have gained increased importance in the prevention and treatment of substance use disorders. Clinical and preclinical research in animal models has identified cognitive phenotypes that might make the individual more prone to initiate and maintain drug use. Specifically, the cognitive traits associated with a vulnerability for drug use or abuse mainly involve frontal symptoms such as increased impulsivity and reduced ability to inhibit

María A. Aguilar (ed.), *Methods for Preclinical Research in Addiction*, Neuromethods, vol. 174,
https://doi.org/10.1007/978-1-0716-1748-9_11, © Springer Science+Business Media, LLC, part of Springer Nature 2022

inappropriate responses—indicative of a "loss of control" over behavior [1, 2]—risk taking and disadvantageous decision making [3], increased novelty response and sensation-seeking [4] and reduced "insight" (i.e., interoception or self-awareness) [5]. In addition to these factors, cognition is impaired by the neurobiological actions of the dependence-inducing drugs. Substances such as alcohol, cannabis, cocaine, methamphetamine, heroin or nicotine cause wide aberrant brain neuroplasticity that compromises the integrity of the learning-related brain regions [6], and a broad range of cognitive functions have been reported to be defective in drug users compared with healthy controls [7–9].

In particular, cocaine abuse has been associated with general cognitive impairment, including attention, impulsivity/compulsivity control and executive processes [7–9]. Focusing on the memory function, persons with cocaine use disorders show defective working memory [7–9], which is the capacity to actively manipulate information *online*. The working memory ability has a strong dependence on the frontal cortex [10], and it also depends on the hippocampus when spatial processing is involved [11]. Moreover, cocaine induces deficits in reference memory [7–9] that involves the short- and long-term acquisition and storage of information. This function widely depends on the hippocampus in the case of declarative memory content, which includes spatial memory (topographical orientation), episodic memory (life events including "what, when and where" aspects) and verbal memory (verbal information, facts, and concepts) [11, 12]. Consistent with the cognitive impairment, chronic cocaine users show neuroplastic changes in memory-related brain regions, such as diminished functional activity and reduced volume and gray matter thickness in frontal cortical areas [13–17] and altered functional activity and interregional connectivity in the hippocampal region [18–20].

The severity and persistence of the neurocognitive impairment induced by cocaine may vary greatly among individuals depending on numerous moderating factors: their premorbid intellectual functioning, the presence of comorbid psychopathology, stressful experiences and the drug use patterns (reviewed in [21]). Regarding this last aspect, higher cocaine does (i.e., grams consumed per week), more prolonged use (years), an earlier age of cocaine use onset or concomitant use of other drugs (polysubstance use) often predict worse neuropsychological performance (reviewed in [21]). Another main variable to consider is the duration of drug abstinence at the time of the cognitive assessment. Longitudinal studies show that 6–12 months of reduced or absent cocaine use lead to a partial or complete recovery in cognitive functions—including working and reference memory—and the associated brain neuroadaptations [9, 16, 17]. This finding indicates that impaired performance in both prefrontal and hippocampus-dependent cognitive domains is maintained by continuous exposure to the drug.

Importantly, the cognitive status has relevant clinical implications in cocaine use disorders because more severe cognitive deficits consistently predict addiction treatment drop-out and relapse in cocaine dependent patients [2, 22–24].

## 1.2 Rodent Models for the Investigation of Cocaine-Induced Cognitive Deficits

### 1.2.1 The "Passive" Chronic Cocaine Administration Model

Considering the above information, it is relevant to investigate the cognitive deficits induced by cocaine exposure, including its modulatory or vulnerability factors and their potential prevention and treatment. Basic research in animal models has traditionally focused more on drug-seeking or taking responses and habits [25], but investigation on the lasting cognitive impairment induced by cocaine exposure has gained increased attention. While there is compelling evidence of cocaine-induced cognitive decline in non-human primates [26–32], here we will focus on the protocols developed in rodents because they are the most easily available and widely used model for biomedical research. Cocaine-induced cognitive impairment has been studied by two main procedures: the cocaine self-administration paradigm [33–41] and the "passive" administration of cocaine by the experimenter [42–53] (Table 1).

In the cocaine self-administration paradigm, which is addressed in other chapters of this book, the rodent assumes an active role and learns to self-administer cocaine infusions by pressing a lever in an operant conditioning chamber. Subsequently, it is tested for cognitive performance, either immediately after ceasing cocaine infusion administration or after a drug withdrawal period of several days or weeks. This paradigm has consistently revealed cocaine-induced cognitive deficits in a variety of prefrontal and hippocampus-dependent domains, including both working and reference memory [33–41]. A notable advantage of the self-administration paradigm is that it allows researchers to study key addiction-related phenomena such as the motivation for the drug, craving, and relapse, all of which are elucidated by the expression of drug seeking/taking responses and may be assessed concomitantly to the cognitive functioning [33, 38]. As a caveat, it is difficult to control for the amount of drug received because this factor is chosen by each animal, and previous interindividual variability (or vulnerability) could possibly affect both the quantities of infused cocaine and the cognitive functions, which are usually more impaired at higher cocaine doses [37, 39]. In any case, a main issue is that this model requires researchers to submit rodents to both complex surgical procedures and extensive training periods in specialized operant chambers in order to self-administer cocaine [54], and neither the required technical expertise nor the equipment may be available in every laboratory.

Given these reasons, the model of "passive" cocaine administration that will be detailed in this chapter has been frequently used to investigate drug-induced cognitive impairment (Table 1). According to this protocol, the rodent forcibly receives a systemic

**Table 1**

Studies using the passive cocaine administration protocol in young or adult rodents to assess the effect of the drug on cognitive performance. Abbreviations: *IP* intraperitoneal, *n.s.* not specified, *PND* postnatal day at the beginning of the experiment (generally coincides with the first day of cocaine administration), *W.P.* drug withdrawal period, that is, days from the last cocaine dose to the beginning of the behavioral testing. Depending on the study, behavioral testing may not always commence with a cognitive test. (*) These experiments administered cocaine in a conditioned place preference apparatus. The rest of the experiments administered cocaine in the home cage. ∨: improved performance in the cocaine-treated animals compared to controls (vehicle-treated); ∧: improved performance in the cocaine-treated animals; =: similar performance of cocaine-treated and control animals

| References | Subjects | Housing | Cocaine treatment | W.P. | Cognitive task(s) | Results |
|---|---|---|---|---|---|---|
| Aguilar et al. (2017) [42] | Male OF1 adult mice (PND 68) | Grouped (4 per cage) | 5 mg/kg (PND 68–69); 15 mg/kg (PND 70–72); 25 mg/kg (PND 75–79); three doses per day; IP | 1 day | NOR (object memory) | ∨ |
| Berardino et al. 2019 [43] | Female CF1 adolescent mice (PND 34–37) | Grouped (4 per cage) | 34 mg/kg cocaine and 4 mg/kg caffeine per day for 10 days, IP | 1 day or 10 days | NOR (object memory) | = 1 day W.P. ∨ 10 days W.P. |
| Davidson et al. 2018 [44] | Male Sprague-Dawley rats (PND 90) | Single | 20 mg/kg per day for 18 days, IP | 3 days | Serial feature negative discrimination test (hippocampal dependent) | ∨ |
| | | | | | Simple discrimination test (hippocampal independent) | = |
| Fole et al. 2015 [45] | Male LEW and F344 adult (PND 56) and adolescent rats (PND 32) | Grouped (5 per cage) | 20 mg/kg per day for 13 days, IP | Adults: 0 days Adolescents: 23 days | NPR (place memory) | ∨ LEW adolescent rats = other groups |
| Gong et al. 2019 [46] | Male C57BL6J and BALB/CJ mice (PND 90) | Grouped (5 per cage) | 5 mg/kg per day, for 10 days (resting period of 2 days after 5 administrations), IP | 0 days | NOR (object memory) Social Recognition test (social recognition memory) | ∨ ∨ C57BL/6J mice = BALB/CJ mice |

| | | | | | | |
|---|---|---|---|---|---|---|
| Krueger et al. 2009 [47] | Male C57BL/6J adult (PND 93) | n.s. | 30 mg/kg per day, for 14 days, IP | 0 days | Instrumental conditioning (acquisition of instrumental response) | < |
| | | | | | Reversal learning (cognitive flexibility) | > |
| | | | | | Three-choice serial reaction time task (attention) | = |
| | | | | 14 days | Delayed matching to position task (working memory) | > |
| Ladrón de Guevara-Miranda et al. 2017 [48] | Male C57BL/6J adult mice (PND 63) | Grouped (3–4 per cage) | 20 mg/kg per day for 12 days | 44 days | NOR (object memory) | > |
| | | | | | NPR (place memory) | > |
| | | | | | Conditioned Place Preference test (acquisition of drug-contextual associations) | > |
| | | | | | Y maze (continuous spatial working memory) | > |
| Ledesma et al. 2017 [49] | Male OF1 adult (PND 55) or juvenile (PND 21) | Grouped (4 per cage) | Ascending doses of 5 mg/kg (2 days), 15 (3 days followed by 2 days of abstinence) and 25 mg/kg (5 days). Three injections per day, IP (juvenile: PND 68–79; adults: PND 102–113) | 1 day | NOR (object memory) | > |
| Mañas-Padilla et al. 2020 [73] | Male C57BL/6J mice (PND 84) | Single | 20 mg/kg per day for 14 days (*) | 28 days | NOR (object memory) | > |
| | | | | | NPR (place memory) | > |
| | | | | | Water maze task (spatial memory acquisition) | > |
| | | | | | Water maze task (spatial memory retention) | = |
| | | | | | Water maze task (cognitive flexibility) | = |
| | | | | | | > |

(continued)

**Table 1**
**(continued)**

| References | Subjects | Housing | Cocaine treatment | W.P. | Cognitive task(s) | Results |
|---|---|---|---|---|---|---|
| | | | | | Water maze task (spatial working memory) | |
| Mendez et al. (2008) [50] | Male Long-Evans rats (300–325 g) | Single | 30 mg/kg per day for 14 days, IP | 93 days | Water maze task (spatial memory acquisition) | = |
| | | | | | Water maze task (spatial memory retention) | > |
| Morisot et al. 2014 [51] | Male C57BL/6J × 129 mice (3–6 months old) | Grouped | 5 mg/kg (day 1); 10 mg/kg (days 2–3); 15 mg/kg (days 4–6); 20 mg/kg (7–10); administered every 12 hours (twice per day). One additional dose (20 mg/kg) on day 11. | 0, 14, 28, 35 or 42 days | NOR (object memory) | ∨ 0, 14, 28 days W.P. = 35, 42 days W.P. |
| Preston et al. 2019 [52] | Male C57BL/6J, (~PND 63–77) | Grouped (5 per cage) | 8 Sessions over days 1–4 and 7–10 with mice receiving cocaine at escalating doses: 4,8,16,24 mg/kg and 16,24,32,32 mg/kg respectively; one dose per day, IP (*) | 21 days | Radial arm maze (continuous spatial working memory training) | = |
| | | | | | Modified radial arm maze task (delayed spatial working memory training) | > |
| Santucci et al. 2010 [53] | Male Long-Evans rat (PND 30) | Grouped | 10 or 20 mg/kg, for 7–8 days, subcutaneous. | 62 days | Object discrimination training (associative stimulus-reward memory) | ∨ (both cocaine doses) |

injection of cocaine, usually in its home-cage, at the doses and timing scheduled by the experimenter. This protocol has been typically used in male rodents submitted to administration of high daily doses of cocaine (reaching 20–30 mg/kg/day) usually during a period of 10–14 consecutive days (Table 1). Although this protocol is simple, it has been demonstrated to induce persistent neurocognitive impairment in both rats and mice. These alterations affect performance in a wide variety of tests (Table 1), which also reveal concomitant neurobiological alterations. For example, cognitively impaired animals submitted to the passive cocaine administration model show abnormal functional activity in brain limbic regions— that seems reduced in the prefrontal cortex [43] but aberrantly increased in hippocampal and temporal regions [48, 51]—, abundant hippocampal neuroadaptations (affecting its synaptic plasticity, i.e., long-term depression [44, 45], neurochemical content [48], and adult hippocampal neurogenesis [55]) and increased serum corticosterone [46] compared to drug-naïve mice.

*1.2.2 Models for the Study of Working and Reference Memory: Continuous Spontaneous Alternation and Object and Place Recognition*

Numerous behavioral paradigms in rodents may be used to model the cognitive domains that are altered by cocaine, such as attention, impulsivity, decision-making, working, and reference memory [56–58]. In this chapter, we will explain methods to assess cocaine-induced impairment of spatial working memory in the continuous spontaneous alternation behavior (SAB) task, as well as on nonspatial and spatial reference memory retention in the novel object and novel place recognition paradigms (NOR and NPR, respectively) in mice.

Continuous SAB is based on the innate tendency of rodents to perform a thorough exploration of novel environments. Therefore, when they are exposed to a "T"- or "Y"-shaped maze, they will spontaneously try to visit the three different arms consecutively in an alternate pattern (e.g., arm "B," arm "A," arm "C"), so none of the arms should remain unexplored more than the others [59]. While there are several tasks that rely on the SAB, the one that we will explain here is the "continuous" variant, meaning that the animal is placed in the apparatus and allowed to freely and "continuously" explore the three arms at its own will for several minutes, performing as many alternations as possible [59]. A rodent must remember which arms were entered previously—and thus which arm must be visited next—in order to achieve a triplet of three different arms consecutively explored, an ability that is assumed to engage spatial working memory [59–61]. Performance in this task may be prevented by disrupting the prefrontal cortex ([62], although there are some divergent reports [63, 64]) and in animal models of hippocampal dysfunction, probably due to its spatial component [65–67].

The NOR and NPR paradigms also exploit the innate tendency of rodents to explore novelty, either for objects or places, to assess short- or long-term reference memory [68]. In the standard versions of these tasks, the animal is allowed to freely explore a set of two identical objects during a sample session. After a retention interval, one of the objects is replaced by an identical copy in its previous location (i.e., a familiar and static object), but the second object of the pair is either replaced by a novel, unknown object (in the NOR task) or by another copy of the familiar object displaced to a novel position in the maze (in the NPR task; also called "novel object location"). Therefore, these two paradigms assess both nonspatial ("what") and spatial ("where") components of episodic-like memory. Performance in both the NOR and NPR tasks has been traditionally linked to the hippocampus, although their neurobiological mechanisms are dissociated. It is generally accepted that the NPR entails more hippocampal dependency than the NOR [69, 70]. Thus, the latter relies mainly on other temporal lobe structures such as the perirhinal cortex [68, 70], and it requires the hippocampus for long-term memory retention intervals or when the contextual and spatial features of the testing environment entail certain complexity (reviewed in [68, 71, 70]).

The rationale for selecting the continuous SAB and the NOR and NPR paradigms for this chapter is that they evaluate three different forms of memory that are impaired in rodents submitted to the passive chronic cocaine administration model (Table 1, and additional unpublished studies from our research group). Specifically, the NOR task has been frequently applied in several laboratories. These studies reported deficits in the cocaine-withdrawn mice at both short (1 min to 2 h) and long (24 h) memory retention intervals (Tables 1 and 2). Importantly, these tasks also stand out for their methodological simplicity. Unlike many other memory paradigms, for example, associative fear memory, operant learning, spatial navigation in the water maze, hole-board or radial arm mazes [56, 58], the tasks described in this chapter rely on the rodents" spontaneous exploratory behavior and thus do not require experimenters to submit the animals to aversive stimuli or previous food restriction procedures as a motivation to perform the task. Furthermore, animals are not submitted to extensive training protocols, a factor that allows a rapid memory assessment (Subheading 3). Finally, it is also worth mentioning that these tests are performed in simple dry mazes (Y-shaped or squared) and require neither complex apparatuses—such as operant conditioning chambers or a water pool—nor any other expensive or sophisticated material (Subheading 2).

**Table 2**
Novel object recognition (NOR) and novel place recognition (NPR) protocols used to study the cognitive-impairing effects of cocaine. Abbreviations: *n.s.* not specified, *W.P.* drug withdrawal period. ∨: impaired performance in the cocaine-treated animals compared to controls (vehicle-treated); =: similar performance of cocaine-treated and control animals

| References | Sample session duration | Memory retention interval (ITI) | Test session duration | Definition of object exploration | Total time exploring objects | Object or place memory |
|---|---|---|---|---|---|---|
| NOR | | | | | | |
| Aguilar et al. (2017) [42] | 3 min | 1 min | 3 min | Intentional contact of the mouse's snout or front paws with the object from a distance of 2 cm or less | n.s. | ∨ |
| Berardino et al. (2019) [43] | 10 min | 24 h | 5 min | Orientation of the head toward an object with the nose 1 cm or less around it | n.s. | = 1 day W.P ∨ 10 days W. P. |
| Gong et al. (2019) [46] | 5 min | 15 min | 5 min | Directing the nose toward the object at a distance of 1 cm and/or touching the object with its nose or paws | ∨ C57BL/6J mice = BALB/CJ mice | ∨ C57BL/6J mice = BALB/CJ mice |
| Ladrón de Guevara-Miranda et al. (2017) [48] | 10 min | 24 h | 10 min | Touching an object with its nose or forepaws | = | ∨ |
| Ledesma et al. (2017) [49] | 3 min | 1 min | 3 min | Intentional contact of the mouse's snout or front paws with the object from a distance of 2 cm or less | n.s. | ∨ |
| Mañas-Padilla et al. (unpublished) | 10 min | 24 h | 10 min | Actively touching an object with its nose or its forepaws; or pointing the nose toward the object at a distance of 0.5 cm or less. | = | ∨ |

(continued)

**Table 2**
**(continued)**

| References | Sample session duration | Memory retention interval (ITI) | Test session duration | Definition of object exploration | Total time exploring objects | Object or place memory |
|---|---|---|---|---|---|---|
| Morisot et al. (2014) [51] | 10 min | 15 min to 2 h | 5 min | Directing the nose within 0.5 cm of the object while looking at, sniffing, or touching it, excluding accidental contacts with it (backing into, standing on the object, etc.) | ∨ 0 days W.P. = 14, 28, 35, 42 days W.P. | ∨ 0, 14, 28 d W.P. = 35, 42 d W.P. |
| NPR | | | | | | |
| Fole et al. (2015) [45] | 5 min | 30 min; 24 h | 5 min | A single interaction was counted when the animal approached the object at a distance of 2 cm or less | n.s. | = ITI 30 min ∨ ITI 24 h (only in LEW adolescent rats, not in adult rats or F344 rats) |
| Ladrón de Guevara-Miranda et al. (2017) [48] | 10 min | 24 h (after NOR test) | 10 min | Touching an object with its nose or fore paws | = | ∨ |
| Mañas-Padilla et al. (unpublished) | 10 min | 24 h (after NOR test) | 10 min | Actively touching an object with its nose or its forepaws; or pointing the nose toward the object at a distance of 0.5 cm or less | = | ∨ |

## 2 Materials

### 2.1 Animals and Housing Conditions

The following aspects should be considered regarding the animals used in the study:

- It is strongly recommended to perform the experiment in rodents of a similar sex, age, and strain, unless the variability induced by these factors is an intended focus of the study. Additionally, when it is possible to identify the litter of each animal (such as in case of in-house breeding), littermates must be equally distributed among the experimental treatments to control for interlitter variations [72]. In our laboratory, we have tested the passive chronic cocaine administration protocol described in this chapter in young male C57BL/6J mice, acquired from Janvier Labs (Le Genest-Saint-Isle, France), that started receiving cocaine either at 9 [48] or 12 weeks of age ([73], Table 1).

- As a minimum, two experimental conditions are required to obtain conclusions: a group of cocaine-treated mice, and a comparable group of vehicle-treated mice as drug naïve controls.

- Considering that behavioral data may involve quite high interindividual variability, a minimum of approximately ten mice per experimental treatment (corresponding with a statistical power $\geq 0.8$ or $\beta \leq 0.2$) will be necessary to achieve statistically significant results in all the three memory tasks described in this chapter.

- Mice are housed under standard conditions (temperature $22\ °C \pm 2\ °C$; 12-h light–dark cycle; lights on at 08:00) with ad libitum access to water and food during the experiment. In case of acquiring the mice from an external supplier, they must arrive at the animal facility at least 1–2 weeks before starting the experiments to allow proper quarantine and acclimation.

- While this protocol has been successfully applied either to grouped or single-housed animals (Table 1), the social housing conditions (number of animals per cage), the size of the cages and the elements they contain (e.g., nesting material) entail environmental enrichment factors that might critically influence results. Therefore, they must be as similar as possible for all animals in the study. If mice are group housed, mice sharing the same cage should ideally be assigned to different experimental treatments (e.g., cocaine or vehicle administration).

- In either the grouped or the single housing condition, providing nesting material is recommended for welfare, and it may attenuate intermale aggression [74] or alleviate the impact of social isolation, respectively [75].

- Every mouse should be assigned a particular and unique identification code (e.g., "cocaine 1," "vehicle 1," "cocaine 2," etc.). This design will allow experimenters to accurately monitor each animal through the experiment. Furthermore, it will allow them to establish correlations among the different evaluated behavioral and/or neurobiological measures. In the case of group housing, many rodent identification systems—such as ear punching—are available and commonly used [76].

### 2.2 Materials for Passive Cocaine Administration

The protocol of passive cocaine administration in mice (Fig. 1a) requires the following materials:

- Cocaine hydrochloride in powder form must be acquired from an authorized supplier (e.g., Alcaliber S.A., Madrid, Spain) and stored at controlled room temperature or at 4 °C, protected from exposure to light. The quantity of powder cocaine required per day may be calculated in advance according to the cocaine dose to be administered and to the number and weight of the animals that should receive the drug.[1] At the experimenter's convenience, powder cocaine may be previously weighted and aliquoted into appropriate containers for daily use (Fig. 1b, c). A high precision milligram balance will be required to weigh the drug.

- Normal saline solution (sterile physiological saline; 0.9% NaCl), stored at controlled room temperature, is used as the vehicle solution to dissolve cocaine. Saline will be administered intraperitoneally to mice at a volume of 10 ml/kg. As explained before, if the number and approximate weight of the animals is known, the volume of physiological saline required may be calculated in advance[2] and aliquoted in appropriate containers (e.g., plastic centrifuge tubes stored at controlled room temperature) ready for daily administration.

- We typically use 0.5 ml injection volume syringes (MYJECTOR; Terumo) with a 0.01 ml graduation and 30 × 1.2″ (0.3 mm × 12 mm) needle gauges for administration of both cocaine and vehicle solutions [77] (Fig. 1b).

- Materials to weigh the mice include an appropriate scale (weighing capacity that covers a range of 0–50 g and a readability of at least 0.1 g) and a small, light box to place the animals for weighing (Fig. 1b). The box may contain disposable tissue paper that can be easily replaced when dirty.

---

[1] As an example, let us consider an experiment using 20 adult mice weighting 25–30 g each, half of which ($n = 10$) will receive cocaine at a dose of 20 mg/kg/day and the rest will be assigned to the control vehicle group. Therefore: Ten mice receiving cocaine × 30 g weight for each (to avoid being short, it is advisable to use their maximum weight for calculations) = 300 g is the total body mass of the mice to be administered daily. If the desired dose is 20 mg of cocaine per 1000 g of weight (i.e., 20 mg/kg dose), then 6 mg of powder cocaine will be required for 300 g of weight by a simple rule of three. In conclusion, 6 mg (0.006 g) of powder cocaine will be required for each daily administration according to this example.

[2] Following the previous example, the group of ten cocaine-treated mice, with a total mass of 300 g, will use a total of 3 ml (0.01 ml/g) of saline solution per administration day (the solution in which the cocaine will be diluted). In this case, it is also necessary to prepare saline solution for the group of vehicle-treated mice that will receive a daily saline administration at a similar volume (10 mg/kg) to the cocaine-treated animals.

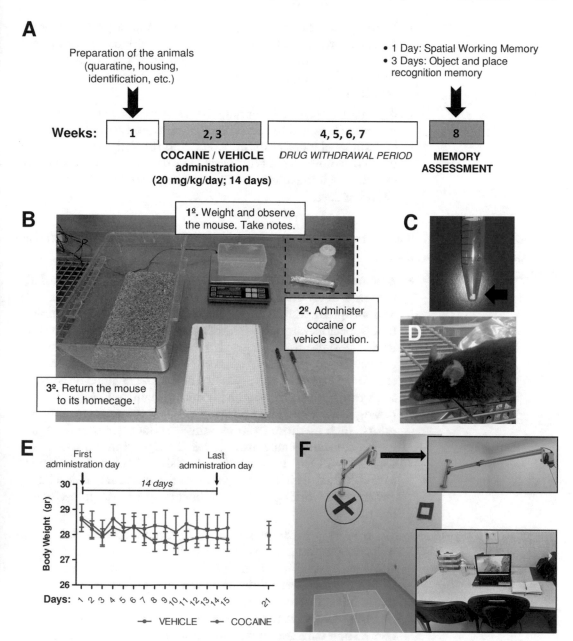

**Fig. 1** Materials and protocol for passive chronic cocaine administration and subsequent cognitive assessment. (**a**) Standard schedule of the protocol of passive cocaine or vehicle administration followed by memory assessment. Details such as the cocaine administration regimen, the duration of the drug withdrawal period or the inclusion of additional behavioral tasks may vary depending on the aims of the study. (**b**) Example of materials and brief description of the procedure required for cocaine and vehicle administration. (**c**) Powder cocaine aliquoted for one daily use in a 15 ml polypropylene centrifuge tube. On the administration day, cocaine may be diluted by adding the saline solution directly to the tube. Powder cocaine is stored at a controlled temperature (4 °C) and protected from light. (**d**) A mouse treated with a 14-day chronic cocaine administration protocol detailed in this chapter. Cocaine-treated mice are not expected to show any external signs of pain or impaired health. (**e**) Body weight of mice administered with cocaine (20 mg/kg/day; $n = 22$) or

- Writing materials (e.g., permanent markers, a pen or a notebook) are required to label the tubes that contain the different solutions and to annotate the weight of the animals and any incident that might occur during the drug administration process (Fig. 1b).

- A previously planned schedule that clearly depicts the identification code assigned to each animal, their experimental treatment, its individual identification mark(s) and the daily order in which it will be administered is required. The daily order of cocaine and vehicle administration may be counterbalanced on alternate days.

### 2.3 Materials for Cognitive Assessment

The memory tasks explained here will require general materials required for behavioral testing:

- A suitable space for behavioral testing should be carefully prepared in advance. Memory assessment should take place in an appropriate testing environment, a room isolated from noises that is spacious enough to contain the apparatuses, the recording devices and the animal cages. It should also contain a seat for the experimenter, preferably next to a small table for a notebook and/or a computer, and separated at least ~1.5 m from the maze so as to not disturb the animals during their test (Fig. 1f). The testing room should be well ventilated (avoiding air currents and odors, such as those from cleaning products or perfume) and with a constant temperature ($22 \pm 2$ °C) and illumination. We usually perform the behavioral tasks detailed here under 100–200 lux, which is equivalent to the illumination level usually used for working areas, avoiding any source of intense bright light.

  Importantly, the mazes must be evenly illuminated. Light reflections or intense shadows may be avoided by either displacing the mazes or modifying the light sources in the room. Zones with different light intensities or brightness may induce the animals to develop spatial preferences during testing [78] that may confound the memory-related measures. Furthermore, contrasting dark and illuminated zones may make recording the data, either by an observer or a video-tracking software, difficult.

**Fig. 1** (continued) vehicle solution ($n = 20$) for 14 consecutive days (days 1–15) and 1 week after (day 21). Cocaine may induce a slight and transitory weight loss. In the example, this effect did not result in a statistically significant effect [repeated measures ANOVA on days 1–15 ("cocaine × day"): effect for "cocaine": $F_{(1, 40)} = 0.275$, $p = 0.603$; "day": $F_{(14, 560)} = 5.294$, $p < 0.000$; "cocaine × day": $F_{(14, 560)} = 1.639$, $p = 0.065$]. (**f**) Example of a setting for behavioral assessment. Black spatial cues are hung on the walls. Data are expressed as mean ± standard error of the mean

The continuous SAB and the NPR tasks entail a spatial component, and thus the walls surrounding the testing apparatus should be distinct from one another in order to facilitate spatial orientation. This may be achieved by placing one distinctive contextual cue on each wall (Fig. 1f). The contextual cues are typically made from black cardboard (to allow contrast from the background) cut in different geometric shapes (e.g., a triangle, a circle, a set of vertical bars, a cross, etc.). Other elements in the room, including the video recording materials, furniture or even the experimenter, may also act as extra-maze contextual cues and should not change positions during behavioral testing. A folding screen or a curtain near the testing maze—with a contextual cue attached—could be used as a room divider to hide certain environmental elements from the animals, such as the whereabouts of the experimenter or the computer. The reader may consult (Fig. 1f) other publications [60, 79] for examples of appropriate room settings. We recommend that the behavioral testing room or environment does not resemble the one where cocaine was previously administered, to avoid retrieval of drug-related memories that may result in anxiety and/or drug *craving* feelings in the cocaine-withdrawn mice during testing.

- Every session must be video recorded for subsequent *off-line* analysis, preferably in zenith view (Figs. 2b and 3d), by an overhead digital camera (Fig. 1f). Depending on the setting, a tripod, extension cords or adjustable plastic clamps may be required to securely hang the camera from the ceiling or from a horizontal support (Fig. 1f). The camera should be firmly fixed in place for recording.

- Identification signs. Carboard cards or a whiteboard are needed to write both the code of the animal that is currently being tested and the testing session that is being performed (e.g., "cocaine 1" mouse, "habituation session"; *see* Subheading 3.3).

- One or several storage devices (a laptop or USB disks) are required for storing the video files.

- A laboratory timer or a similar device is required to monitor the duration of the sessions. This device will preferably have no sound so as to not to disturb the animals.

- Materials for cleaning the mazes should include a sprayer dispenser bottle containing a solution of 30% ethanol diluted in water. Disposable paper towels are required to clean the mazes and remove any odor cues after each session.

- A previously planned schedule that clearly depicts the daily order in which mice will be tested is necessary. For behavioral assessment, a mouse from each experimental treatment is tested in alternation (e.g., vehicle-treated mouse, cocaine-treated mouse,

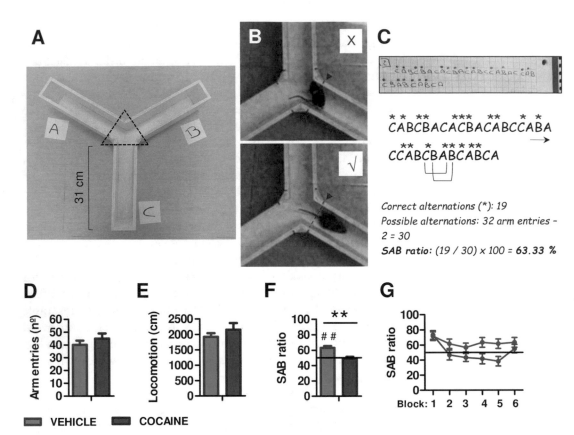

**Fig. 2** Assessment of the effects of cocaine on spatial working memory (continuous spontaneous alternation behavior [SAB]). (**a**) The Y-maze apparatus used for this task. Please note that the three different arms are clearly identified (A, B, and C). The imaginary dotted triangle indicates the region considered as the neutral central area. (**b**) Arm entries may be scored observationally. To facilitate scoring, we recommend delimiting the entrance of the arm by visual landmarks that are visible by the experimenter but not by the animal (indicated by red arrows). An arm entry is annotated only when the four paws of the mouse have clearly crossed this imaginary boundary, as shown in the photograph below. (**c**) For an easy quantification of the total number of correct alternations, we annotate the complete sequence of arm entries and subsequently add a color mark above every letter that starts a correct alternation (e.g., CBA but not BAB). We then count the total number of color marks. The example shows the calculation of the SAB ratio for a session of one animal. (**d–g**) Sample data from one experiment ($n = 12$ mice per treatment) using the protocol described in this chapter. The cocaine-treated mice were similar to the controls in exploratory parameters (**d, e**), but they were impaired in their continuous SAB ratio (**f**). The SAB ratio may also be split into "blocks." In (**g**), we calculated the SAB ratio in "blocks" of every five possible alternations, until a total of 30 alternations. Difference between groups: **$p < 0.000$; Difference versus chance performance: ##$p < 0.000$. Chance performance is set at 50%. Data are expressed as mean ± standard error of the mean

vehicle-treated mouse, cocaine-treated mouse and so on). This design will help control for potentially confounding variables—such as the total time of habituation to the experimental room or the influence of circadian rhythms—between the experimental treatments. In the case of the NOR and the NPR tasks, the schedule will also show the types of objects required for each trial and their position in the maze (e.g., Fig. 3c).

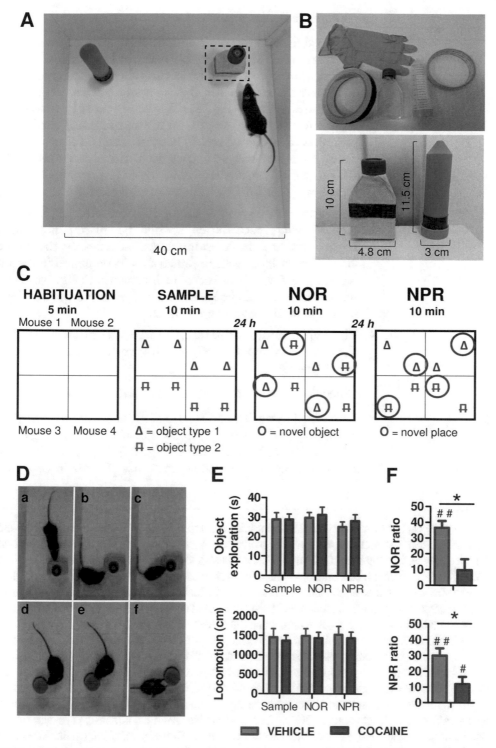

**Fig. 3** Assessment of the effects of cocaine on reference memory for objects and places. (**a**) Photograph of the open field apparatus displaying two object types; a mouse is approaching one of the objects. Object explorations do not always imply that the mouse is directly touching the object because interactions that

Specific materials required for the continuous SAB task:

- Y-shaped maze. The continuous SAB is performed on a Y-maze with three arms of the same size that are placed at 120 degrees to each other and converge into a central triangular area. In mice, appropriate measures for the arms are 31 cm long × 8 cm wide × 15 cm high (Fig. 2a). The maze must be made from a light but durable material with a water-resistant surface (e.g., plexiglass or lacquered wood) because it will be cleaned multiple times. Its color must be chosen so it will contrast visually with the fur of the animals that will be tested (e.g., mazes in white or light-gray color for C57BL6J mice with black fur).

- Each arm must be named after a different letter (A, B, or C). A card showing each letter should be placed next to the corresponding arm. A landmark that is visible in the recorded videos but hidden from the mice should be placed in each arm separated 3.5 cm from the arms intersection (Fig. 2a, b). This marker will act as an imaginary boundary that separates the arm from a neutral central zone (Fig. 2a, b).

- The main measure of this task, the SAB, may be easily analysed in the recorded videos by a trained observer who uses only pen and paper (Fig. 2c).

Specific materials required for the NOR and NPR tasks:

- Open-field maze. The NOR and NPR task will be performed in a square open-field maze, typically measuring 40 cm long × 40 cm wide × 40 cm high for mice (Fig. 3a). Variants of this maze frequently include four separate arenas (80 cm long × 80 cm

---

**Fig. 3** (continued) occur at a distance of 0.5–2 cm from the object itself (the dashed line that the mouse seems about to cross with its nose) are typically scored as explorations (Table 2). (**b**) Two different object types for mice may be fabricated from a 50 ml cell culture flask and a 50 ml conical centrifuge tube. The tube is covered with material from a blue nitrile glove and the flask is covered with masking tape in light yellow to provide them with a different color and texture. A strip of black duct tape is added to both objects to facilitate their visual contrast over the light-colored maze background. (**c**) Schedule for testing four mice. Importantly, a mouse must be tested always in the same arena; and the arenas must be counterbalanced across cocaine- and vehicle-treated animals. Both the type of novel object (type 1 or 2) and the novel object and place in the maze are also counterbalanced across mice. The duration of the habituation session is 5 min, while the sample, NOR and NPR sessions last for 10 min. The habituation and sample session are performed on day 1, and the NOR and NPR sessions are performed on days 2 and 3. (**d**) Behaviors typically considered as object exploration. The mouse may sniff the object while having its four paws on the floor (*a*, *b*, and *d*) or while rearing supported on the object (*c* and *e*). Merely touching the object with the forepaws (*f*) is also frequently considered as object exploration, although there is not a consensus (Table 2). (**e, f**) Sample data ($n = 20$ mice per treatment) using the protocol described in this chapter. (**e**) Cocaine-treated mice were similar to controls in their exploratory activity throughout the test. (**f**) Cocaine treated mice were impaired in both memory measures. Difference between groups: *$p < 0.05$; difference versus chance performance: ##$p < 0.000$, #$p < 0.05$. Chance performance is set as zero. Data are expressed as mean ± standard error of the mean

wide $\times$ 40 cm high maze) that will allow the experimenter to test four mice at once (Fig. 3c). As stated for the Y-maze, it is important that the surface of this maze is washable and provides sufficient visual contrast to clearly distinguish the animals, even their nose tip, to assess object exploration (Fig. 3d).

- Objects. While the NPR task uses two identical copies of an object, at least two different objects (e.g., object type "1" and "2") are needed for the NOR task (Fig. 3b, c). We typically use objects made from laboratory expendables that have a similar volume but different geometric shape (Fig. 3b). It is mandatory that these two object types are clearly distinguishable from one other and hold a similar intrinsic appeal for the mice [79, 80]. For example, rodents will inherently prefer objects that they could climb on over nonaffordable objects [81], objects that are more complex (i.e., textured objects, with irregular patterns and increased number of elements [78]) or bigger in size over smoother, simpler, and smaller objects. It is important to rule out all these confounding factors, so the preference in exploring one object over the other in the NOR test could only be attributed to novelty.[3] Multiple copies of the same object are required, considering that the NOR and the NPR sessions are performed with an identical copy of the familiar object but not with the familiar object itself (Subheading 3).

- Adhesive putty or double-sided tape is useful to fix the objects to the maze floor because they should neither be tipped nor displaced by the animals.

- An instrument to register the duration of exploration of each object is recommended. A laboratory timer, a chronometer or a similar device operated manually or through a keyboard will allow the experimenter to quantify the total time the mouse explores each object. Freely available software for the analysis of animal behavior may facilitate this register and also provide other relevant variables such as the frequency of explorations [82]. Commercial software packages may also allow automatic registration of object exploration and the simultaneous assessment of spatiotemporal variables such as the total locomotion or the time spent in different maze zones. Alternatively, in absence of any of these devices, it is possible to quantify object exploration by registering the start and stop times of object investigation as they are displayed by the time stamps on the video [79].

---

[3] If the experimenter is unsure about whether the two object types chosen for the NOR task are intrinsically preferred equally by their mice, this factor should be confirmed in a pilot study. For this validation, naïve animals—of the same sex, strain and age than the ones that will be used in the experiment—will perform one habituation session to the maze and then a sample session containing one copy of each object type, whose position in the maze will be counterbalanced across mice. If the time spent exploring the two object types is similar in the group of mice, these object types have equal intrinsic motivational properties.

# 3   Methods

### 3.1   Procedure for Chronic Passive Administration of Cocaine or Vehicle

Cocaine (e.g., 20 mg/kg/day) and vehicle administrations are performed once daily starting at approximately 09:30 a.m. In our experiments, we used a schedule of 12 [48] or 14 [73] consecutive administration days (Fig. 1a).

1. Carry the animals to the room where the administrations will be performed. Allow them 5–10 min for habituation.

2. Freshly dilute powder cocaine in saline solution, for a dose of 20 mg/kg in a volume of 10 ml/kg (*see* Subheading 2.2 for a more detailed explanation). Appropriately label the tube ("COCAINE") and protect this solution from light. Powder cocaine is readily dissolved when saline solution is added to its container after gently manual shaking. A magnetic stirrer is not necessary.

3. Label the container of drug-free saline ("VEHICLE") so that it is clearly distinguishable from the container of the cocaine solution.

4. Identify the mouse that should be administered first, according to your schedule. Weigh it and observe its general appearance (posture, fur condition, external wounds—e.g., due to fighting with cage companions—reflexes and other indicators of pain or stress [83]). Annotate its weight and any other information relevant to its health status. Across the administration days, mice in the cocaine-treated group will usually experience a slight weight loss and/or a delayed weight gain compared to their vehicle counterparts (Fig. 1e). However, their general health should not be impaired (Fig. 1f). We have not observed that this cocaine administration regime increases aggression among male cagemates. Observing wounded animals would indicate excessive aggressive behavior, and thus single housing may be considered.

5. Intraperitoneally administer a volume of 10 ml/kg of the "vehicle" or "cocaine" solution depending on the mouse's experimental treatment (Fig. 1b). The 10 mg/kg administration volume for intraperitoneal injections is consistent with good laboratory practices [84], and it allows a rapid calculation of the volume needed for each particular animal.[4] For repeated intraperitoneal injections, we recommend to alternate daily the side of the body used for administration (to avoid repeatedly puncturing the same area).

---

[4] Given that 10 mg/kg equals 0.01 ml/g of body weight, a mouse weighing 26 g will receive a volume of 0.26 ml. Decimals are rounded to the nearest hundredth (e.g., a mouse weighting 28.6 g will receive 0.29 ml).

6. Return the mouse to its home cage. In the case of group-housed animals, the administered mouse may be returned to an auxiliary cage until all of their cagemates have been administered the appropriate treatment to avoid confusion (Fig. 1b). In the cocaine-administered mice, increased vertical and/or horizontal activity should be observed briefly after drug administration (due to the psychostimulant and hyperlocomotor effects of cocaine).

7. Syringes and needles must be replaced as much as required. Special care must be taken so that mice in the cocaine-treated group never share any needle or syringe with the vehicle-treated animals, to avoid the latter to be exposed to any quantity of the drug.

### 3.2 Withdrawal from Chronic Cocaine

The cocaine withdrawal period extends from the last drug administration to the behavioral testing (Fig. 1a). Different withdrawal durations may influence the intensity of the cognitive deficits (Table 1) as well as the presence of concomitant behavioral alterations (i.e., anxiety due to drug craving). In this chapter, we propose a drug withdrawal period of ~4 weeks. With this period, we observed lasting cognitive deficits in absence of anxiety-like or hyperactivity responses in the cocaine-treated animals that could obscure cognitive assessment ([73], Table 1).

During extended cocaine withdrawal, animals should remain undisturbed but examined for general health and briefly handled twice per week by the experimenter. In this way, they will remain habituated to handling and should reduce their anxiety during subsequent behavioral testing. Depending on the requirements of the study, the cocaine withdrawal period may also be used to apply pharmacological or environmental interventions that aim to influence the cognitive deficits.

### 3.3 Memory Assessment

In our laboratory, the behavioral assessment is performed during the light-phase (usually starting at 09:00). A 24-h interval or longer should separate two different behavioral tasks to avoid any interference in cognitive performance due to task switching, or a reduced motivation to explore due to extensive daily assessment. Considering this factor, the continuous SAB may be performed on the first day of testing, while the different sessions required for the NOR and NPR tasks may be performed on the three subsequent days (Fig. 1a).

#### 3.3.1 Procedure for the Continuous SAB Test

The continuous SAB is evaluated by a single Y-maze session.

1. Transfer all the mice to the behavioral assessment room and leave them undisturbed for at least 20 min.

2. Check that the room setting is correct (temperature, ventilation, video recording devices, etc.) and that the Y-maze is

correctly placed under the camera. The *zoom* must be the maximum allowed that ensures that the entire surface of the maze is recorded.

3. Identify the mouse that should be tested first, according to your schedule, and write the corresponding identification sign (e.g., cardboard with the code of the animal and the testing day). Remember to test cocaine- and vehicle-treated mice alternatively.

4. Start recording the video. Be sure that the corresponding identification sign is recorded at the beginning of the trial by briefly placing it under the camera.

5. Gently pick the mouse by its tail and place it on a small box or a surface that allows grasping (e.g., a cage grid or the forearm of the experimenter covered by a coat or a tissue) to transfer it to the Y-maze.

6. Gently release the mouse in the center of the starting arm, facing the central zone. The starting arm may be the same for all animals or may be counterbalanced.

7. Go to your seat and start the timer. Allow the mouse to freely explore the Y-maze for 6 min. Annotate any incident during testing (e.g., an unexpected noise).

8. When the time is over, stop the video recording.

9. Return the mouse to its home cage.

10. Carefully clean the maze and let the surface dry.

11. Repeat the procedure with the next mouse.

12. After the testing day is finished, transfer all the video files from the camera to one or more storage devices.

### 3.3.2 Procedure for the NOR and NPR Tests

The object recognition tasks will include four testing sessions performed over 3 days: habituation and sample sessions (day 1), NOR session (day 2) and NPR session (day 3; Fig. 3c). The memory retention interval will be 24 h to assess long-term reference memory, although cocaine-induced deficits have been reported at shorter delays (Table 2).

*Day 1: habituation and sample sessions*

1. Transfer all the mice to the behavioral assessment room and leave them undisturbed for at least 20 min.

2. Check that the room setting is correct (temperature, ventilation, video recording devices, etc.) and that the open field maze is correctly placed under the camera. The *zoom* must be the maximum allowed so that the entire surface of the maze is recorded. The open field should be empty, with no objects, for the habituation session.

3. Identify the mouse (or mice if using an open field with four arenas) that should be tested first, according to your schedule, and write the corresponding identification sign (with the code of the animal(s) and the testing day and session, i.e., "habituation").

4. Start recording the video. Be sure that the identification sign is recorded at the beginning of the trial by briefly placing it under the camera.

5. Gently pick the mouse by its tail and place it on a small box or a surface that allows grasping (e.g., a cage grid or the forearm of the experimenter covered by a coat or a tissue) to transfer it to the open field maze.

6. Gently release the mouse in the center of the arena. Repeat this process with the remaining three mice if four animals are tested at once. Remember to test cocaine- and vehicle-treated mice alternatively or during the same session (in the case of simultaneous assessment of several mice).

7. Once all mice are in the maze, go to your seat and start the timer. Allow the mice to freely explore the empty arena for 5 min. Annotate any incident during testing (e.g., an unexpected noise). This habituation session will allow the mouse or mice to become familiar with the apparatus.

8. When the time is over, stop the video recording.

9. Return the mouse or mice to their corresponding home cage.

10. Carefully clean the maze and let the surface dry.

11. Repeat the procedure with the next mouse (or group of mice) until all mice have been tested.

12. Prepare the open field maze for the "sample" session. Place two identical copies of one of the objects (which will be designed as the "familiar" object) equidistant from two adjacent corners, separated ~7 cm from the walls (Fig. 3c). We recommend that the type of object designated as "familiar" as well as the placements of the objects in the maze are counterbalanced across mice of both treatments (Fig. 3c).

13. The "sample" session must begin at least 90 min after the habituation session. Test all the mice in the "sample" session as described in **steps 3–10**. Test the mice in the same order as they were tested in the "habituation" session. In the case of a maze with multiple compartments, make sure that each mouse is always tested in the same arena. Importantly, this sample session must last 10 min to ensure sufficient exploration of the objects.

14. Carefully clean the maze, including the objects it contains, and let the surface dry. Continue testing the remaining mice.

15. After the testing day is finished, transfer all the video files from the camera to one or more storage devices.

*Day 2: NOR session*

This test is performed the day after the habituation and sample sessions (~24 h retention interval). In the open field, replace one of the familiar objects with an identical copy. Replace the other familiar object by a copy of the novel object and place it in the previous position (Fig. 3c). Perform a 10 min session for all the mice as described in the previous steps.

*Day 3: NPR session*

This test is performed the day after the NOR session (~24 h retention interval). Replace the familiar object with an identical copy. Replace the novel object with another copy of the familiar object, but displace this object to the vertically opposite corner. In this way, two copies of the familiar object ("static" and "displaced") will be located in diagonally opposite corners (Fig. 3c). Perform a 10 min session for all the mice as described in the previous steps.

### 3.4 Observational Registry and Analysis of Behavioral Data

Observational analysis of the memory related-variables is simple and cost-free, and thus it is the most common choice among laboratories. Observational registries should be performed by one or more trained observers who are preferably blind to the treatment conditions. Importantly, unexperienced observers should undergo a previous training, by registering a set of sessions (e.g., ten different sessions) several times until a consistent criteria is achieved. An accordance >90% (assessed by Pearson's correlation or by intraclass correlation coefficient [85]) will indicate a sufficient intraobserver reproducibility. These registers must also be checked for accordance (interobserver variability) with the criteria of an experienced observer.

#### 3.4.1 Spatial Working Memory in the Y-Maze

The continuous SAB will be registered observationally using paper and pen.

- Start playing the video. Skip the initial part of the video where the sign identifying the animal is recorded to perform a registry blind to the experimental treatment group.

- As soon as the mouse is released in the maze, annotate the letter corresponding to the starting arm (e.g., C). For 6 min, score an arm entry by annotating the corresponding arm code (e.g., A, B, or C) each time the mouse, coming from the center zone, enters the arm by crossing the previously defined boundary with its four paws (Fig. 2b). If a mouse leaves the arm and reenters it (i.e., crosses the arm boundary to enter the center zone and then enters the arm again), this maneuver is scored as a new entry in this arm.

- After the session is registered, observe the sequence of letters and score one correct alternation for each of the three different letters that appear consecutively (Fig. 2c). For example, the sequence of arm entries "CABCBABACBAB" includes ten possible alternations, six of which are performed correctly: CAB (1), ABC (1), BCB (0), CBA (1), BAB (0), ABA (0), BAC (1), ACB (1), CBA (1), and BAB (0); AB does not count as a possible alternation because one arm entry is lacking to complete a triplet.

  Relevant measures for this task are:

- *Total number of arm entries* as a measure of exploratory activity, and to calculate the total number of possible alternations, defined as: (total number of arm entries − 2).

- *Continuous SAB ratio*, determined by the equation: [(total number of correct alternations/total number of possible alternations) × 100]. This value is as a measure of spatial working memory.

- Other measures that—optionally—may be assessed in this task are: total locomotion, total time spent in each arm, percent of arm perseverations (reentries) or continuous SAB split into blocks of 4 to 5 triplets each to study its change across the session.

  Once the measures are calculated for each mouse, it is possible to perform the required statistical comparisons:

- For a measure of exploratory activity, compare the total number of arm entries between groups. We have typically not found differences in this exploration parameter or in total locomotion for cocaine-treated mice after a drug withdrawal period of at least 4 weeks (Fig. 2d, e).

- For each treatment group, compare their SAB ratio with chance performance. This comparison may be performed by a one-sample *t*-test. Mathematically, the chance of alternating three different arms in a three-arm maze is 22% [86]. However, we have observed that mice tested in our Y-Maze (either cocaine- or vehicle-treated) will tend to enter any of the two arms ahead of them as they are leaving an explored arm. Thus, the event of turning around and reentering the arm from where they just left is possible but rare for a mouse. Considering that arm reentries are unlikely in our mice, we have set our chance performance level at a 50% correct SAB [86]. In our experiments, the vehicle-treated mice usually show a SAB ratio $\geq 60\%$, which is significantly greater than chance (one sample *t* test: $t_{(11)} = 5.500$, $p < 0.000$; for data in Fig. 2f); while the cocaine-withdrawn mice may perform at chance level (50%; Fig. 2f) [48].

- Compare the SAB ratio between the two groups. This is the key comparison to reveal a spatial working memory deficit in the cocaine-withdrawn mice: $t$ test for independent samples: $t_{(11)} = -4.225$, $p < 0.000$ for data in Fig. 2f; repeated measures analysis of variance (ANOVA) per session block ("cocaine × block") effect for "cocaine": $F(1, 22) = 16.784$, $p < 0.000$; "day": $F(5, 110) = 3.809$, $p = 0.003$ for data in Fig. 2g.

- Importantly, if the groups are not equal in their level of exploratory activity (i.e., locomotion or total number of arm entries), it is possible that exploratory activity could have influenced the SAB [60]. For example, abundant arm entries may gradually increase the working memory load due to proactive interference. As an attempt to reduce the potential influence of exploratory activity on the SAB, the SAB ratio may be calculated for a similar number of arm entries in all animals (e.g., for the first 20 or 30 arm entries).

*3.4.2 Reference Memory for Objects and Places*

The NOR and NPR tasks will be registered observationally, usually using a time-counter device (*see* Subheading 2) to score the time the mouse spends exploring each object.

- Start playing the video. Skip the initial part of the video where the sign identifying the animal is recorded to achieve a register that is blind to the experimental treatment group.

- The observational registration may commence once the mouse is released in the open-field and last for 10 min. Even in the event of testing several mice at once, every mouse must be observed individually.

- Typically, the experimenter must start the time counter when the animal begins one exploration and stop it when the exploration is finished. Explorations may occur in a fraction of a second (i.e., starting and stopping the counter almost immediately) or they may last for several seconds.

- Object exploration is defined as the mouse sniffing the object by touching it with its nose or pointing its nose toward an object that is less than 1–2 cm away [69–71, 79, 80] (Fig. 3a, d). We and other authors also score as exploration the behavior of actively touching the object with the forepaws [35, 48, 81, 87–89] (Table 2). This phenomenon usually occurs when the mouse rears supported on the object; frequently, the mouse looks at the object and/or sniffs it while rearing (Fig. 3d). Other behaviors are never considered as object exploration, for example, climbing the object, biting/chewing the object, resting near the object, standing on the top of the object, jumping supported on the object, or touching the object with the body or tail (Table 2).

Relevant measures for this task are as follows.

- *Total time of object exploration* (i.e., sum of time exploring both objects); as a measure of motivation to explore the objects. This measure should be reported for the sample session as well as for the NOR and NPR sessions.

- *Object memory ratio*, determined in the NOR session by the equation: [(time exploring the novel object − time exploring the familiar object)/total time exploring both objects) × 100]. This values measures long-term reference memory (24-h retention interval) for objects.

- *Place memory ratio*, determined in the NPR session by the equation: (time exploring the displaced object − time exploring the static object)/total time exploring both objects) × 100. This value is a measure of long-term reference memory (24-h retention interval) for locations.

- Other measures that—optionally—may be assessed in this task are: total locomotion, frequency of explorations and mean duration of the explorations. Furthermore, the habituation session in the empty open-field may be used for a detailed assessment of exploratory and anxiety-like behavior, for example, by quantifying thigmotaxis and behaviors such as rearing or grooming [90].

Once the measures are calculated for each mouse, one can perform the required statistical comparisons.

- Compare the total time of object exploration between groups across sessions because altered exploration of objects may affect memory-related measures (Fig. 3e). Other exploratory parameters such as locomotion may also be analysed (Fig. 3e).

- In the NOR and NPR sessions, for each treatment group compare each memory ratio with chance performance. According to their formulation, the ratio indicates how much time (in %) the mouse prefers the novel or the displaced object over the familiar and static one. Therefore, chance performance (i.e., absence of preference) corresponds to a value of zero. The control, vehicle-treated mice should display positive NOR and NPR ratios (typically $\geq 25–30\%$), which are significantly different from zero (one sample $t$-test for NOR: $t_{(19)} = 8.407$, $p < 0.000$; for NPR: $t_{(19)} = 6.512$, $p < 0.000$; for data in Fig. 3f). These data would confirm a preference for the novel object or place (i.e., a memory for the familiar ones). The cocaine-treated mice may perform by chance (NOR data in Fig. 3f), data that would indicate the absence of novelty discrimination, or over chance (NPR data in Fig. 3f: $t_{(19)} = 2.586$, $p = 0.018$). Negative values significantly less than zero are uncommon in mice as they would indicate a manifest aversion for novelty.

- Compare each memory ratio between groups. This is the key comparison to reveal an object or a place recognition memory deficit in the cocaine-withdrawn mice ($t$ test for independent samples: NOR: $t_{(38)} = -3.278$, $p = 0.002$; for NPR: $t_{(38)} = -2.757$, $p = 0.009$; for data in Fig. 3f).

## 4 Notes

In this chapter, we described the model of passive chronic cocaine administration in mice and three popular and affordable behavioral paradigms to study the impairing effects of cocaine on different forms of memory (spatial working memory and long-term reference memory for objects and places). These memories are supported by distinct neurobiological pathways.

- The memory paradigms detailed here may be integrated in a more extensive behavioral test battery [48, 88] to assess the effects of cocaine on additional emotional or cognitive functions. Most of the behavioral assessment batteries abide by performing the tasks in order of increasing invasiveness [91–93]. Tasks based on unconditioned anxiety and spontaneous exploration are performed first because they may be the most affected by previous experience. Furthermore, they may also provide an opportunity for the animal to become familiar with the testing environment, a phenomenon that might facilitate subsequent cognitive assessment. On the contrary, the more invasive tasks that involve exposure to highly aversive and stressful stimuli, require deprivation of water or food and/or involve an extended training are performed at latter stages [91–93]. An extended behavioral assessment could be planned as follows: (1) noninvasive tasks for anxiety-like behavior (the elevated plus maze, the light–dark box, etc.); (2) memory tasks based on spontaneous exploration (e.g., continuous SAB and the object/place memory paradigms explained in this chapter); (3) highly stressful or demanding tasks (i.e., the forced swimming test, the tail suspension test, fear conditioning, spatial navigation in the water maze, hole-board or radial arm mazes, etc.).

- It is mandatory that the rodent population selected for the study, when tested in drug naïve conditions, is a consistently good performer (i.e., above chance level; Subheading 3) in the cognitive tasks in which the effects of cocaine will be evaluated. In rodent populations that—due to their strain, sex, age, or any other individual or environmental circumstance—have difficulties in performing a cognitive task in drug-free conditions, the impairing effect of cocaine may not be detected [45].

- Considering this possibility, the behavioral protocols reported here may be adapted according to the study population. For example, if the 24 h NOR and NPR tasks result too difficult for the control drug-naïve animals, shorter memory retention intervals may facilitate performance. Other normal, drug-naïve rodents may progressively loss motivation for novelty as the testing session progresses. In this case, the "Object" and "Place" memory ratios may be calculated only for the first 1–5 min of the session when animals are inclined to perform the task [35, 70]. In the Y-Maze, the continuous SAB ratio may also be calculated for a certain number of the initial alternations [48].

Other variants of the memory tasks described here may be used when appropriate. For example, *delayed* SAB introduces a memory retention interval that separates arm choices. This modification enhances the long-term memory component and hippocampal dependency [59, 94]. Variants of the NOR and NPR tasks assess memory for the object presentation order or place [95]. Increasing the similarity between the objects and/or their spatial positions [96] or the memory retention interval [68] enhances the difficulty and hippocampal dependence, while a shorter retention interval and more distinctive objects to be discriminated may facilitate the task. Adaptation of the behavioral protocols to rats, would also require a modification of the maze measures.

- Another issue that might confound memory assessment based on spontaneous exploration is that some animals may show a low exploratory activity of the maze arms or objects, independently of their drug treatment. A low total exploratory activity may strongly bias the result of the calculated memory ratios. In this case, the duration of the testing session may be increased so the rodent will perform a sufficient number of arm entries or object explorations for analysis. Alternatively, if this problem only affects one or a few animals, they may be excluded from the study.

- While useful for studying cocaine-induced cognitive decline, the passive cocaine administration model as described in this chapter will not allow experimenters to observe the expression of voluntary cocaine seeking or taking responses. Nonetheless, it is possible to increase the information provided by this model, for example, by pairing the administrations with specific contextual cues, such as in a conditioned place preference paradigm [52]. This design would allow the experimenter to study cocaine reward or the memory for cocaine-stimuli associations in addition to cognitive impairment [52].

- Despite the consistent effects of the passive cocaine administration model to induce cognitive impairment, it is important to note that most studies have applied this protocol to young male rodents (Table 1). Considering this fact, using this protocol to study the influence of gender or age as vulnerability factors on cocaine-induced memory impairment would entail an interesting and unexplored research area. The impact of different cocaine doses or administration schedules may also be investigated. Moreover, this model may be applied to pregnant females to study the effect of prenatal cocaine exposure on the cognitive status of the offspring (reviewed in [97]).

- The passive cocaine administration model seems to be adequate to test pharmacological or environmental interventions on the vulnerability for, prevention and/or treatment of the drug-induced cognitive deficits. For example, an environmental enrichment protocol, as described in another chapter of this book, may be applied before, during, and/or after cocaine administration. In this case, additional experimental groups should be added to the study (e.g., Vehicle–no treatment; Cocaine–no treatment; Cocaine+Treatment; and, optionally, Vehicle+Treatment).

## Acknowledgments

This research was funded by grants from the Spanish Ministry of Economy and Competitiveness (MINECO, Agencia Estatal de Investigación [AEI]) cofounded by the European Research Development Fund (FEDER, UE; PSI2015-73156-JIN to E.C-O.; PSI2017-82604R to L.J.S.). The authors M.C.M-P., S.G-R., and F.A-G. hold predoctoral grants from the Spanish Ministry of Science, Innovation and Universities (FPU17/00276 to M.C.M-P.; PRE2018-085673 to F.A-G.; and FPU18/00941 to S.G-R.)
The authors acknowledge the IBIMA and University of Malaga's Animal Facility and their staff for their valuable assistance during the behavioral experiments and maintenance of the mice.

## References

1. Jupp B, Caprioli D, Dalley JW (2013) Highly impulsive rats: modelling an endophenotype to determine the neurobiological, genetic and environmental mechanisms of addiction. Dis Model Mech 6(2):302–311

2. Verdejo-Garcia A, Lawrence AJ, Clark L (2008) Impulsivity as a vulnerability marker for substance-use disorders: review of findings from high-risk research, problem gamblers and genetic association studies. Neurosci Biobehav Rev 32(4):777–810

3. Balogh KN, Mayes LC, Potenza MN (2013) Risk-taking and decision-making in youth: relationships to addiction vulnerability. J Behav Addict 2(1)

4. Belin D, Deroche-Gamonet V (2012) Responses to novelty and vulnerability to cocaine addiction: contribution of a multi-

symptomatic animal model. Cold Spring Harb Perspect Med 2(11)

5. Goldstein RZ, Craig AD, Bechara A, Garavan H, Childress AR, Paulus MP, Volkow ND (2009) The neurocircuitry of impaired insight in drug addiction. Trends Cogn Sci 13 (9):372–380

6. Sampedro-Piquero P, Santin LJ, Castilla-Ortega E (2019) Aberrant brain neuroplasticity and function in drug addiction: a focus on learning-related brain regions. In: Palermo S, Morese R (eds) Behavioral neuroscience. IntechOpen, London. Available from: https://www.intechopen.com/online-first/aberrant-brain-neuroplasticity-and-function-in-drug-addiction-a-focus-on-learning-related-brain-regi

7. Sampedro-Piquero P, Ladron de Guevara-Miranda D, Pavon FJ, Serrano A, Suarez J, Rodriguez de Fonseca F, Santin LJ, Castilla-Ortega E (2019) Neuroplastic and cognitive impairment in substance use disorders: a therapeutic potential of cognitive stimulation. Neurosci Biobehav Rev 106:23–48

8. Spronk DB, van Wel JH, Ramaekers JG, Verkes RJ (2013) Characterizing the cognitive effects of cocaine: a comprehensive review. Neurosci Biobehav Rev 37(8):1838–1859

9. Vonmoos M, Hulka LM, Preller KH, Minder F, Baumgartner MR, Quednow BB (2014) Cognitive impairment in cocaine users is drug-induced but partially reversible: evidence from a longitudinal study. Neuropsychopharmacology 39(9):2200–2210

10. Funahashi S (2017) Working memory in the prefrontal cortex. Brain Sci 7(5)

11. Wirt RA, Hyman JM (2017) Integrating spatial working memory and remote memory: interactions between the medial prefrontal cortex and hippocampus. Brain Sci 7(4):43

12. Squire LR (1992) Memory and the hippocampus: a synthesis from findings with rats, monkeys, and humans. Psychol Rev 99 (2):195–231

13. Ersche KD, Barnes A, Jones PS, Morein-Zamir S, Robbins TW, Bullmore ET (2011) Abnormal structure of frontostriatal brain systems is associated with aspects of impulsivity and compulsivity in cocaine dependence. Brain 134(Pt 7):2013–2024

14. Fein G, Di Sclafani V, Meyerhoff DJ (2002) Prefrontal cortical volume reduction associated with frontal cortex function deficit in 6-week abstinent crack-cocaine dependent men. Drug Alcohol Depend 68(1):87–93

15. Hester R, Garavan H (2004) Executive dysfunction in cocaine addiction: evidence for

discordant frontal, cingulate, and cerebellar activity. J Neurosci 24(49):11017–11022

16. Hirsiger S, Hanggi J, Germann J, Vonmoos M, Preller KH, Engeli EJE, Kirschner M, Reinhard C, Hulka LM, Baumgartner MR, Chakravarty MM, Seifritz E, Herdener M, Quednow BB (2019) Longitudinal changes in cocaine intake and cognition are linked to cortical thickness adaptations in cocaine users. Neuroimage Clin 21:101652

17. Parvaz MA, Moeller SJ, d'Oleire Uquillas F, Pflumm A, Maloney T, Alia-Klein N, Goldstein RZ (2017) Prefrontal gray matter volume recovery in treatment-seeking cocaine-addicted individuals: a longitudinal study. Addict Biol 22(5):1391–1401

18. Adinoff B, Gu H, Merrick C, McHugh M, Jeon-Slaughter H, Lu H, Yang Y, Stein EA (2015) Basal hippocampal activity and its functional connectivity predicts cocaine relapse. Biol Psychiatry 78(7):496–504

19. Ding X, Lee SW (2013) Cocaine addiction related reproducible brain regions of abnormal default-mode network functional connectivity: a group ICA study with different model orders. Neurosci Lett 548:110–114

20. Tau GZ, Marsh R, Wang Z, Torres-Sanchez T, Graniello B, Hao X, Xu D, Packard MG, Duan Y, Kangarlu A, Martinez D, Peterson BS (2014) Neural correlates of reward-based spatial learning in persons with cocaine dependence. Neuropsychopharmacology 39 (3):545–555

21. Mahoney JJ (2019) Cognitive dysfunction in individuals with cocaine use disorder: potential moderating factors and pharmacological treatments. Exp Clin Psychopharmacol 27 (3):203–214

22. Aharonovich E, Hasin DS, Brooks AC, Liu X, Bisaga A, Nunes EV (2006) Cognitive deficits predict low treatment retention in cocaine dependent patients. Drug Alcohol Depend 81 (3):313–322

23. Fox HC, Jackson ED, Sinha R (2009) Elevated cortisol and learning and memory deficits in cocaine dependent individuals: relationship to relapse outcomes. Psychoneuroendocrinology 34(8):1198–1207

24. Ruiz Sanchez de Leon JM, Pedrero Perez E, Llanero Luque M, Rojo Mota G, Olivar Arroyo A, Bouso Saiz JC, Puerta Garcia C (2009) Neuropsychological profile in cocaine addiction: issues about addict's social environment and predictive value of cognitive status in therapeutic outcomes. Adicciones 21 (2):119–132

25. Everitt BJ (2014) Neural and psychological mechanisms underlying compulsive drug seeking habits and drug memories—indications for novel treatments of addiction. Eur J Neurosci 40(1):2163–2182

26. Gould RW, Gage HD, Nader MA (2012) Effects of chronic cocaine self-administration on cognition and cerebral glucose utilization in Rhesus monkeys. Biol Psychiatry 72 (10):856–863

27. Kangas BD, Doyle RJ, Kohut SJ, Bergman J, Kaufman MJ (2019) Effects of chronic cocaine self-administration and N-acetylcysteine on learning, cognitive flexibility, and reinstatement in nonhuman primates. Psychopharmacology 236(7):2143–2153

28. Kromrey SA, Gould RW, Nader MA, Czoty PW (2015) Effects of prior cocaine self-administration on cognitive performance in female cynomolgus monkeys. Psychopharmacology 232(11):2007–2016

29. Liu S, Heitz RP, Sampson AR, Zhang W, Bradberry CW (2008) Evidence of temporal cortical dysfunction in rhesus monkeys following chronic cocaine self-administration. Cereb Cortex 18(9):2109–2116

30. Melamed JL, de Jesus FM, Aquino J, Vannuchi CRS, Duarte RBM, Maior RS, Tomaz C, Barros M (2017) Differential modulatory effects of cocaine on marmoset monkey recognition memory. Prog Brain Res 235:155–176

31. Porter JN, Gurnsey K, Jedema HP, Bradberry CW (2013) Latent vulnerability in cognitive performance following chronic cocaine self-administration in rhesus monkeys. Psychopharmacology 226(1):139–146

32. Porter JN, Olsen AS, Gurnsey K, Dugan BP, Jedema HP, Bradberry CW (2011) Chronic cocaine self-administration in rhesus monkeys: impact on associative learning, cognitive control, and working memory. J Neurosci 31 (13):4926–4934

33. Bechard AR, LaCrosse A, Namba MD, Jackson B, Knackstedt LA (2018) Impairments in reversal learning following short access to cocaine self-administration. Drug Alcohol Depend 192:239–244

34. Briand LA, Flagel SB, Garcia-Fuster MJ, Watson SJ, Akil H, Sarter M, Robinson TE (2008) Persistent alterations in cognitive function and prefrontal dopamine D2 receptors following extended, but not limited, access to self-administered cocaine. Neuropsychopharmacology 33(12):2969–2980

35. Briand LA, Gross JP, Robinson TE (2008) Impaired object recognition following prolonged withdrawal from extended-access cocaine self-administration. Neuroscience 155 (1):1–6

36. Fijal K, Nowak E, Leskiewicz M, Budziszewska B, Filip M (2015) Working memory deficits and alterations of ERK and CREB phosphorylation following withdrawal from cocaine self-administration. Pharmacol Rep 67(5):881–889

37. George O, Mandyam CD, Wee S, Koob GF (2008) Extended access to cocaine self-administration produces long-lasting prefrontal cortex-dependent working memory impairments. Neuropsychopharmacology 33 (10):2474–2482

38. Gobin C, Schwendt M (2017) The effects of extended-access cocaine self-administration on working memory performance, reversal learning and incubation of cocaine-seeking in adult male rats. J Addict Prev 5(1)

39. Gobin C, Shallcross J, Schwendt M (2019) Neurobiological substrates of persistent working memory deficits and cocaine-seeking in the prelimbic cortex of rats with a history of extended access to cocaine self-administration. Neurobiol Learn Mem 161:92–105

40. Kantak KM, Udo T, Ugalde F, Luzzo C, Di Pietro N, Eichenbaum HB (2005) Influence of cocaine self-administration on learning related to prefrontal cortex or hippocampus functioning in rats. Psychopharmacology 181 (2):227–236

41. Saddoris MP, Carelli RM (2014) Cocaine self-administration abolishes associative neural encoding in the nucleus accumbens necessary for higher-order learning. Biol Psychiatry 75 (2):156–164

42. Aguilar MA, Ledesma JC, Rodriguez-Arias M, Penalva C, Manzanedo C, Minarro J, Arenas MC (2017) Adolescent exposure to the synthetic Cannabinoid WIN 55212-2 modifies cocaine withdrawal symptoms in adult mice. Int J Mol Sci 18(6):1326

43. Berardino BG, Fesser EA, Belluscio LM, Gianatiempo O, Pregi N, Canepa ET (2019) Effects of cocaine base paste on anxiety-like behavior and immediate-early gene expression in nucleus accumbens and medial prefrontal cortex of female mice. Psychopharmacology 236(12):3525–3539

44. Davidson TL, Hargrave SL, Kearns DN, Clasen MM, Jones S, Wakeford AGP, Sample CH, Riley AL (2018) Cocaine impairs serial-feature negative learning and blood-brain barrier integrity. Pharmacol Biochem Behav 170:56–63

45. Fole A, Martin M, Morales L, Del Olmo N (2015) Effects of chronic cocaine treatment

during adolescence in Lewis and Fischer-344 rats: novel location recognition impairment and changes in synaptic plasticity in adulthood. Neurobiol Learn Mem 123:179–186

46. Gong D, Zhao H, Liang Y, Chao R, Chen L, Yang S, Yu P (2019) Differences in cocaine- and morphine-induced cognitive impairments and serum corticosterone between C57BL/6J and BALB/cJ mice. Pharmacol Biochem Behav 182:1–6

47. Krueger DD, Howell JL, Oo H, Olausson P, Taylor JR, Nairn AC (2009) Prior chronic cocaine exposure in mice induces persistent alterations in cognitive function. Behav Pharmacol 20(8):695–704

48. Ladron de Guevara-Miranda D, Millon C, Rosell-Valle C, Perez-Fernandez M, Missiroli M, Serrano A, Pavon FJ, Rodriguez de Fonseca F, Martinez-Losa M, Alvarez-Dolado M, Santin LJ, Castilla-Ortega E (2017) Long-lasting memory deficits in mice withdrawn from cocaine are concomitant with neuroadaptations in hippocampal basal activity, GABAergic interneurons and adult neurogenesis. Dis Model Mech 10(3):323–336

49. Ledesma JC, Aguilar MA, Gimenez-Gomez P, Minarro J, Rodriguez-Arias M (2017) Adolescent but not adult ethanol binge drinking modulates cocaine withdrawal symptoms in mice. PLoS One 12(3):e0172956

50. Mendez IA, Montgomery KS, LaSarge CL, Simon NW, Bizon JL, Setlow B (2008) Long-term effects of prior cocaine exposure on Morris water maze performance. Neurobiol Learn Mem 89(2):185–191

51. Morisot N, Le Moine C, Millan MJ, Contarino A (2014) CRF(2) receptor-deficiency reduces recognition memory deficits and vulnerability to stress induced by cocaine withdrawal. Int J Neuropsychopharmacol 17(12):1969–1979

52. Preston CJ, Brown KA, Wagner JJ (2019) Cocaine conditioning induces persisting changes in ventral hippocampus synaptic transmission, long-term potentiation, and radial arm maze performance in the mouse. Neuropharmacology 150:27–37

53. Santucci AC, Rabidou D (2011) Residual performance impairments in adult rats trained on an object discrimination task subsequent to cocaine administration during adolescence. Addict Biol 16(1):30–42

54. Kmiotek EK, Baimel C, Gill KJ (2012) Methods for intravenous self administration in a mouse model. J Vis Exp 70:e3739

55. Castilla-Ortega E, Ladron de Guevara-Miranda D, Serrano A, Pavon FJ, Suarez J, Rodriguez de Fonseca F, Santin LJ (2017) The impact of cocaine on adult hippocampal neurogenesis: potential neurobiological mechanisms and contributions to maladaptive cognition in cocaine addiction disorder. Biochem Pharmacol 141:100–117

56. Buccafusco JJ (2009) Methods of Behavior Analysis in Neuroscience. Taylor & Francis Group, LLC, Boca Raton FL

57. El Massioui N, Lamirault C, Yague S, Adjeroud N, Garces D, Maillard A, Tallot L, Yu-Taeger L, Riess O, Allain P, Nguyen HP, von Horsten S, Doyere V (2016) Impaired Decision Making and Loss of Inhibitory-Control in a Rat Model of Huntington Disease. Front Behav Neurosci 10:204

58. Rodriguiz RM, Wetsel WC (2006) Assessments of Cognitive Deficits in Mutant Mice. In: Levin ED, Buccafusco JJ (eds) Animal Models of Cognitive Impairment. Taylor & Francis Group, LLC, Boca Raton FL

59. Hughes RN (2004) The value of spontaneous alternation behavior (SAB) as a test of retention in pharmacological investigations of memory. Neurosci Biobehav Rev 28(5):497–505

60. Miedel CJ, Patton JM, Miedel AN, Miedel ES, Levenson JM (2017) Assessment of Spontaneous Alternation, Novel Object Recognition and Limb Clasping in Transgenic Mouse Models of Amyloid-beta and Tau Neuropathology. J Vis Exp 123:55523

61. Prieur EA, Jadavji NM (2019) Assessing Spatial Working Memory Using the Spontaneous Alternation Y-maze Test in Aged Male Mice. Bio-protocol 9(3):e3162

62. Murray AJ, Woloszynowska-Fraser MU, Ansel-Bollepalli L, Cole KL, Foggetti A, Crouch B, Riedel G, Wulff P (2015) Parvalbumin-positive interneurons of the prefrontal cortex support working memory and cognitive flexibility. Sci Rep 5:16778

63. Deacon RM, Penny C, Rawlins JN (2003) Effects of medial prefrontal cortex cytotoxic lesions in mice. Behav Brain Res 139 (1–2):139–155

64. Pooters T, Laeremans A, Gantois I, Vermaercke B, Arckens L, D'Hooge R (2017) Comparison of the spatial-cognitive functions of dorsomedial striatum and anterior cingulate cortex in mice. PLoS One 12(5):e0176295

65. De Bundel D, Schallier A, Loyens E, Fernando R, Miyashita H, Van Liefferinge J, Vermoesen K, Bannai S, Sato H, Michotte Y, Smolders I, Massie A (2011) Loss of system x (c)- does not induce oxidative stress but decreases extracellular glutamate in hippocampus and influences spatial working memory and

limbic seizure susceptibility. J Neurosci 31 (15):5792–5803

66. Faizi M, Bader PL, Saw N, Nguyen TV, Beraki S, Wyss-Coray T, Longo FM, Shamloo M (2012) Thy1-hAPP(Lond/Swe+) mouse model of Alzheimer's disease displays broad behavioral deficits in sensorimotor, cognitive and social function. Brain Behav 2(2):142–154

67. Hidaka N, Suemaru K, Takechi K, Li B, Araki H (2011) Inhibitory effects of valproate on impairment of Y-maze alternation behavior induced by repeated electroconvulsive seizures and c-Fos protein levels in rat brains. Acta Med Okayama 65(4):269–277

68. Cohen SJ, Stackman RW Jr (2015) Assessing rodent hippocampal involvement in the novel object recognition task. A review. Behav Brain Res 285:105–117

69. Barker GR, Warburton EC (2011) When is the hippocampus involved in recognition memory? J Neurosci 31(29):10721–10731

70. Winters BD, Forwood SE, Cowell RA, Saksida LM, Bussey TJ (2004) Double dissociation between the effects of peri-postrhinal cortex and hippocampal lesions on tests of object recognition and spatial memory: heterogeneity of function within the temporal lobe. J Neurosci 24(26):5901–5908

71. Denny CA, Burghardt NS, Schachter DM, Hen R, Drew MR (2012) 4- to 6-week-old adult-born hippocampal neurons influence novelty-evoked exploration and contextual fear conditioning. Hippocampus 22 (5):1188–1201

72. Lazic SE, Essioux L (2013) Improving basic and translational science by accounting for litter-to-litter variation in animal models. BMC Neurosci 14:37

73. Mañas-Padilla MC, Gil-Rodríguez S, Sampedro-Piquero P, Ávila-Gámiz F, Rodríguez de Fonseca F, Santín LJ, Castilla-Ortega E (2021) Remote memory of drug experiences coexists with cognitive decline and abnormal adult neurogenesis in an animal model of cocaine-altered cognition. Addict Biol 26: e12886

74. Van Loo PL, Van Zutphen LF, Baumans V (2003) Male management: coping with aggression problems in male laboratory mice. Lab Anim 37(4):300–313

75. Kulesskaya N, Rauvala H, Voikar V (2011) Evaluation of social and physical enrichment in modulation of behavioural phenotype in C57BL/6J female mice. PLoS One 6(9): e24755

76. Wang L (2005) A primer on rodent identification methods. Lab Anim (NY) 34(4):64–67

77. JoVE Science Education Database (2019) Lab animal research. Compound administration I. JoVE, Cambridge

78. Hughes RN (1997) Intrinsic exploration in animals: motives and measurement. Behav Process 41(3):213–226

79. Denninger JK, Smith BM, Kirby ED (2018) Novel object recognition and object location behavioral testing in mice on a budget. J Vis Exp 141:10.3791/58593

80. Gulinello M, Mitchell HA, Chang Q, Timothy O'Brien W, Zhou Z, Abel T, Wang L, Corbin JG, Veeraragavan S, Samaco RC, Andrews NA, Fagiolini M, Cole TB, Burbacher TM, Crawley JN (2018) Rigor and reproducibility in rodent behavioral research. Neurobiol Learn Mem 165:106780

81. Heyser CJ, Chemero A (2012) Novel object exploration in mice: not all objects are created equal. Behav Process 89(3):232–238

82. Friard O, Gamba M (2016) BORIS: a free, versatile open-source event-logging software for video/audio coding and live observations. Methods Ecol Evol 7:1325–1330

83. Burkholder T, Foltz C, Karlsson E, Linton CG, Smith JM (2012) Health evaluation of experimental laboratory mice. Curr Protoc Mouse Biol 2:145–165

84. Diehl KH, Hull R, Morton D, Pfister R, Rabemampianina Y, Smith D, Vidal JM, van de Vorstenbosch C (2001) A good practice guide to the administration of substances and removal of blood, including routes and volumes. J Appl Toxicol 21(1):15–23

85. Popovic ZB, Thomas JD (2017) Assessing observer variability: a user's guide. Cardiovasc Diagn Ther 7(3):317–324

86. Holcomb LA, Gordon MN, Jantzen P, Hsiao K, Duff K, Morgan D (1999) Behavioral changes in transgenic mice expressing both amyloid precursor protein and presenilin-1 mutations: lack of association with amyloid deposits. Behav Genet 29(3):177–185

87. Albasser MM, Chapman RJ, Amin E, Iordanova MD, Vann SD, Aggleton JP (2010) New behavioral protocols to extend our knowledge of rodent object recognition memory. Learn Mem 17(8):407–419

88. Ladron de Guevara-Miranda D, Pavon FJ, Serrano A, Rivera P, Estivill-Torrus G, Suarez J, Rodriguez de Fonseca F, Santin LJ, Castilla-Ortega E (2016) Cocaine-conditioned place preference is predicted by previous anxiety-like behavior and is related to an increased number of neurons in the basolateral amygdala. Behav Brain Res 298(Pt B):35–43

89. Prado Lima MG, Schimidt HL, Garcia A, Dare LR, Carpes FP, Izquierdo I, Mello-Carpes PB (2018) Environmental enrichment and exercise are better than social enrichment to reduce memory deficits in amyloid beta neurotoxicity. Proc Natl Acad Sci U S A 115(10): E2403–E2409

90. Sestakova N, Puzserova A, Kluknavsky M, Bernatova I (2013) Determination of motor activity and anxiety-related behaviour in rodents: methodological aspects and role of nitric oxide. Interdiscip Toxicol 6(3):126–135

91. Lad HV, Liu L, Paya-Cano JL, Parsons MJ, Kember R, Fernandes C, Schalkwyk LC (2010) Behavioural battery testing: evaluation and behavioural outcomes in 8 inbred mouse strains. Physiol Behav 99(3):301–316

92. McIlwain KL, Merriweather MY, Yuva-Paylor LA, Paylor R (2001) The use of behavioral test batteries: effects of training history. Physiol Behav 73(5):705–717

93. Wolf A, Bauer B, Abner EL, Ashkenazy-Frolinger T, Hartz AM (2016) A Comprehensive Behavioral Test Battery to Assess Learning and Memory in 129S6/Tg2576 Mice. PLoS One 11(1):e0147733

94. Aggleton JP, Hunt PR, Rawlins JN (1986) The effects of hippocampal lesions upon spatial and non-spatial tests of working memory. Behav Brain Res 19(2):133–146

95. DeVito LM, Eichenbaum H (2010) Distinct contributions of the hippocampus and medial prefrontal cortex to the "what-where-when" components of episodic-like memory in mice. Behav Brain Res 215(2):318–325

96. Ces A, Burg T, Herbeaux K, Heraud C, Bott JB, Mensah-Nyagan AG, Mathis C (2018) Age-related vulnerability of pattern separation in C57BL/6J mice. Neurobiol Aging 62:120–129

97. Garcia-Pardo MP, De la Rubia Orti JE, Aguilar Calpe MA (2017) Differential effects of MDMA and cocaine on inhibitory avoidance and object recognition tests in rodents. Neurobiol Learn Mem 146:1–11

# Chapter 12

# Transcranial Direct Current Stimulation to Reduce Addiction-Related Behaviors in Mice

Stéphanie Dumontoy, Adeline Etievant, Andries Van Schuerbeek, and Vincent Van Waes

## Abstract

Transcranial direct current stimulation (tDCS) is a neuromodulation method used in humans to increase or decrease cortical excitability in a noninvasive and painless manner. A weak electric current flows between two electrodes, an anode and a cathode, placed on the scalp. This technique has gained considerable interest recently as a tool for the treatment of several psychiatric disorders, including depression and addiction. However, the mechanisms underlying its beneficial effects remain poorly understood, requiring further investigations in human as well as animal models. In Besançon (Laboratoire de Recherches Intégratives en Neurosciences et Psychologie Cognitive, Université Bourgogne Franche-Comté), we have developed a mouse model of tDCS to study its mechanisms of action, using a stimulation protocol similar to those in clinical trials. In this chapter, we will describe this model and its different variants (i.e., tDCS on awake restrained mice; tDCS on anesthetized mice; tDCS in freely moving mice).

**Key words** Neuromodulation, tDCS, Addiction, Frontal cortex, Translational research, Mice

## 1 Introduction

Transcranial direct current stimulation (tDCS) is an innovative, noninvasive, and safe neuromodulation technique. It induces long-lasting changes in cortical excitability via a weak constant electric current that flows between two electrodes: an anode (+) and a cathode (−). It is thought that the current under the anode causes membrane depolarization of the neurons, which brings their membrane potentials closer to the threshold of excitability, making it more likely to generate an action potential. On the other hand, the cathode induces hyperpolarization, which decreases neuronal excitability and therefore corresponds to inhibition of the neurons [1]. However, this description is oversimplified and many parameters influence the effect of the current on the neurons, such as orientation of the neurons in the electric field, cell morphology, the

María A. Aguilar (ed.), *Methods for Preclinical Research in Addiction*, Neuromethods, vol. 174,
https://doi.org/10.1007/978-1-0716-1748-9_12, © Springer Science+Business Media, LLC, part of Springer Nature 2022

cortical area considered, and the length and intensity of stimulation [2].

Electrode position on the skull is selected according to the expected outcome. The dorsolateral prefrontal cortex (DLPFC) is usually targeted to decrease depression and addiction-related symptoms because it is involved in various cognitive and behavioral functions, such as decision-making and impulsivity, and is connected to many cortical and subcortical regions. In the context of addiction, dependence leads to alterations in DLPFC function. A decrease in activity of the DLPFC might be at the origin of inappropriate responses to drugs, such as the reduction of inhibitory behavior or relapse phenomena [3, 4].

For a decade, tDCS has been used experimentally to treat several psychiatric disorders, including depressive symptoms [5, 6]. Recently, modulation of cortical excitability by tDCS has gained interest in the scientific community as a mean of decreasing maladaptive behavior in drug-dependent patients [7]. Several teams have studied the effects of tDCS in patients suffering from addiction to alcohol [8–19], nicotine [20–28], cocaine [13, 29–32], marijuana [33], methamphetamines [34–36], and heroin [37, 38]. These studies generally report a reduction in the craving and in symptoms associated with withdrawal (depression, relapse, etc.), but discrepant results are also sometimes reported [17, 25].

The findings are encouraging because current available treatment options for drug cessation remain limited and are associated with poor long-term success rates [39]. In this context, electric noninvasive stimulations appear to be attractive techniques worthy of further investigation to reduce addictive symptoms [40]. Moreover, tDCS presents several advantages: it is noninvasive, affordable, easy-to-use, portable, painless, safe, and well tolerated.

Despite its growing popularity, the neurobiological mechanisms underlying the effects of tDCS on brain function and behavior remain poorly understood. For example, pharmacological studies in humans suggested that several neurotransmission systems might be implicated in the effects of tDCS, such as the dopaminergic system [41, 42] and the glutamate/gamma-aminobutyric acid (GABA) balance (i.e., excitation/inhibition) [43]. Other data suggest that the effect of tDCS might involve long-term potentiation (LTP)-like plasticity through N-methyl D-aspartate (NMDA) receptors [44]. Various other mechanisms are also under investigation, such as the modulation of cerebral blood flow by tDCS [45, 46] and the involvement of brain-derived neurotrophic factor (BDNF) [47–49] or glial cells (e.g., astrocytes) [50].

Several animal models of tDCS have been developed for translational studies. The majority of the laboratories used rodents (as exemplified in this chapter), but tDCS has also been tested in domestics cats, where the two electrodes were located on the animal's head [51], and in rabbits, with the stimulation electrode surgically fixed to the skull and the counterelectrode placed on one ear [52]. Our tDCS model in mice is based on the seminal model of Nitsche and collaborators [53, 54], with similar stimulation parameters to those used in clinical trials. Mice receive a 20-min stimulation twice a day for 5 consecutive days at an intensity of 0.2 mA (cf. 2 mA for humans). With this protocol, we have previously shown in mice that repeated tDCS prevents abnormal behaviors associated with abstinence from chronic nicotine consumption [54] and produces long-lasting attenuation of cocaine-induced behavioral responses and gene regulation in corticostriatal circuits [55].

Three variants of tDCS will be presented in this chapter: tDCS on awake restrained mice; tDCS on anesthetized mice; and tDCS in freely moving mice.

## 2 Materials

### 2.1 Animals

- Swiss or C57BL-6/J male or female mice (Janvier, France), adolescents (5 weeks old) or adults (8 weeks old).

The animals are housed in groups of four or five per cage (before the surgery) under standard laboratory conditions of humidity (30–70%), temperature (19–25 °C), and a 12/12-h light–dark cycle (lights on at 7:00 am), with food and water available *ad libitum*.

### 2.2 Surgery

#### 2.2.1 Anesthesia

- Ketamine hydrochloride (Clorketam® 1000, Vetoquinol, France).
- Xylazine (Rompun® 2%, Bayer, France).
- Sterile saline injectable solution (NaCl 0.9%; Baxter SA, Belgium or Cooper, France).
- Physiological serum or Tears Naturale® (Alcon, Belgium) to protect the eye.
- Stereotaxic apparatus (Stoelting, USA).

#### 2.2.2 Fixation of the Stimulation Electrode Holder

- Stereotaxic arm adaptor (Dixi Medical, France).
- Cotton swap.
- Electrode holder: tubular plastic jack, internal diameter 2.1 mm (DIXI Medical, France) (as shown in Fig. 1).
- Glass ionomer dental cement (GC® Fuji I, GC Corporation, Japan).
- Warming pad (Thermacage, Datesand Technologies, UK).

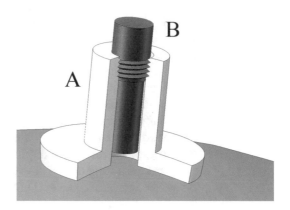

**Fig. 1** An electrode holder (internal diameter 2.1 mm, A) is surgically fixed onto the skull with a coating of glass ionomer cement. Then the stimulation electrode (anode, B), covered with a gold layer, is screwed into the electrode holder previously filled with saline. Only the saline solution is in contact with the skull

*2.2.3  Postsurgical Care*

- Chlortetracycline hydrochloride antibacterial (Aureomycin Evans 3%, Cooper, France).

- Glycyrrhizin acid anti-itch (P.O.12® Enoxolone 2%, Sanofi, France).

- Paracetamol 10% (Doliprane, Sanofi, France) diluted in drinking water.

- Povidone-iodine 10% (Iso-Betadine® Dermicum, Meda Pharma NV, Belgium).

**2.3  Stimulation**

- Stimulation electrode: anode, 3.5 mm$^2$, covered with a gold layer (DIXI Medical, France) (as shown Fig. 1).

- Reference electrode: 4.5–8.75 cm$^2$, rectangle (Physiomed Elektromedizin AG, Germany).

- Conductive gel (Ten20®, Weaver and Company, USA or ReegaPha 6101, MEI, France).

- DC-Stimulator Plus (NeuroConn, Germany).

- Open-tES stimulator [56].

- Reference electrode: rubber-plate electrode.

*2.3.1  For Restrained Animals*

- Restraint box: homemade (as described in Subheading 3.3.1 and Fig. 3).

*2.3.2  For Anesthetized Animals*

- Isoflurane (1000 mg/kg Isoflo®, Zoetis Belgium SA, Belgium).
- Jacket (homemade).

<table>
<tr><td>

*2.3.3 For Freely Moving Animals*

</td><td>

- Screws 0–80 × 3/32 N (Bilaney Consultants GmbH, Germany).
- Tissue adhesive (3 M Vetbond, 3 M Deutschland GmbH Consumer Health Care, Germany).
- Commutator system (Two-Channel Commutator SL2C/SB; Bilaney Consultants GmbH, Germany) (as described in Subheading 3.3.3 and Fig. 5).
- Bipolar cable: with spring covering 305-000 and bipolar cable 305-491/2 (Bilaney Consultants GmbH, Germany).
- Paper tape (tesa®, tesa SE, Germany).

</td></tr>
</table>

# 3  Methods

## 3.1  Surgery

### 3.1.1  Anesthesia

Before surgery, mice are allowed 1 week of acclimation in the animal facility, during which they are repeatedly handled. An electrode holder is then surgically fixed onto the skull 1 week before the start of the experiments.

Animals are anesthetized with an intraperitoneal injection of a fresh mixture of ketamine hydrochloride and xylazine (80 mg/kg and 12 mg/kg, respectively) diluted in NaCl 0.9%. If your stereotaxic device is equipped, mice can alternatively be anesthetized with isoflurane 3.5% in an induction chamber for 2 min and anesthesia can be maintained on the stereotaxic frame (isoflurane 1.5–2%).

The anesthetized mice are placed onto the stereotaxic apparatus. A drop of physiological serum or artificial tears is then applied to each eye to avoid corneal drying during surgery. The skin of the anesthetized animal is cleaned with a cotton swab soaked in 70% ethanol.

### 3.1.2  Electrode Holder Fixation

The skull is exposed by making a sagittal incision starting from the supraorbital region. The incision site must be approximately 1–2 cm in length, large enough to receive the electrode holder. The edges of the skin are separated using two surgical clamps. The skull is scraped slightly with the tip of a scalpel to create micro scratches in order to allow better grip of the cement. The cranium is dried using a cotton swab (to remove blood and conjunctive tissue).

For the freely moving protocol, two small holes (diameter approximately 1 mm) are drilled into the skull posterior to the Bregma (left–right) and nylon mounting screws (0–80 × 3/32 N) are fastened in the holes to provide optimal fixation of the electrode holder. Additionally, a drop of tissue adhesive is placed on/around the screws. This step is not necessary for restrained or anesthetized animals.

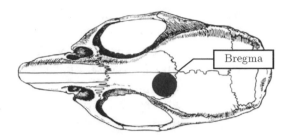

**Fig. 2** The center of the electrode holder (red point) is positioned over the left frontal cortex 1 mm rostral and 1 mm left of Bregma (the point at which the coronal suture intersects the sagittal suture). Adapted from Paxinos and Franklin [57]

The center of the electrode holder is positioned over the left frontal cortex 1 mm anterior to the coronal suture and 1 mm left of the sagittal suture (Fig. 2). The electrode holder is placed on the stereotaxic arm (with an adaptor from DIXI Medical, Besançon) and carefully positioned against the skull (close contact is necessary to prevent the dental cement entering under it).

Glass ionomer dental cement is prepared according to the manufacturer's instructions (GC® Fuji I, GC Corporation, Japan). The standard powder to liquid ratio is 1.8/1.0 g (i.e., one level scoop of powder to two drops of liquid). Thin layers of glass ionomer cement are applied on the lower part of the electrode holder. Layers must form a hill-shaped structure for better grip of the electrode holder. The screw thread inside the electrode holder must be completely clean (total absence of cement). Once all the layers are dry, the stereotaxic arm must be carefully removed while holding the electrode holder using pliers. The animal is then gently removed from the stereotaxic apparatus and placed on a warming pad until fully awake.

*3.1.3 Postsurgical Care*

Antibacterial and anti-itch pomades are applied around the incision site (Aureomycin and P.O.12). Paracetamol is administered orally to the mice during the first 24 h following surgery (10 mg diluted in 100 ml of drinking water). Additionally, the electrode holder can be filled with a povidone–iodine (10%, topical use) liquid solution, which acts as an antiseptic to avoid infection at the skull surface.

After surgery, each animal can recover for 1 week before undergoing tDCS. During this period, and also for the remainder of the experiments, mice are placed in individual cages to prevent them from removing or mutually damaging the electrode holder.

*3.2 tDCS Setup*

Make sure that the battery of the tDCS stimulator (DC-Stimulator Plus or Open-tES) is fully charged before starting the stimulation session. Parameters are set to a 20-min duration and 0.2 mA current intensity. Stimulation starts with a 10 s linear fade-in (gradual increase in intensity) and ends with a 10 s fade-out (gradual decrease in intensity) in order to avoid sudden changes in the current. The ten stimulation sessions are carried out over a period of 5 consecutive days ($2 \times 20$ min/day), with an interstimulation interval of 3 h minimum (intra-day) (*see* **Note 1**).

**3.3 Stimulations**

Animals are placed in the experimental room 30 min before stimulation in order to acclimatize.

The electrode holder is filled with saline solution to establish a contact area of 3.5 mm$^2$ toward the skull. The stimulation electrode (anode) is screwed into the electrode holder. A larger conventional rubber-plate electrode (cathode) serves as the counterelectrode and is placed onto the ventral thorax (there are several mounting possibilities for the cathode; *see* below) (*see* **Note 3**). This setting prevents the bypassing of current (shunting effect) that would occur in the case of two juxtaposed encephalic electrodes in mice. Electrodes are then connected to the tDCS stimulator, the anode to the positive terminal and the cathode to the negative terminal. The resistance must be kept below 90 k$\Omega$ throughout the stimulation session. If the resistance becomes higher than 90 k$\Omega$, the direct current stimulator stops automatically (security mode). Sham-stimulated mice are subjected to the same procedure (surgeries and electrode montage) but the current is not delivered (*see* **Note 2**).

**3.3.1 Animal Restrained During Stimulation**

Animals are awake and restrained during tDCS to prevent possible interactions between the tDCS effects and anesthetic drugs. During the stimulation, a custom-made restraining box is used to hold the cathode against the ventral thorax. Conductive gel is added between the electrode and the chest. The animal is restrained by placing a foam-covered lid on its back, and its tail is fixed on the table with an adhesive bandage (Fig. 3).

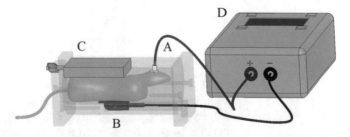

**Fig. 3** Animal restrained during stimulation: the anode (contact area 3.5 mm$^2$, A) is screwed into the electrode holder over the left frontal cortex and the cathode (rubber-plate electrode, 8.75 cm$^2$, B) is placed onto the ventral thorax using the restraint box (C); a $2 \times 20$ min/day constant current of 0.2 mA is applied transcranially using a direct current stimulator (D) for 5 consecutive days, with a linear fade-in–fade-out of 10 s

*3.3.2 Animal Anesthetized During Stimulation*

Animals are under light anesthesia with isoflurane during stimulation. Each mouse is individually anesthetized with isoflurane 3.5% in an induction chamber for 2 min. When the mouse falls asleep, it is removed from the chamber and the cathode (surrounded by a moistened pad and covered with conductive gel) is quickly placed over the ventral thorax using a jacket (homemade). The mouse is then placed back into the induction chamber. When it sleeps deeply, the isoflurane concentration is decreased to 1.5% and the anode is screwed into the electrode holder (Fig. 4).

*3.3.3 Freely Moving Animal During Stimulation*

Mice are briefly anesthetized with isoflurane in an induction chamber before the start of the tDCS procedure. Subsequently, the cathode is placed onto the ventral thorax and fixed using paper tape. Good conductance is obtained by adding conductive paste and a drop of saline solution on the reference electrode. During stimulation, mice are awake and able to freely move in their home cage owing to the commutator system (Fig. 5).

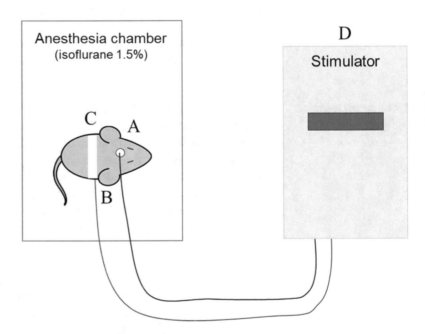

**Fig. 4** Animal anesthetized during stimulation: the anode (contact area 3.5 mm², A) is screwed into the electrode holder over the left frontal cortex and the cathode (rubber-plate electrode, 4.5 cm², B) is placed onto the ventral thorax using a jacket (C); a 2 × 20 min/day constant current of 0.2 mA is applied transcranially using a direct current stimulator (D) for 5 consecutive days, with a linear fade-in–fade-out of 10 s

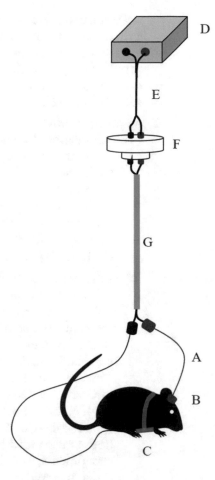

**Fig. 5** Freely moving animal during stimulation: the anode (contact area 3.5 mm², A) is screwed into the electrode holder (B) over the left frontal cortex and the cathode (rubber-plate electrode, 4.5 cm², C) is fixed onto the ventral thorax using paper tape; electrodes are connected to the direct current stimulator (D) using bipolar cable (E), a commutator system (F) and bipolar cable with spring covering (G); a 2 × 20 min/day constant current of 0.2 mA is applied transcranially for 5 consecutive days, with a linear fade-in–fade-out of 10 s

## 4   Notes

1. Variations of the parameters/protocols of stimulation

   The different parameters of stimulation can be modified as needed: the type of stimulation (single vs. repeated), duration (from 3 to 20 min), intensity (from 0.025 to 0.2 mA), and polarity (anodal/cathodal) [58].

   With regard to the type of stimulation, a single stimulation (1 × 20 min at 0.2 mA) in anesthetized mice seems to be as effective as ten stimulations (2 × 20 min/day at 0.2 mA during 5 consecutive days) to alleviate behavioral despair in the forced swimming test. However, this beneficial effect fades faster with a single stimulation compared to ten stimulations [58].

Alternatively, other types of stimulation can be used, such as pulsed [59] or oscillatory [60] tDCS, transcranial alternating current stimulation [61], or transcranial random noise stimulation [62].

2. Major problems.

    (a) Stimulation does not start or stop (resistance is too high)

      It may be necessary to add additional saline solution to the electrode holder and screw the electrode well to the bottom. It also may be relevant to verify the cathode (add conductive gel, press well over the thorax). Finally, as a last resort and after the latter interventions, check the connecting cables (conductance). If there is cement in the electrode holder the animal cannot be stimulated.

    (b) Loss of the electrode holder

      When the electrode is screwed into the electrode holder, the head stage can accidentally come off from the skull. If this happens during stimulation, the animal cannot be reintegrated into the group of stimulated animals and will be excluded from the experiment. If the concern occurs before or after stimulation, the animal can be anesthetized using isoflurane and the electrode holder can be fixed again using dental cement (as described in Subheading 3.1.2). It is important to house mice individually after surgery until the end of the experiments, especially in cases of repeated stimulation, in order to avoid the risk of mutual damage or loss of the electrode holder.

    (c) Loss of contact of reference electrode with the thorax

      When using a restraint box, the mouse can occasionally detach its tail from the adhesive tape and manage to get out of the box. If it is a sham-stimulated animal, just put it back in the box and fix its tail again using a new paper tape. For a stimulated mouse, the current cuts off due to lack of contact of the reference electrode with the thorax, in which case the animal must quickly be put back in the restraint box and stimulation restarted from that point. Time lost needs to be taken into account and added to the remaining time.

      When freely moving, a mouse can reach to detach the electrode from the thorax, cutting off the current due to absence of contact of the reference electrode with the thorax. In this case, stimulation is paused and the mouse is placed in an induction chamber filled with isoflurane for brief anesthesia. Next, the reference electrode can be replaced on the ventral thorax and fixed using paper tape. When placed back in its home cage, stimulation can be continued.

3. Advantages and disadvantages of the different protocols of stimulation

The major difference between the three protocols described here is the emotional/wake state of the mice during stimulation. tDCS in awake mice (restrained or freely moving) is closer to what is done in humans. This could be very important considering that the effects of tDCS are very likely to be state dependent. Moreover, there is no interaction in this case between tDCS and the anesthetic agent [63].

The major disadvantage of using the restraint box is that it is stressful for the mice, which might interfere with the behavioral and neurobiological effects of tDCS. This stress is avoided when mice are anesthetized but the isoflurane may interact with the stimulation [63]. A disadvantage of freely moving stimulation is that it is a heavier protocol to set up and has a more complicated assembly. Furthermore, although stress is reduced it cannot be completely eliminated in this protocol.

# 5   Conclusion

The face validity of the animal model of tDCS presented above was confirmed in a former study [54]. Indeed, we have shown that the behavioral parameters affected by electrical stimulations in clinical studies were also affected in our animal model (e.g., depression related behaviors, visuospatial and working memory, and craving for nicotine). Interestingly, the parameters that are not affected in human (e.g., basal levels of stress and anxiety) were not affected in mice either. Using this animal model, we also recently demonstrated that tDCS decreased the craving for cocaine [55], and reduced the relapse in an animal model of alcohol oral self-administration (unpublished data). These data give additional evidence for a beneficial effect of tDCS to reduce drug consumption in dependent patients. Altogether, our animal model of tDCS has a translational value and will be useful to further investigate the neurobiological mechanisms underlying the effect of tDCS on addiction-related behaviors. It will accelerate the development of an innovative, safe, and easy-to-use technique that could be used in human to treat addiction-related behaviors, in combination with other approaches such as pharmacological treatments and cognitive behavioral therapies.

# References

1. Stagg CJ, Nitsche MA (2011) Physiological basis of transcranial direct current stimulation. Neuroscientist 17:37–53

2. Philip NS, Nelson B, Frohlich F et al (2017) Low-intensity transcranial current stimulation in psychiatry. Am J Psychiatry 174:628–639

3. Garavan H, Hester R (2007) The role of cognitive control in cocaine dependence. Neuropsychol Rev 17:337–345

4. Koob GF, Volkow ND (2009) Neurocircuitry of addiction. Neuropsychopharmacology 35:217–238

5. Brunoni AR, Ferrucci R, Fregni F et al (2012) Transcranial direct current stimulation for the treatment of major depressive disorder: a summary of preclinical, clinical and translational findings. Prog Neuro-Psychopharmacol Biol Psychiatry 39:9–16

6. Nitsche MA, Boggio PS, Fregni F et al (2009) Treatment of depression with transcranial direct current stimulation (tDCS): a review. Exp Neurol 219:14–19

7. Ekhtiari H, Tavakoli H, Addolorato G et al (2019) Transcranial electrical and magnetic stimulation (tES and TMS) for addiction medicine: a consensus paper on the present state of the science and the road ahead. Neurosci Biobehav Rev 104:118–140

8. Boggio PS, Sultani N, Fecteau S et al (2008) Prefrontal cortex modulation using transcranial DC stimulation reduces alcohol craving: a double-blind, sham-controlled study. Drug Alcohol Depend 92:55–60

9. Holla B, Biswal J, Ramesh V et al (2020) Effect of prefrontal tDCS on resting brain fMRI graph measures in alcohol use disorders: a randomized, double-blind, sham-controlled study. Prog Neuro-Psychopharmacol Biol Psychiatry 102:109950

10. Klauss J, Penido Pinheiro LC, Silva Merlo BL et al (2014) A randomized controlled trial of targeted prefrontal cortex modulation with tDCS in patients with alcohol dependence. Int J Neuropsychopharmacol 17:1793–1803

11. Klauss J, Anders QS, Felippe LV et al (2018) Multiple sessions of transcranial direct current stimulation (tDCS) reduced craving and relapses for alcohol use: a randomized placebo-controlled trial in alcohol use disorder. Front Pharmacol 9:716

12. Nakamura-Palacios EM, de Almeida Benevides MC, da Penha Zago-Gomes M et al (2012) Auditory event-related potentials (P3) and cognitive changes induced by frontal direct current stimulation in alcoholics according to Lesch alcoholism typology. Int J Neuropsychopharmacol 15:601–616

13. Nakamura-Palacios EM, Lopes IBC, Souza RA et al (2016) Ventral medial prefrontal cortex (vmPFC) as a target of the dorsolateral prefrontal modulation by transcranial direct current stimulation (tDCS) in drug addiction. J Neural Transm (Vienna) 123:1179–1194

14. da Silva MC, Conti CL, Klauss J et al (2013) Behavioral effects of transcranial direct current stimulation (tDCS) induced dorsolateral prefrontal cortex plasticity in alcohol dependence. J Physiol Paris 107:493–502

15. den Uyl TE, Gladwin TE, Wiers RW (2015) Transcranial direct current stimulation, implicit alcohol associations and craving. Biol Psychol 105:37–42

16. den Uyl TE, Gladwin TE, Wiers RW (2016) Electrophysiological and behavioral effects of combined transcranial direct current stimulation and alcohol approach bias retraining in hazardous drinkers. Alcohol Clin Exp Res 40:2124–2133

17. den Uyl TE, Gladwin TE, Rinck M et al (2017) A clinical trial with combined transcranial direct current stimulation and alcohol approach bias retraining. Addict Biol 22:1632–1640

18. Vanderhasselt M-A, Allaert J, De Raedt R et al (2020) Bifrontal tDCS applied to the dorsolateral prefrontal cortex in heavy drinkers: influence on reward-triggered approach bias and alcohol consumption. Brain Cogn 138:105512

19. Wietschorke K, Lippold J, Jacob C et al (2016) Transcranial direct current stimulation of the prefrontal cortex reduces cue-reactivity in alcohol-dependent patients. J Neural Transm (Vienna) 123:1173–1178

20. Boggio PS, Liguori P, Sultani N et al (2009) Cumulative priming effects of cortical stimulation on smoking cue-induced craving. Neurosci Lett 463:82–86

21. Falcone M, Bernardo L, Ashare RL et al (2016) Transcranial direct current brain stimulation increases ability to resist smoking. Brain Stimul 9:191–196

22. Falcone M, Bernardo L, Wileyto EP et al (2019) Lack of effect of transcranial direct current stimulation (tDCS) on short-term smoking cessation: results of a randomized, sham-controlled clinical trial. Drug Alcohol Depend 194:244–251

23. Fecteau S, Agosta S, Hone-Blanchet A et al (2014) Modulation of smoking and decision-making behaviors with transcranial direct

current stimulation in tobacco smokers: a preliminary study. Drug Alcohol Depend 140:78–84

24. Fregni F, Liguori P, Fecteau S et al (2008) Cortical stimulation of the prefrontal cortex with transcranial direct current stimulation reduces cue-provoked smoking craving: a randomized, sham-controlled study. J Clin Psychiatry 69:32–40

25. Kroczek AM, Häußinger FB, Rohe T et al (2016) Effects of transcranial direct current stimulation on craving, heart-rate variability and prefrontal hemodynamics during smoking cue exposure. Drug Alcohol Depend 168:123–127

26. Meng Z, Liu C, Yu C et al (2014) Transcranial direct current stimulation of the frontal-parietal-temporal area attenuates smoking behavior. J Psychiatr Res 54:19–25

27. Xu J, Fregni F, Brody AL et al (2013) Transcranial direct current stimulation reduces negative affect but not cigarette craving in overnight abstinent smokers. Front Psych 4:112

28. Aronson Fischell S, Ross TJ, Deng Z-D et al (2020) Transcranial direct current stimulation applied to the dorsolateral and ventromedial prefrontal cortices in smokers modifies cognitive circuits implicated in the nicotine withdrawal syndrome. Biol Psychiatry Cogn Neurosci Neuroimaging 5:448–460

29. Batista EK, Klauss J, Fregni F et al (2015) A randomized placebo-controlled trial of targeted prefrontal cortex modulation with bilateral tDCS in patients with crack-cocaine dependence. Int J Neuropsychopharmacol 18: pyv066

30. Conti CL, Nakamura-Palacios EM (2014) Bilateral transcranial direct current stimulation over dorsolateral prefrontal cortex changes the drug-cued reactivity in the anterior cingulate cortex of crack-cocaine addicts. Brain Stimul 7:130–132

31. Conti CL, Moscon JA, Fregni F et al (2014) Cognitive related electrophysiological changes induced by non-invasive cortical electrical stimulation in crack-cocaine addiction. Int J Neuropsychopharmacol 17:1465–1475

32. Gorini A, Lucchiari C, Russell-Edu W et al (2014) Modulation of risky choices in recently abstinent dependent cocaine users: a transcranial direct-current stimulation study. Front Hum Neurosci 8:661

33. Boggio PS, Zaghi S, Villani AB et al (2010) Modulation of risk-taking in marijuana users by transcranial direct current stimulation (tDCS)

of the dorsolateral prefrontal cortex (DLPFC). Drug Alcohol Depend 112:220–225

34. Shahbabaie A, Golesorkhi M, Zamanian B et al (2014) State dependent effect of transcranial direct current stimulation (tDCS) on methamphetamine craving. Int J Neuropsychopharmacol 17:1591–1598

35. Shahbabaie A, Hatami J, Farhoudian A et al (2018) Optimizing electrode montages of transcranial direct current stimulation for attentional bias modification in early abstinent methamphetamine users. Front Pharmacol 9:907

36. Shahbabaie A, Ebrahimpoor M, Hariri A et al (2018) Transcranial DC stimulation modifies functional connectivity of large-scale brain networks in abstinent methamphetamine users. Brain Behav 8:e00922

37. Wang Y, Shen Y, Cao X et al (2016) Transcranial direct current stimulation of the frontal-parietal-temporal area attenuates cue-induced craving for heroin. J Psychiatr Res 79:1–3

38. Sharifi-Fardshad M, Mehraban-Eshtehardi M, Shams-Esfandabad H et al (2018) Modulation of drug craving in crystalline-heroin users by transcranial direct stimulation of dorsolateral prefrontal cortex. Addict Health 10:173–179

39. O'Brien CP (2008) Evidence-based treatments of addiction. Philos Trans R Soc Lond Ser B Biol Sci 363:3277–3286

40. Fraser PE, Rosen AC (2012) Transcranial direct current stimulation and behavioral models of smoking addiction. Front Psych 3

41. Nitsche MA, Lampe C, Antal A et al (2006) Dopaminergic modulation of long-lasting direct current-induced cortical excitability changes in the human motor cortex. Eur J Neurosci 23:1651–1657

42. Borwick C, Lal R, Lim LW et al (2020) Dopamine depletion effects on cognitive flexibility as modulated by tDCS of the dlPFC. Brain Stimul 13:105–108

43. Krause B, Márquez-Ruiz J, Cohen Kadosh R (2013) The effect of transcranial direct current stimulation: a role for cortical excitation/inhibition balance? Front Hum Neurosci 7:602

44. Liebetanz D, Nitsche MA, Tergau F et al (2002) Pharmacological approach to the mechanisms of transcranial DC-stimulation-induced after-effects of human motor cortex excitability. Brain 125:2238–2247

45. Iyer PC, Rosenberg A, Baynard T et al (2019) Influence of neurovascular mechanisms on response to tDCS: an exploratory study. Exp Brain Res 237:2829–2840

46. Wachter D, Wrede A, Schulz-Schaeffer W et al (2011) Transcranial direct current stimulation induces polarity-specific changes of cortical blood perfusion in the rat. Exp Neurol 227:322–327

47. Lu Y, Christian K, Lu B (2008) BDNF: a key regulator for protein synthesis-dependent LTP and long-term memory? Neurobiol Learn Mem 89:312–323

48. Podda MV, Cocco S, Mastrodonato A et al (2016) Anodal transcranial direct current stimulation boosts synaptic plasticity and memory in mice via epigenetic regulation of Bdnf expression. Sci Rep 6:22180

49. Fritsch B, Reis J, Martinowich K et al (2010) Direct current stimulation promotes BDNF-dependent synaptic plasticity: potential implications for motor learning. Neuron 66:198–204

50. Monai H, Hirase H (2018) Astrocytes as a target of transcranial direct current stimulation (tDCS) to treat depression. Neurosci Res 126:15–21

51. Schweid L, Rushmore RJ, Valero-Cabré A (2008) Cathodal transcranial direct current stimulation on posterior parietal cortex disrupts visuo-spatial processing in the contralateral visual field. Exp Brain Res 186:409–417

52. Márquez-Ruiz J, Leal-Campanario R, Sánchez-Campusano R et al (2012) Transcranial direct-current stimulation modulates synaptic mechanisms involved in associative learning in behaving rabbits. Proc Natl Acad Sci U S A 109:6710–6715

53. Liebetanz D, Klinker F, Hering D et al (2006) Anticonvulsant effects of transcranial direct-current stimulation (tDCS) in the rat cortical ramp model of focal epilepsy. Epilepsia 47:1216–1224

54. Pedron S, Monnin J, Haffen E et al (2014) Repeated transcranial direct current stimulation prevents abnormal behaviors associated with abstinence from chronic nicotine consumption. Neuropsychopharmacology 39:981–988

55. Pedron S, Beverley J, Haffen E et al (2017) Transcranial direct current stimulation produces long-lasting attenuation of cocaine-induced behavioral responses and gene regulation in corticostriatal circuits. Addict Biol 22:1267–1278

56. Pedron S, Dumontoy S, Dimauro J et al (2020) Open-tES: an open-source stimulator for transcranial electrical stimulation designed for rodent research. PLoS One 15:e0236061

57. Paxinos G, Franklin KBJ (2007) The mouse brain in stereotaxic coordinates. Academic, San Diego

58. Peanlikhit T, Van Waes V, Pedron S et al (2017) The antidepressant-like effect of tDCS in mice: a behavioral and neurobiological characterization. Brain Stimul 10:748–756

59. Datta A, Dmochowski JP, Guleyupoglu B et al (2013) Cranial electrotherapy stimulation and transcranial pulsed current stimulation: a computer based high-resolution modeling study. NeuroImage 65:280–287

60. Mizrak E, Kim K, Roberts B et al (2018) Impact of oscillatory tDCS targeting left prefrontal cortex on source memory retrieval. Cogn Neurosci 9:194–207

61. Antal A, Boros K, Poreisz C et al (2008) Comparatively weak after-effects of transcranial alternating current stimulation (tACS) on cortical excitability in humans. Brain Stimul 1:97–105

62. Moret B, Donato R, Nucci M et al (2019) Transcranial random noise stimulation (tRNS): a wide range of frequencies is needed for increasing cortical excitability. Sci Rep 9:15150

63. Gersner R, Kravetz E, Feil J et al (2011) Long-term effects of repetitive transcranial magnetic stimulation on markers for neuroplasticity: differential outcomes in anesthetized and awake animals. J Neurosci 31:7521–7526

# Chapter 13

## Manipulating Reconsolidation to Weaken Drug Memory

### Amy L. Milton

### Abstract

The long-term prevention of relapse is one of the most challenging aspects of treating drug addiction. Relapse can be precipitated by environmental cues previously associated with the drug high in a pavlovian manner. These drug-associated conditioned stimuli (CSs) become maladaptive emotional memories that interact with instrumental drug-seeking and drug-taking memories to enhance drug use, and these memories can persist for a lifetime.

However, memories are not immutable once formed, and it has become clear over the past few decades that under certain conditions, memories can become once again susceptible to disruption, requiring "reconsolidation" to persist in the brain. This reconsolidation process has therefore been identified as a potential target for disrupting the maladaptive emotional memories that contribute to relapse risk. This chapter considers how reconsolidation-based approaches can be applied to animal analogues of drug-seeking behavior, and some future research directions that will be important for translating these approaches to large-scale clinical populations.

**Key words** Memory, Reconsolidation, Addiction, Pro-abstinence, Anti-relapse, Rat, Translational

---

## 1 Introduction

Arguably, the prevention of relapse to drug-seeking behavior is one of the most difficult aspects of treating drug addiction. While detoxifying individuals and promoting drug abstinence can often be achieved in the short term, the risks of relapse remain high for years, if not decades, after an individual's last drug use [1].

A highly influential account of why relapse risk remains so high when the physiological withdrawal effects of drug have abated postulates that drug addiction hijacks normal reward learning mechanisms to produce "maladaptive emotional memories" [2]. Seeking and consuming of rewards (including drugs of abuse) relies upon the formation of instrumental memories; the actions that lead to the presentation of the reward ("drug-seeking") and to its consumption ("drug-taking"). These instrumental memories can be goal-directed, "action-outcome" memories where an action is made because a reward is desired [3]. With extensive

María A. Aguilar (ed.), *Methods for Preclinical Research in Addiction*, Neuromethods, vol. 174,
https://doi.org/10.1007/978-1-0716-1748-9_13, © Springer Science+Business Media, LLC, part of Springer Nature 2022

experience ("overtraining"), a less predictable correlation between the action and the outcome, or under the influence of psychostimulants [4], habitual, "stimulus-response" memories form [3]. The transition from goal-directed to habitual behavioral control is associated with changes in dependence of the memory on different striatal regions, and if coupled with reduced prefrontal cortical inhibition, can ultimately lead to compulsive drug-seeking behavior in a proportion of individuals [5].

Drug-seeking and drug-taking behavior are not executed in isolation, but occur in particular contexts and in the presence of specific, predictive environmental stimuli that become associated with the drug high in a pavlovian manner [6]. These drug-associated pavlovian conditioned stimuli (CSs) can interact with instrumental memories to support and enhance drug-seeking behavior in at least three psychologically and neurobiologically separable ways [6]. Pavlovian CS-drug memories are extremely persistent and resistant to extinction [7], and potent precipitators of relapse [1, 8]. Therefore, if it was possible to disrupt pavlovian CS-drug memories, it may also be possible to reduce the risk of relapse to drug-seeking in the longer term.

The prospect of disrupting old and well-established drug-associated memories was not considered feasible until the rediscovery of the process of "memory reconsolidation" in the early 2000s. Originally documented as "cue-dependent amnesia" in the 1960s [9, 10], the discovery that pavlovian fear memories require protein synthesis in the amygdala to reconsolidate after certain conditions of retrieval [11] reinvigorated the field. Shortly after this initial demonstration of reconsolidation, the notion that memory reconsolidation could present a target for disrupting memories that contribute to mental health disorders—such as posttraumatic stress disorder and drug addiction—emerged.

Since its rediscovery, much research has focused on characterizing the mechanisms that underlie the reconsolidation process, mostly by comparison to the initial consolidation of memories. Memory reconsolidation consists of several constituent processes (Fig. 1). First, the memory destabilizes via a process that depends upon activation of specific neurotransmitter receptors [12–16], activation of protein phosphatases [17] and protein degradation [18]. The destabilization process is distinct from memory retrieval [14, 19, 20] and for pavlovian memories, appears to be dependent upon a "violation of expectations" between what is expected, and what actually occurs [21–24]. This requirement is consistent with the hypothesis that reconsolidation has evolved to provide a mechanism for memory updating [25]. Once destabilized, the memory is considered to be in the unstable and transient "active state" [26], requiring restabilization to the stable "inactive state" in order to persist. This restabilization process—sometimes confusingly also referred to as "reconsolidation"—depends upon activation of

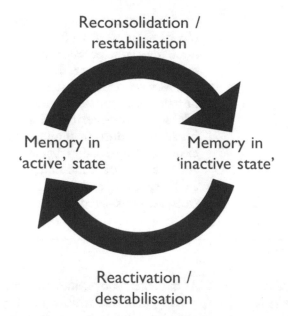

**Fig. 1** Subcomponents of memory reconsolidation. Following their initial consolidation, memories are stored in an "inactive" state that is stable and persistent. Under certain conditions—procedurally termed "reactivation"—the memory becomes destabilized and enters the transient and unstable "active" state. In order to persist, the memory must undergo a "restabilization" process (sometimes confusingly also termed "reconsolidation")

specific neurotransmitter receptors [27–29], protein kinases [30–32], immediate early genes [33, 34] and protein synthesis [35–37]. Restabilization appears to be completed within 4–6 h after the memory is destabilized, thus providing a reasonably long therapeutic window for the manipulation of old and well-established memories.

The notion that reconsolidation could be exploited to provide a novel form of treatment for mental health disorders like drug addiction has received a great deal of attention in the preclinical literature (*see* [6, 38–40] for review) and has been successfully used in some small-scale human trials [41–46]. It is true that there are a number of challenges that need to be addressed for this approach to be used at a large scale—for example, managing the individual variability in drug use histories and how this will influence the conditions under which the memory becomes unstable [47], and the interaction of pavlovian and instrumental memories [40, 48]. However, the evidence indicates that old, well-established drug memories *can* be disrupted in rats [34] and humans [41–46]. Old and strong memories are more resistant to destabilization than younger or weaker memories [49] but they do reconsolidate—just under different parameters of reactivation [50].

There are many open questions in the field of drug memory reconsolidation. Broadly speaking, these are all addressed with the same general experimental approach in the animal literature: animals are trained such that they learn the memory of interest, the memory is reactivated at a later time point under the influence of a putative amnestic agent, and subsequently the animals are tested for relapse to drug-seeking behavior. There are specific variants of this broad procedure, including the type of behavioral test used to examine drug-seeking behavior, the drug of abuse investigated, and the type of amnestic agent used to manipulate the memory. These specifics, and other relevant issues such as appropriate controls, are considered in more detail in the next section.

## 2    Methods

### 2.1    Selection of Memory Type to Target for Reconsolidation Blockade

As discussed above, addiction is a complex disorder in which drug-seeking behavior is influenced by a number of different types of memory. The memories that can be targeted for reconsolidation blockade will depend upon the choice of behavioral procedure used to investigate drug-seeking behavior (*see* **Note 1**).

The majority of rodent studies attempting to disrupt the reconsolidation of drug memories relevant to addiction have used either the conditioned place preference (CPP) procedure, or drug self-administration procedures. CPP is rapidly trained (4–8 sessions) and robust, though it requires administration of drug by the experimenter rather than drug self-administration by the animal. By contrast, drug self-administration procedures often require more extensive training (typically 3–10 sessions, though training can occur over a longer period if overtrained memories are of interest). Self-administration of drugs such as cocaine or heroin usually require surgical implantation of a chronic, indwelling intravenous catheter so that animals can be connected to a syringe of drug via a sterile line during the conditioning sessions (though for an example of oral cocaine self-administration, *see* [51]). Alcohol can be self-administered orally by animals, though usually following a procedure to habituate animals to its aversive taste (*see* **Note 2**). Ultimately the choice of training procedure used to produce the formation of a drug memory will depend on the type of memory that is of experimental interest.

There are many variants of instrumental drug self-administration tasks that allow the contribution of different memories to relapse to be assessed (Fig. 2). If pavlovian CS-drug memories are of interest, then procedures such as acquisition of a new instrumental response for conditioned reinforcement (ANR) allows the conditioned reinforcing properties of a CS to be studied in isolation. Similarly, behavioral tasks such as pavlovian-instrumental transfer (PIT), pavlovian conditioned approach (or autoshaping)

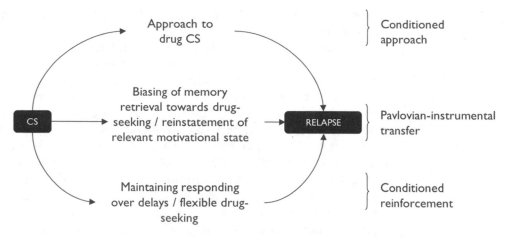

**Fig. 2** The "three routes to relapse" by which drug-associated CSs can influence drug-seeking behavior. (Reproduced, with permission, from [6])

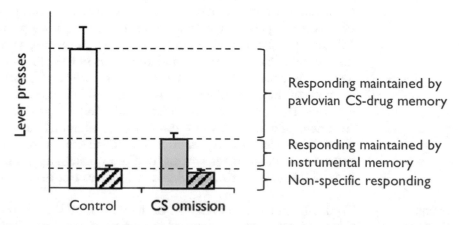

**Fig. 3** Drug-seeking is supported by both pavlovian and instrumental memories in many behavioral tasks. If animals are trained on drug self-administration procedures and subsequently tested for relapse to drug-seeking behavior, it is possible to determine the contribution of pavlovian and instrumental memories to drug-seeking with the use of appropriate control conditions. An inactive lever (striped bars) that has no programmed consequence serves as a control for nonspecific responding. A group in which a drug-associated CS is omitted ("CS omission") allows responding maintained by the instrumental memory to be determined. Any enhancement in drug-seeking observed in the control group can be attributed to the presence of the pavlovian CS

or second-order schedules of reinforcement (Fig. 2) allow the study of specific influences of pavlovian CSs on drug-seeking behavior. These procedures are useful for demonstrating the disruption of pavlovian CS-drug memories, as the dependence of the behavior on a specific psychological process (e.g., conditioned reinforcement) provides behavior that is readily impaired if the memory supporting the psychological process is disrupted. However, the strength of these tasks—that the behavior is "psychologically pure"—is also a limitation, as necessarily tasks such as ANR or PIT do not fully capture the full range of ways that a CS-drug memory can influence relapse behavior.

It can often be advantageous to combine "psychologically pure" animal learning theory-inspired behavioral tasks with parallel experiments using tasks of greater translational relevance. Reinstatement tasks—whether following extinction of instrumental responding for drug in the "extinction–reinstatement" procedure, or following a period of drug abstinence without instrumental extinction—have been most commonly used in the literature, though second-order schedules of conditioned reinforcement have also been investigated. Although these tasks depend upon many psychological processes for performance of the behavior, including both pavlovian and instrumental memories, with appropriate control conditions it is possible to determine which memories within an associative network have been disrupted. For example, using control groups in which the CS is omitted at test, or by comparing responding during the presence and absence of a CS within the same animals, it is possible to distinguish the impact of a disrupted pavlovian memory on behavior that is supported by a still-intact instrumental memory (Fig. 3). Similarly, studies of instrumental memory reconsolidation [52–54] usually omit pavlovian CSs from the behavioral procedures to allow clearer demonstrations of the manipulation on instrumental behavior. Ultimately, a strong test of the reconsolidation blockade approach for the treatment of addiction will require the use of animal analogues of compulsive drug-seeking, such as the "3-criteria" model [55] or demonstrations of drug-seeking that are resistant to punishment [56, 57]. However, to date these models have not been tested.

## 2.2 Selection of Amnestic Agent and Route of Administration

To disrupt the reconsolidation of drug-associated memories, two things are required; a behavioral procedure effective at destabilizing the memory, and a manipulation that will prevent the memory from restabilizing and reentering the stable inactive state (Fig. 1). The majority of studies investigating drug memory reconsolidation have used pharmacological methods to prevent the restabilization of the memory, with very few using optogenetic [58, 59] or chemogenetic [60–62] strategies. Other studies have used behavioral methods of memory disruption, including the "extinction within the reconsolidation window" or "retrieval–extinction" procedure [41, 42]. However, it remains to be determined whether all instances of the "retrieval–extinction" effect depend upon the disruption of reconsolidation—rather than the facilitation of extinction [63]—and so these will not be considered further here.

The choice of amnestic agent to disrupt drug-associated memory reconsolidation will depend upon the specific research question of interest. Two widely used amnestic agents are the β-adrenergic receptor antagonist propranolol, and the NMDA subtype of glutamate receptor antagonist dizocilpine (MK-801); however, a wide range of molecules previously shown to be involved in synaptic plasticity and memory consolidation have been targeted, including protein synthesis inhibitors [64], immediate early genes [33, 34], protein kinases [30–32, 65, 66], and retrograde signaling molecules [67]. The route of administration will depend upon both the specific research question and the amnestic agent selected. Some amnestic agents, including protein synthesis inhibitors such as anisomycin, have a high mortality rate when administered systemically. Others, such as oligodeoxynucleotides targeting immediate early genes, require direct access to neuronal membranes. Consequently, these amnestic agents are given intracerebrally. Intracerebral administration is also appropriate when the contribution of specific neural regions to memory reconsolidation informs the research question. As for other pharmacological studies, systemic administration of amnestic agents provides less information on neural circuitry underlying any observed memory disruption but is potentially more translatable to clinical populations.

Regardless of the pharmacological target of the amnestic agent, habituation to the administration procedure is strongly advised to avoid any risk of treatment administration being used as a discriminative stimulus (*see* **Note 3**).

**2.3  Relative Timing of Memory Reactivation and Administration of Amnestic Agent**

The timing of amnestic agent administration relative to memory reactivation has been a subject of controversy within the reconsolidation literature, with some [68] arguing that amnestic agents can *only* be administered after memory reactivation, in order to avoid confounding effects on memory retrieval. Others [69] maintain that although post-reactivation treatments are preferable, in practice amnestic agents sometimes do need to be administered prior to memory reactivation. This may be because the route of administration or type of amnestic agent leads to a delay in the peak amnestic effect, or because the target of the amnestic agent has a time-limited role in the reconsolidation process. Where amnestic agents have acute side effects that complicate interpretation of the memory reactivation session—for example, the hyperactivity induced by administration of the NMDA receptor antagonist MK-801, which can interfere with the expression of conditioned freezing in the reactivation of fear memory [70]—then short-term memory tests can be of use. These tests, which are typically performed 3–4 h after the reactivation session, occur after the acute effects of the amnestic agent have worn off, but before the emergence of any memory deficit induced by the targeting of memory reconsolidation.

**2.4  Choice of Memory Reactivation Protocol**

Alongside the choice of amnestic agent, the choice of the procedure for memory reactivation is one of the most challenging aspects of experiments aiming to disrupt memory reconsolidation. The field does not yet have a full understanding of the parameters that are maximally effective at inducing the destabilization of a memory (rather than its simple retrieval) without entering the "limbo" in which the original memory becomes stable once again, before engaging the new learning mechanisms associated with extinction [71–73]. There has been debate over whether very old and strong memories reconsolidate [74]. However, it has been shown that CS-drug memories still reconsolidate when 27 days have elapsed between the end of training and reactivation [34], indicating that old and strong memories may simply require different reactivation parameters to new memories. Until there is a validated biomarker for CS-drug memory destabilization, the design of memory reactivation sessions relies primarily on parameters shown to be effective in previous research.

There appear to be critical differences in the parameters that underlie pavlovian and instrumental memory reconsolidation. For pavlovian memories, the "violation of expectations" required to induce memory destabilization [21] is usually achieved by omitting the CS during a truncated reactivation session. For CPP studies, this is often through a brief memory retention test in which animals are allowed to express their preference across the compartments of the chamber. For self-administration studies, we and others have previously observed that reexposing animals to approximately 10% of the CS-drug exposures they received during training, in the absence of drug reinforcement, is sufficient to induce memory destabilization [27, 70, 75]. Where drug-associated CSs have been presented contingent on instrumental responding during training, then reexposure should also be contingent on responding during the memory reactivation session [76, 77]. The reactivation session is usually truncated compared to the usual training session; this may be an important factor in inducing memory destabilization in its own right [78] and is usually necessary due to short-term increases in the rate of responding observed in animals that are not receiving reinforcement.

The reactivation parameters required to destabilize instrumental memories appear to differ from those that induce lability of pavlovian memories. Instrumental response-drug memories have been less extensively investigated than CS-drug memories, but appear to depend critically upon the reinforcement of responding during the memory reactivation session [52, 53]. This may account for findings showing that disruption of a pavlovian CS-drug memory (through a nonreinforced reactivation session, as described above) leaves the instrumental memory intact [27]. A consistent finding is that even though responding needs to be contingently reinforced to induce instrumental memory destabilization, a change in the reinforcement contingency is also necessary—and this should be to a more volatile or unpredictable schedule. For example, animals trained to self-administer cocaine on a continuous reinforcement (fixed ratio 1, or FR1) schedule show a subsequent reduction in responding for drug if they receive MK-801 prior to a session in which the memory is reactivated using a variable, but not fixed, ratio schedule [52, 53]. The extent to which volatility, in addition to "violation of expectations" contributes to instrumental memory reconsolidation requires further investigation.

Control groups are always important, regardless of the exact parameters used to reactivate the memory or the specific memory targeted for disruption. Different controls are required depending on the outcome of the intervention. When amnesia is observed at the subsequent memory retention test, it is important to include control groups treated with the amnestic agent in the absence of an effective memory reactivation session. These serve to rule out nonspecific reductions of the treatment, independent of memory destabilization. Often these are "nonreactivated" controls (*see* **Note 4**) but "delayed" controls—which undergo the memory reactivation procedure but receive the amnestic agent outside of the reconsolidation window—can also be used.

Those who favor a "memory integration" view of memory reconsolidation [79] would argue that when an amnestic effect is observed, it is also advisable to include controls that exclude the possibility of state-dependent memory retrieval. This would require groups that receive administration of the same amnestic agent prior to the test session, in order to determine whether the memory is observed in groups that received the same agent (vehicle or drug treatment) prior to reactivation and test, but not groups that received different treatments prior to reactivation and test. Nonspecific, state-dependent effects have been observed for drugs such as systemic protein synthesis inhibitors, which induce a range of side effects [79], but for other treatments these same effects have not been reported.

Different control groups are necessary if a treatment fails to produce amnesia at a subsequent test. Besides demonstrating that the dose used is sufficient to produce effects on memory (e.g., [80]), it is also necessary to demonstrate that the reactivation session was sufficient to induce destabilization of the memory. As memory retrieval and memory destabilization can be dissociated, until there is a marker for memory destabilization, the most straightforward control group to address this point is a group that receive an established amnestic agent. These "positive controls" establish the effectiveness of the memory reactivation procedure, ruling out ineffective memory destabilization as an explanation for a lack of amnestic effect.

**2.5  Memory Retention Tests**

The final phase of memory reconsolidation experiments is testing the memory to reveal any deficit. The choice of specific memory test will depend on the type of memory targeted, though in the context of drug memories it will usually involve assessing an aspect of drug-seeking behavior. Considering that the ultimate aim of translational research targeting drug memory reconsolidation is a long-term reduction in relapse risk, then ideally tests should be conducted repeatedly and in different situations (e.g., tests of CS-induced relapse, stress-induced relapse, drug-primed relapse) to determine whether the memory returns. Behaviorally, it is not possible to show that a memory has been erased [81], though future research may be able to use engram-labelling techniques [82] to address this issue.

## 3    Notes

1. Regardless of the behavioral procedure used, it is important to maintain a level of consistency across the training, reactivation and test phases of the experiment. Animals should be trained and tested at approximately the same time of day, and in the same chambers, unless a context change is part of the memory reactivation procedure. Although there has not yet been a full parametric analysis of the conditions under which memories destabilize, it has been established that a "violation" of what the animal expects and what actually occurs is critical for inducing memory destabilization [21]. Careful attention to the reliability of different cues—including circadian and interoceptive cues—provides greater experimental control over the degree to which expectations are violated during the memory reactivation procedure.

2. Several different procedures have been used to habituate rats to the aversive taste of ethanol prior to oral self-administration, including saccharin fading [83] and intermittent two-bottle choice [84]. Some have suggested that habituation to the taste of ethanol is unnecessary to establish instrumental self-administration of ethanol [85, 86]. We have previously observed that saccharin fading may alter the dynamics of alcohol memory reconsolidation [87] and that animals that have not been habituated to the taste of ethanol do not readily consume it despite instrumentally working for it [87]. For this reason, we now use two-bottle choice as our standard method of habituation.

3. Habituation of animals to the procedure used to administer the amnestic agent is strongly advised, particularly if the administration is prior to the memory reactivation session. The reasons for this are twofold; first, to avoid any undue stress on the animals on the day of the reactivation session, and second, so that the treatment cannot be used to distinguish between normal training and the memory reactivation session. The memory reactivation session, particularly for pavlovian memories, is often not reinforced. Consequently, if treatment was only given prior to the reactivation session, then the injection

or infusion used to administer the treatment could be used by the animal as a discriminative cue that the session would not be reinforced. This may bias the animal toward new learning and consolidation of a new memory in the reactivation session, rather than reconsolidation of the previously learned memory. In the absence of treatment administration—for example, at the subsequent memory retention test—this would lead to a failure to observe amnesia due to ineffective memory destabilization, rather than an ineffective amnestic agent. For a similar reason, where experiments are being conducted in teams it is also advisable to avoid any change in experimenter between training, reactivation and test sessions.

4. Where nonreactivated controls are used, it is important to ensure that the memory of interest is not reactivated accidentally while administering the amnestic treatment. Anecdotally, animals may use different cues to trigger memory reactivation, including the method used to transfer them from the home cage to the testing apparatus, or the presence of the experimenter associated with the training procedure. Ideally, common factors between transporting animals to receive drug injections and transporting animals to the training sessions should be avoided for nonreactivated groups.

## 4  Conclusions

There remain unanswered questions in drug memory reconsolidation research, which can be addressed using the approaches detailed above. These primarily relate to the translation of reconsolidation-based approaches to human clinical populations, where drug self-administration histories are not standardized (as they are to some degree in animal research) and the drug-associated cues that trigger relapse may not be so readily identifiable.

More research is required to determine the parameters of reconsolidation and extinction of drug-associated memories, particularly when the age and strength of these memories differs. As noted above, it appears that old and well-established cocaine memories do reconsolidate [34], but it may be that the transition between memory retrieval, reconsolidation, limbo, and extinction [71, 72] change in an age- and strength-dependent manner [50]. Individual differences may also affect these parameters [47]. Therefore, a reliable and noninvasive marker of memory destabilization would be highly valuable in translating reconsolidation-based approaches to the clinical situation.

Another key area of research is the identification of amnestic agents or approaches that are well-tolerated in humans. The β-adrenergic receptor antagonist propranolol has received much attention in this respect, primarily because it is safe to use in humans, and has been done so in studies investigating the reconsolidation of fear memories in healthy participants [19, 23, 24, 88, 89] and patients with posttraumatic stress disorder [90–92] and specific phobia [93]. However, propranolol has not been as effective in human studies of drug memory reconsolidation, where some studies have reported reductions in drug craving and physiological responses to drug-associated CSs [46, 94, 95], while others have found no effect [96] or even an enhancement of craving [97]. However, these negative studies did not include "positive controls" to demonstrate that the parameters of the memory reactivation were sufficient to destabilize the memory (as described above). It is also worth noting that propranolol produces mixed effects in animal models of drug-seeking too, with some but not all of the "three routes to relapse" [6] being disrupted by the administration of systemic propranolol at reactivation [28, 75, 98, 99]. NMDA receptor antagonists produce larger amnestic effects when given in conjunction with memory reactivation in animals [100], but pose a greater challenge to clinical translation. Identification of amnestic agents that can be administered to humans with a minimum of side effects will also be key in the translational of reconsolidation-based approaches to drug-addicted patients.

Despite the remaining questions, disrupting maladaptive drug-associated memories to reduce the long-term risk of relapse in drug-addicted patients would represent a transformative approach in the treatment of addiction. The further development and optimization of reconsolidation-based interventions will benefit markedly from a translational and backtranslational approach, and from basic research identifying the underlying mechanisms that govern the transition between memory retrieval and memory destabilization.

# References

1. Gawin FH, Kleber HD (1992) Clinician's guide to cocaine addiction: theory, research and treatment. The Guildford Press, New York, NY, pp 33–52

2. Everitt BJ, Dickinson A, Robbins TW (2001) The neuropsychological basis of addictive behaviour. Brain Res Rev 36:129–138

3. Dickinson A (1985) Actions and habits: the development of behavioural autonomy. Philos Trans R Soc Lond B 308:67–78

4. Nelson A, Killcross S (2006) Amphetamine exposure enhances habit formation. J Neurosci 26:3805–3812

5. Everitt BJ, Robbins TW (2005) Neural systems of reinforcement for drug addiction: from actions to habits to compulsion. Nat Neurosci 8:1481–1489

6. Milton AL, Everitt BJ (2010) The psychological and neurochemical mechanisms of drug memory reconsolidation: implications for the treatment of addiction. Eur J Neurosci 31:2308–2319

7. Di Ciano P, Everitt BJ (2004) Conditioned reinforcing properties of stimuli paired with self-adminstered cocaine, heroin or sucrose: implications for the persistence of addictive behavior. Neuropharmacology 47:202–213

8. de Wit H, Stewart J (1981) Reinstatement of cocaine-reinforced responding in the rat. Psychopharmacology 75:134–143

9. Schneider AM, Sherman W (1968) Amnesia: a function of the temporal relation of footshock to electroconvulsive shock. Science 159:219–222

10. Misanin JR, Miller RR, Lewis DJ (1968) Retrograde amnesia produced by electroconvulsive shock after reactivation of a consolidated memory trace. Science 160:554–555

11. Nader K, Schafe GE, LeDoux JE (2000) Fear memories require protein synthesis in the amygdala for reconsolidation after retrieval. Nature 406:722–726

12. Ferrer Monti RI et al (2016) An appetitive experience after fear memory destabilization attenuates fear retention: involvement GluN2B-NMDA receptors in the basolateral amygdala complex. Learn Mem 23:465–478

13. Holehonnur R et al (2016) Increasing the GluN2A/GluN2B ratio in neurons of the mouse basal and lateral amygdala inhibits the modification of an existing fear memory trace. J Neurosci 36:9490–9504

14. Milton AL et al (2013) Double dissociation of the requirement for GluN2B- and GluN2A-containing NMDA receptors in the destabilization and restabilization of a reconsolidating memory. J Neurosci 33:1109–1115

15. Reichelt AC, Exton-McGuinness MT, Lee JLC (2013) Ventral tegmental dopamine dysregulation prevents appetitive memory destabilization. J Neurosci 33:14205–14210

16. Suzuki A, Mukawa T, Tsukagoshi A, Frankland PW, Kida S (2008) Activation of LVGCCs and CB1 receptors required for destabilization of reactivated contextual fear memories. Learn Mem 15:426–433

17. Yu Y-J, Huang C-H, Chang C-H, Gean P-W (2016) Involvement of protein phosphatases in the destabilization of methamphetamine-associated contextual memory. Learn Mem 23:486–493

18. Jarome TJ, Werner CT, Kwapis JL, Helmstetter FJ (2011) Activity dependent protein degradation is critical for the formation and stability of fear memory in the amygdala. PLoS One 6:e24349

19. Sevenster D, Beckers T, Kindt M (2012) Retrieval per se is not sufficient to trigger reconsolidation of human fear memory. Neurobiol Learn Mem 97:338–345

20. Santoyo-Zedillo M, Rodriguez-Ortiz CJ, Chavez-Marchetta G, Bermudez-Rattoni F, Balderas I (2014) Retrieval is not necessary to trigger reconsolidation of object recognition memory in the perirhinal cortex. Learn Mem 21:452–456

21. Pedreira ME, Pérez-Cuesta LM, Maldonado H (2004) Mismatch between what is expected and what actually occurs triggers memory reconsolidation or extinction. Learn Mem 11:579–585

22. Fernández RS, Boccia MM, Pedreira ME (2016) The fate of memory: reconsolidation and the case of prediction error. Neurosci Biobehav Rev 68:423–441

23. Sevenster D, Beckers T, Kindt M (2013) Prediction error governs pharmacologically induced amnesia for learned fear. Science 339:830–833

24. Sevenster D, Beckers T, Kindt M (2014) Prediction error demarcates the transition from retrieval, to reconsolidation, to new learning. Learn Mem 21:580–584

25. Lee JLC, Nader K, Schiller D (2017) An update on memory reconsolidation updating. Trends Cogn Sci 21:531. https://doi.org/10.1016/j.tics.2017.1004.1006

26. Lewis DJ (1979) Psychobiology of active and inactive memory. Psychol Bull 86:1054–1083

27. Milton AL, Lee JLC, Butler VJ, Gardner RJ, Everitt BJ (2008) Intra-amygdala and systemic antagonism of NMDA receptors prevents the reconsolidation of drug-associated memory and impairs subsequently both novel and previously acquired drug-seeking behaviors. J Neurosci 28:8230–8237

28. Milton AL et al (2012) Antagonism at NMDA receptors, but not β-adrenergic receptors, disrupts the reconsolidation of pavlovian conditioned approach and instrumental transfer for ethanol-associated conditioned stimuli. Psychopharmacology 219:751–761

29. Vengeliene V, Olevska A, Spanagel R (2015) Long-lasting effect of NMDA receptor antagonist memantine on ethanol-cue association and relapse. J Neurochem 135:1080–1085

30. Valjent E, Corbillé AG, Bertran-Gonzelez J, Hervé D, Girault JA (2006) Inhibition of ERK pathway or protein synthesis during reexposure to drugs of abuse erases previously learned place preference. Proc Natl Acad Sci U S A 103:2932–2937

31. Sanchez H, Quinn JJ, Torregrossa MM, Taylor JR (2010) Reconsolidation of a cocaine-associated stimulus requires amygdalar protein kinase A. J Neurosci 30:4401–4407

32. Rich MT et al (2015) Effects of amygdalar CaMKII activity on extinction and reconsolidation of a cocaine-associated memory. Drug Alcohol Depend 146:e80

33. Lee JLC, Di Ciano P, Thomas KL, Everitt BJ (2005) Disrupting reconsolidation of drug

memories reduces cocaine seeking behavior. Neuron 47:795–801

34. Lee JLC, Milton AL, Everitt BJ (2006) Cue-induced cocaine seeking and relapse are reduced by disruption of drug memory reconsolidation. J Neurosci 26:5881–5887

35. Bernardi RE, Lattal KM, Berger SP (2007) Anisomycin disrupts a contextual memory following reactivation in a cocaine-induced locomotor activity paradigm. Behav Neurosci 121:156–163

36. Robinson MJF, Franklin KBJ (2007) Effects of anisomycin on consolidation and reconsolidation of a morphine-conditioned place preference. Behav Brain Res 178:146–153

37. von der Goltz C et al (2009) Cue-induced alcohol seeking behaviour is reduced by disrupting the reconsolidation of alcohol-related memories. Psychopharmacology 205:389–397

38. Torregrossa MM, Corlett PR, Taylor JR (2011) Aberrant learning and memory in addiction. Neurobiol Learn Mem 96:609–623

39. Torregrossa MM, Taylor JR (2013) Learning to forget: manipulating extinction and reconsolidation processes to treat addiction. Psychopharmacology 226:659–672

40. Exton-McGuinness MTJ, Milton AL (2018) Reconsolidation blockade for the treatment of addiction: challenges, new targets, and opportunities. Learn Mem 25:492–500

41. Xue Y-X et al (2012) A memory retrieval-extinction procedure to prevent drug craving and relapse. Science 336:241–245

42. Germeroth LJ et al (2017) Effect of a brief memory updating intervention on smoking behavior: a randomized clinical trial. JAMA Psychiatry 74:214–223

43. Das RK, Lawn W, Kamboj SK (2015) Rewriting the valuation and salience of alcohol-related stimuli via memory reconsolidation. Transl Psychiatry 5:e645

44. Das RK, Walsh K, Hannaford J, Lazzarino AI, Kamboj SK (2018) Nitrous oxide may interfere with the reconsolidation of drinking memories in hazardous drinkers in a prediction-error-dependent manner. Eur Neuropsychopharmacol 28:828–840

45. Das RK et al (2019) Ketamine can reduce harmful drinking by pharmacologically rewriting drinking memories. Nat Commun 10:5187

46. Xue Y-X et al (2017) Effect of selective inhibition of reactivated nicotine-associated memories with propranolol on nicotine craving. JAMA Psychiatry 74:224–232

47. Kuijer EJ, Ferragud A, Milton AL (2020) Retrieval-extinction and relapse prevention: rewriting maladaptive drug memories? Front Behav Neurosci 14:23. https://doi.org/10.3389/fnbeh.2020.00023

48. Vousden GH, Milton AL (2017) The chains of habits: too strong to be broken by reconsolidation blockade? Curr Opin Behav Sci 13:158–163

49. Suzuki A et al (2004) Memory reconsolidation and extinction have distinct temporal and biochemical signatures. J Neurosci 24:4787–4795

50. Reichelt AC, Lee JLC (2013) Appetitive Pavlovian goal-tracking memories reconsolidate only under specific conditions. Learn Mem 20:51–60

51. Miles FJ, Everitt BJ, Dickinson A (2003) Oral cocaine seeking by rats: action or habit? Behav Neurosci 117:927–938

52. Exton-McGuinness MTJ, Patton RC, Sacco LB, Lee JLC (2014) Reconsolidation of a well-learned instrumental memory. Learn Mem 21:468–477

53. Exton-McGuinness MTJ, Lee JLC (2015) Reduction in responding for sucrose and cocaine reinforcement by disruption of memory reconsolidation. eNeuro 2:e0009-0015.2015

54. Exton-McGuinness MTJ, Drame ML, Flavell CR, Lee JLC (2019) On the resistance to relapse to cocaine-seeking following impairment of instrumental cocaine memory reconsolidation. Front Behav Neurosci 13:242

55. Deroche-Gamonet V, Belin D, Piazza PV (2004) Evidence for addiction-like behavior in the rat. Science 305:1014–1017

56. Pelloux Y, Everitt BJ, Dickinson A (2007) Compulsive drug seeking by rats under punishment: effects of drug taking history. Psychopharmacology 194:127–137

57. Giuliano C et al (2018) Evidence for a long-lasting compulsive alcohol seeking phenotype in rats. Neuropsychopharmacology 43:728–738

58. Rich MT, Huang YH, Torregrossa MM (2019) Plasticity at thalamo-amygdala synapses regulates cocaine-cue memory formation and extinction. Cell Rep 26:1010–1020

59. Rich MT, Huang YH, Torregrossa MM (2020) Calcineurin promotes neuroplastic changes in the amygdala associated with weakened cocaine-cue memories. J Neurosci 40:1344. https://doi.org/10.1523/JNEUROSCI.0453-1519.2019

60. Liu C et al (2018) Retrieval-induced upregulation of Tet3 in pyramidal neurons of the dorsal hippocampus mediates cocaine-associated memory reconsolidation. Int J Neuropsychopharmacol 21:255–266

61. Yuan K et al (2020) Basolateral amygdala is required for reconsolidation updating of heroin-associated memory after prolonged withdrawal. Addict Biol 25:e12793. https://doi.org/10.1111/adb.12793

62. Xue Y-X et al (2017) Selective inhibition of amygdala neuronal ensembles encoding nicotine-associated memories prevents nicotine seeking and relapse. Biol Psychiatry 82:781. https://doi.org/10.1016/j.biopsych.2017.1004.1017

63. Cahill EN, Milton AL (2019) Neurochemical and molecular mechanisms underlying the retrieval-extinction effect. Psychopharmacology 236:111. https://doi.org/10.1007/s00213-00018-05121-00213

64. Milekic MH, Brown SD, Castellini C, Alberini CM (2006) Persistent disruption of an established morphine conditioned place preference. J Neurosci 26:3010–3020

65. Miller CA, Marshall JF (2005) Molecular substrates for retrieval and reconsolidation of cocaine-associated contextual memory. Neuron 47:873–884

66. Rich MT, Torregrossa MM (2015) CaMKII inhibition affects cocaine-cue memory processes to attenuate reinstatement. Drug Alcohol Depend 156:e189–e190

67. Itzhak Y, Anderson KL (2007) Memory reconsolidation of cocaine-associated context requires nitric oxide signaling. Synapse 61:1002–1005

68. Schiller D, Phelps EA (2011) Does reconsolidation occur in humans? Front Behav Neurosci 5:24

69. Finnie PSB, Nader K (2012) The role of metaplasticity mechanisms in regulating memory destabilization and reconsolidation. Neurosci Biobehav Rev 36:1667–1707

70. Lee JLC, Milton AL, Everitt BJ (2006) Reconsolidation and extinction of conditioned fear: inhibition and potentiation. J Neurosci 26:10051–10056

71. Merlo E, Milton AL, Goozée ZY, Theobald DEH, Everitt BJ (2014) Reconsolidation and extinction are dissociable and mutually exclusive processes: behavioral and molecular evidence. J Neurosci 34:2422–2231

72. Merlo E, Milton AL, Everitt BJ (2018) A novel retrieval-dependent memory process revealed by the arrest of ERK1/2 activation

in the basolateral amygdala. J Neurosci 38:3199–3207

73. Cassini LF, Flavell CR, Amaral OB, Lee JLC (2017) On the transition from reconsolidation to extinction of contextual fear memories. Learn Mem 24:392–399

74. Alberini CM, Milekic MH, Tronel S (2006) Mechanisms of memory stabilization and de-stabilization. Cell Mol Life Sci 63:999–1008

75. Milton AL, Lee JLC, Everitt BJ (2008) Reconsolidation of appetitive memories for both natural and drug reinforcement is dependent on β-adrenergic receptors. Learn Mem 15:88–92

76. Lee JLC, Everitt BJ (2008) Reactivation-dependent amnesia for appetitive memories is determined by the contingency of stimulus presentation. Learn Mem 15:390–393

77. Brown TE, Lee BR, Sorg BA (2008) The NMDA antagonist MK-801 disrupts reconsolidation of a cocaine-associated memory for conditioned place preference but not for self-administration in rats. Learn Mem 15:857–865

78. Pedreira ME, Maldonado H (2003) Protein synthesis subserves reconsolidation or extinction depending on reminder duration. Neuron 38:863–869

79. Gisquet-Verrier P et al (2015) Integration of new information with active memory accounts for retrograde amnesia: a challenge to the consolidation/reconsolidation hypothesis? J Neurosci 35:11623–11633

80. Lee JLC, Everitt BJ, Thomas KL (2004) Independent cellular processes for hippocampal memory consolidation and reconsolidation. Science 304:839–843

81. Gold PE, King RA (1974) Retrograde amnesia: storage failure versus retrieval failure. Psychol Rev 81:465–469

82. Josselyn SA, Köhler S, Frankland PW (2015) Finding the engram. Nat Rev Neurosci 16:521–534

83. Tolliver GA, Sadeghi KG, Samson HH (1987) Ethanol preference following the sucrose-fading initiation procedure. Alcohol 5:9–13

84. Carnicella S, Ron D, Barak S (2014) Intermittent ethanol access schedule in rats as a preclinical model of alcohol abuse. Alcohol 48:243–252

85. Augier E et al (2014) Wistar rats acquire and maintain self-administration of 20% ethanol without water deprivation, saccharin/sucrose fading, or extended access training. Psychopharmacology 231:4561–4568

86. Augier E, Dulman RS, Singley E, Heilig M (2017) A method for evaluating the reinforcing properties of ethanol in rats without water deprivation, saccharin fading or extended access training. J Vis Exp (119):53305. https://doi.org/10.3791/53305

87. Puaud M, Ossowska Z, Barnard J, Milton AL (2018) Saccharin fading is not required for the acquisition of alcohol self-administration, and can alter the dynamics of cue-alcohol memory reconsolidation. Psychopharmacology 235:1069–1082

88. Kindt M, Soeter M, Vervliet B (2009) Beyond extinction: erasing human fear responses and preventing the return of fear. Nat Neurosci 12:256–258

89. Soeter M, Kindt M (2015) Retrieval cues that trigger reconsolidation of associative fear memory are not necessarily an exact replica of the original learning experience. Front Behav Neurosci 9:122

90. Kindt M, van Emmerik A (2016) New avenues for treating emotional memory disorders: towards a reconsolidation intervention for posttraumatic stress disorder. Therapeut Adv Psychopharmacol 6:283–295

91. Brunet A et al (2008) Effect of post-retrieval propranolol on psychophysiologic responding during subsequent script-driven traumatic imagery in post-traumatic stress disorder. J Psychiatr Res 42:503–506

92. Brunet A et al (2011) Trauma reactivation under the influence of propranolol decreases posttraumatic stress symptoms and disorder: 3 open-label trials. J Clin Psychopharmacol 31:547–550

93. Soeter M, Kindt M (2015) An abrupt transformation of phobic behavior after a post-retrieval amnestic agent. Biol Psychiatry 78:880–886

94. Saladin ME et al (2014) Post-retrieval propranolol may alter reconsolidation of trauma memory in individuals with PTSD and comorbid alcohol dependence. Drug Alcohol Depend 140:e193

95. Lonergan M et al (2016) Reactivating addiction-related memories under propranolol to reduce craving: a pilot randomized controlled trial. J Behav Ther Exp Psychiatry 50:245–249

96. Pachas GN et al (2015) Single dose propranolol does not affect physiologic or emotional reactivity to smoking cues. Psychopharmacology 232:1619–1628

97. Jobes ML et al (2015) Effects of prereactivation propranolol on cocaine craving elicited by imagery script/cue sets in opioid-dependent polydrug users: a randomized study. J Addict Med 9:491–498

98. Schramm MJW, Everitt BJ, Milton AL (2016) Bidirectional modulation of alcohol-associated memory reconsolidation through manipulation of adrenergic signaling. Neuropsychopharmacology 41:1103–1111

99. Dunbar AB, Taylor JR (2016) Inhibition of protein synthesis but not β-adrenergic receptors blocks reconsolidation of a cocaine-associated cue memory. Learn Mem 23:391–398

100. Das RK, Freeman TP, Kamboj SK (2013) The effects of N-methyl-D-aspartate and β-adrenergic receptor antagonists on the reconsolidation of reward memory: a meta-analysis. Neurosci Biobehav Rev 37:240–255

# Correction to: Methods for Preclinical Research in Addiction

María A. Aguilar

**Correction to:**
**Front Matter and Chapter 8 in: María A. Aguilar (ed.),** *Methods for Preclinical Research in Addiction, Neuromethods, vol. 174,*
https://doi.org/10.1007/978-1-0716-1748-9

The book was inadvertently published with the below mentioned errors.

1. Incorrect affiliation of the author "José Enrique De La Rubia Orti" in the contributors' list in Front Matter and in the metadata of chapter 8.

2. Incorrect affiliation of the authors "María A. Aguilar, Claudia Calpe-López and M. Ángeles Martínez-Caballero" in the metadata of chapter 8.

These errors have now been corrected by updating the affiliations as mentioned below.

1. The affiliation of "José Enrique De La Rubia Orti" has been updated as "Faculty of Nursing, Catholic University of Valencia, Valencia, Spain" in the contributors' list, in Front Matter, and in the metadata of chapter 8.

2. The affiliation of authors "María A. Aguilar, Claudia Calpe-López and M. Ángeles Martínez-Caballero" has been updated as "Unit of Research 'Neurobehavioral Mechanisms and Endophenotypes of Addictive Behavior,' Department of Psychobiology, Faculty of Psychology, University of Valencia, Valencia, Spain; Faculty of Nursing, Catholic University of Valencia, Valencia, Spain" in the metadata of chapter 8.

---

The updated online version of the book can be found at https://doi.org/10.1007/978-1-0716-1748-9 and https://doi.org/10.1007/978-1-0716-1748-9_8

María A. Aguilar (ed.), *Methods for Preclinical Research in Addiction*, Neuromethods, vol. 174,
https://doi.org/10.1007/978-1-0716-1748-9_14, © Springer Science+Business Media, LLC, part of Springer Nature 2022

# INDEX

María A. Aguilar (ed.), *Methods for Preclinical Research in Addiction*, Neuromethods, vol. 174, https://doi.org/10.1007/978-1-0716-1748-9, © Springer Science+Business Media, LLC, part of Springer Nature 2022

Printed in the United States
by Baker & Taylor Publisher Services